# LIQUID CHROMATOGR
# MASS SPECTROMETRY:
# AN INTRODUCTION

**Analytical Techniques in the Sciences (AnTS)**

*Series Editor*: Davie J. Ando, Consultant, Dartford, Kent, UK

A series of open learning/distance learning books which covers all of the major analytical techniques and their application in the most important areas of physical, life and materials sciences.

*Titles Available in the Series*

**Analytical Instrumentation: Performance Characteristics and Quality**
Graham Currell, University of the West of England, Bristol, UK

**Fundamentals of Electroanalytical Chemistry**
Paul M.S. Monk, Manchester Metropolitan University, Manchester, UK

**Introduction to Environmental Analysis**
Roger N. Reeve, University of Sunderland, UK

**Polymer Analysis**
Barbara H. Stuart, University of Technology, Sydney, Australia

**Chemical Sensors and Biosensors**
Brian R. Eggins, University of Ulster at Jordanstown, Northern Ireland, UK

**Methods for Environmental Trace Analysis**
John R. Dean, Northumbria University, Newcastle, UK

**Liquid Chromatography Mass Spectrometry: An Introduction**
R. Ardrey, University of Huddersfield, Huddersfield, UK

**The Analysis of Drugs of Abuse: A Systematic Approach**
Michael D. Cole, Anglia Polytechnic University, Cambridge, UK

*Forthcoming Titles*

**Infrared Spectroscopy: Experimentation and Applications**
Barbara H. Stuart, University of Technology, Sydney, Australia

**Techniques of Organic Mass Spectrometry**
R. Ardrey, University of Huddersfield, Huddersfield, UK

# LIQUID CHROMATOGRAPHY–MASS SPECTROMETRY: AN INTRODUCTION

**Robert E. Ardrey**
*University of Huddersfield, Huddersfield, UK*

John Wiley & Sons, Ltd

Telephone (+44) 1243 779777

Email (for orders and customer service enquiries): cs-books@wiley.co.uk
Visit our Home Page on www.wileyeurope.com or www.wiley.com

Reprinted with corrections September, 2003



***Other Wiley Editorial Offices***

John Wiley & Sons Inc., 111 River Street, Hoboken, NJ 07030, USA

Jossey-Bass, 989 Market Street, San Francisco, CA 94103-1741, USA

Wiley-VCH Verlag GmbH, Boschstr. 12, D-69469 Weinheim, Germany

John Wiley & Sons Australia Ltd, 33 Park Road, Milton, Queensland 4064, Australia

John Wiley & Sons (Asia) Pte Ltd, 2 Clementi Loop #02-01, Jin Xing Distripark, Singapore 129809

John Wiley & Sons Canada Ltd, 22 Worcester Road, Etobicoke, Ontario, Canada M9W 1L1

Wiley also publishes its books in a variety of electronic formats. Some content that appears in print may not be available in electronic books.

***Library of Congress Cataloging-in-Publication Data***

Ardrey, R. E.
Liquid chromatography-mass spectrometry : an introduction / Robert E. Ardrey.
p. cm. – (Analytical techniques in the sciences)
Includes bibliographical references and index.
ISBN 0-471-49799-1 (cloth : alk. paper) – ISBN 0-471-49801-7 (pbk. : alk. paper)
1. Liquid chromatography. 2. Mass spectrometry. I. Title. II. Series.

QP519.9.L55 A73 2003
543′.0894 – dc21 2002028084

***British Library Cataloguing in Publication Data***

A catalogue record for this book is available from the British Library

ISBN 0-471-49799-1 (Cloth)
ISBN 0-471-49801-7 (Paper)

Typeset in 10/12pt Times by Laserwords Private Limited, Chennai, India
Printed and bound in Great Britain by MPG Books Ltd, Bodmin, Cornwall
This book is printed on acid-free paper responsibly manufactured from sustainable forestry in which at least two trees are planted for each one used for paper production.

# Contents

# Series Preface

There has been a rapid expansion in the provision of further education in recent years, which has brought with it the need to provide more flexible methods of teaching in order to satisfy the requirements of an increasingly more diverse type of student. In this respect, the *open learning* approach has proved to be a valuable and effective teaching method, in particular for those students who for a variety of reasons cannot pursue full-time traditional courses. As a result, John Wiley & Sons Ltd first published the Analytical Chemistry by Open Learning (ACOL) series of textbooks in the late 1980s. This series, which covers all of the major analytical techniques, rapidly established itself as a valuable teaching resource, providing a convenient and flexible means of studying for those people who, on account of their individual circumstances, were not able to take advantage of more conventional methods of education in this particular subject area.

Following upon the success of the ACOL series, which by its very name is predominately concerned with Analytical *Chemistry*, the *Analytical Techniques in the Sciences* (AnTS) series of open learning texts has now been introduced with the aim of providing a broader coverage of the many areas of science in which analytical techniques and methods are being increasingly applied. With this in mind, the AnTS series of texts seeks to provide a range of books which will cover not only the actual techniques themselves, but *also* those scientific disciplines which have a necessary requirement for analytical characterization methods.

Analytical instrumentation continues to increase in sophistication, and as a consequence, the range of materials that can now be almost routinely analysed has increased accordingly. Books in this series which are concerned with the *techniques* themselves will reflect such advances in analytical instrumentation, while at the same time providing full and detailed discussions of the fundamental concepts and theories of the particular analytical method being considered. Such books will cover a variety of techniques, including general instrumental analysis, spectroscopy, chromatography, electrophoresis, tandem techniques, electroanalytical methods, X-ray analysis and other significant topics. In addition, books in the series will include the *application* of analytical techniques in areas such as environmental science, the life sciences, clinical analysis, food science, forensic analysis, pharmaceutical science, conservation and archaeology, polymer science and general solid-state materials science.

Written by experts in their own particular fields, the books are presented in an easy-to-read, user-friendly style, with each chapter including both learning objectives and summaries of the subject matter being covered. The progress of the reader can be assessed by the use of frequent self-assessment questions (SAQs) and discussion questions (DQs), along with their corresponding reinforcing or remedial responses, which appear regularly throughout the texts. The books are thus eminently suitable both for self-study applications and for forming the basis of industrial company in-house training schemes. Each text also contains a large amount of supplementary material, including bibliographies, lists of acronyms and abbreviations, and tables of SI Units and important physical constants, plus where appropriate, glossaries and references to literature sources.

It is therefore hoped that this present series of textbooks will prove to be a useful and valuable source of teaching material, both for individual students and for teachers of science courses.

*Dave Ando*
*Dartford, UK*

# Preface

In this book, I have tried to show the way in which high performance liquid chromatography–mass spectrometry (LC–MS) has developed, somewhat slowly it has to be said, into a powerful hybrid analytical technique.

In the first chapter, I have discussed the limitations of high performance liquid chromatography (HPLC) and mass spectrometry when used in isolation and how the combination of the two allows these to be overcome. In this chapter, the effect of combining the two techniques with regard to the individual performance characteristics are explored.

In Chapters 2 and 3, brief descriptions of HPLC and MS are provided. These are not fully comprehensive but are intended to provide a brief description of those aspects of each of the techniques which are pertinent to a consideration of LC–MS.

Seven different LC–MS interfaces are described in Chapter 4, with particular emphasis being placed on their advantages and disadvantages and the ways in which the interface overcomes (or fails to overcome) the incompatibilities of the two techniques. The earlier interfaces are included for historical reasons only as, for example, the moving-belt and direct-liquid-introduction interfaces, are not currently in routine use. The final chapter (Chapter 5) is devoted to a number of illustrative examples of the way in which LC–MS has been used to solve various analytical problems.

I have tried to make it clear that the LC–MS combination is usually more powerful that either of the individual techniques in isolation and that a holistic approach must be taken to the development of methodologies to provide data from which the required analytical information may be obtained. Data analysis is of crucial importance in this respect and for this reason the computer processing of LC–MS data is considered in some detail in both Chapters 3 and 5.

LC–MS is still not used in many laboratories where it would be a cost-effective investment. In order that interested readers can gauge whether they should 'test the water', a number of applications which **illustrate** the range of analyses and the analytical performance that may be obtained from modern LC–MS interfaces have been described. Although your precise application may not appear here, I hope that the descriptions are general enough for the reader to draw parallels with their own work.

*Bob Ardrey*
*University of Huddersfield, UK*

# Acknowledgements

I gratefully acknowledge the assistance of the following, without whom this book would not have been completed.

My colleagues Lindsay Harding (mass spectroscopist), Carl Hall (analytical scientist) and Terry Pearson (chromatographer) from the Department of Chemical and Biological Sciences at The University of Huddersfield for helpful discussions and their criticisms and suggestions made during the writing of the book. Their combined expertise has hopefully meant that the text is appropriate for its intended audience and that the author has not assumed too much prior knowledge on the part of the reader. Their painstaking checking of the manuscript (on a number of occasions) is gratefully recorded. In particular, I must acknowledge the part that Terry Pearson has played in my retaining the little sanity I have as Huddersfield Town FC, our joint passion, have experienced more 'downs' than 'ups' in recent years! I wish him well in his retirement.

Dave Ando, from John Wiley & Sons Ltd, for his constant encouragement from the time of our initial discussions through to copy-editing and proof-reading of the final manuscript, and the hours spent discussing the state of English cricket and the 'downs' and 'ups' (in that order) of Manchester City FC, the latter being *his* passion, *not* mine!

Micromass, the mass spectrometry company, for permission to use their technical literature and application notes and, in particular, Chris Herbert for helpful discussions and access to his computer graphics.

Finally, but not least, my wife, Lesley, for her forbearance and support while the preparation of this book has taken up the majority of my time.

# Abbreviations, Acronyms and Symbols

| | |
|---|---|
| Ala | alanine |
| APCI | atmospheric-pressure chemical ionization |
| Arg | arginine |
| AS | aerospray |
| Asn | asparagine |
| Asp | aspartic acid |
| BPI | base peak intensity |
| CI | chemical ionization |
| CID | collision-induced dissociation |
| CNL | constant neutral loss |
| CVF | cone-voltage fragmentation |
| Cys | cysteine |
| Da | dalton (atomic mass unit) |
| DC | direct current |
| DLI | direct-liquid introduction |
| EI | electron ionization |
| EPA | Environmental Protection Agency (USA) |
| ESA | electrostatic analyser |
| ESI | electrospray ionization |
| eV | electronvolt |
| FAB | fast-atom bombardment |
| Fuc | fucose (a deoxy sugar) |
| Gal | galactose |
| GalNAc | *N*-acetylgalactosamine |
| GC | gas chromatography |

| | |
|---|---|
| GC–MS | gas chromatography – in combination with mass spectrometry |
| GlcNAc | *N*-acetylglucosamine |
| Gln | glutamine |
| Glu | glutamic acid |
| Gly | glycine |
| Hex | hexose (a monosaccharide with six carbon atoms) |
| HexNAc | *N*-acetylhexosamine |
| His | histidine |
| HIV | human immunodeficiency virus |
| HPLC | high performance liquid chromatography |
| i.d. | internal diameter |
| Ile | isoleucine |
| IS | internal standard |
| IUPAC | International Union of Pure and Applied Chemistry |
| kV | kilovolt |
| LC | liquid chromatography |
| LC–FTIR | (high performance) liquid chromatography in combination with Fourier-transform infrared (spectroscopy) |
| LC–MS | (high performance) liquid chromatography in combination with mass spectrometry |
| LC–MS–MS | (high performance) liquid chromatography in combination with tandem mass spectrometry |
| LC–ToF-MS | (high performance) liquid chromatography in combination with time-of-flight mass spectrometry |
| Leu | leucine |
| LOD | limit of detection |
| LOQ | limit of quantitation |
| Lys | lysine |
| $(M - H)^-$ | deprotonated molecular ion |
| $(M + H)^+$ | protonated molecular ion |
| MAGIC | monodisperse aerosol generating interface for chromatography |
| MALDI | matrix-assisted laser desorption ionization |
| MALDI–ToF | matrix-assisted laser desorption ionization with a time-of-flight mass analyser |
| max ent | maximum entropy |
| Met | methionine |
| MID | multiple-ion detection |
| MIKES | mass-analysed ion kinetic energy spectrometry |
| MRM | multiple-reaction monitoring |
| MS | mass spectrometry |

| | |
|---|---|
| MS–MS | mass spectrometry in combination with mass spectrometry (tandem mass spectrometry) |
| $MS^n$ | multiple stages of tandem mass spectrometry |
| NIH | National Institute of Health (USA) |
| NIST | National Institute of Standards and Technology (USA) |
| o.d. | outside diameter |
| ODS | octadecyl silyl stationary phase used in high performance liquid chromatography |
| PAGE | polyacrylamide gel electrophoresis |
| Phe | phenylalanine |
| Pro | proline |
| PSD | post-source decay |
| QC | quality control |
| Q–ToF | quadrupole time-of-flight mass analyser |
| Q–ToF–LC–MS–MS | quadrupole time-of-flight mass analyser in combination with (high performance) liquid chromatography and tandem mass spectrometry |
| RF | radiofrequency |
| RIC | reconstructed ion chromatogram |
| RMM | relative molecular mass |
| RSD | relative standard deviation |
| S/N | signal-to-noise ratio |
| SD | standard deviation |
| SDM | selected-decomposition monitoring |
| SDS–PAGE | sodium dodecyl sulfate–polyacrylamide gel electrophoresis |
| Ser | serine |
| SIM | selected-ion monitoring |
| SIR | selected-ion recording |
| SRM | selected-reaction monitoring |
| TFA | trifluoroacetic acid |
| Thr | threonine |
| TIC | total-ion current |
| ToF | time-of-flight |
| Trp | tryptophan |
| TSP | thermospray |
| Tyr | tyrosine |
| UV | ultraviolet |
| V | volt |
| vol/vol | volume by volume |
| Val | valine |

| | |
|---|---|
| $B$ | magnetic field (magnitude) |
| $CV$ | coefficient of variation |
| $E$ | electrostatic analyser voltage |
| $H$ | chromatographic plate height |
| $k'$ | capacity factor (in high performance liquid chromatography) |
| $L$ | length of chromatographic column |
| $m/z$ | mass-to-charge ratio |
| $MW$ | molecular weight |
| $N$ | number of theoretical plates for a chromatographic column |
| $R$ | resolution – chromatographic or mass spectral; correlation coefficient |
| $R^2$ | coefficient of determination |
| $RE$ | relative error |
| $t_0$ | retention time of non-retained component in high performance liquid chromatography (dead time) |
| $t_{an}$ | retention time of analyte |
| $w$ | chromatographic peak width |
| | |
| $\alpha$ | chromatographic selectivity |
| $\lambda_{max}$ | wavelength of maximum absorption in a UV spectrum |

# About the Author

**Robert E. Ardrey, B.Sc., Ph.D.**

Bob Ardrey obtained a first degree in Chemistry from the University of Surrey where he went on to obtain his doctorate studying the chemistry of *trans*-2,3-dichloro-1,4-dioxan and the stereochemistry of its reaction products using primarily mass spectrometry and nuclear magnetic resonance spectroscopy. He then carried out post-doctoral research at King's College, London, into the development of emitters for field-desorption mass spectrometry.

He then joined the Central Research Establishment of the Home Office Forensic Science Service (as it then was) at Aldermaston where he developed thermogravimetry–MS, pyrolysis-MS, GC–MS and LC–MS methodologies for the identification of analytes associated with crime investigations. It was here that his interest in LC–MS began with the use of an early moving-belt interface. This interest continued during periods of employment with two manufacturers of LC–MS equipment, namely Kratos and subsequently Interion, the UK arm of the Vestec Corporation of Houston, Texas, the company set up by Marvin Vestal, the primary developer of the thermospray LC–MS interface.

In 1990, Bob set up a mass spectrometry consultancy which he ran until becoming a Senior Lecturer in Analytical Chemistry within the Department of Chemical and Biological Sciences at the University of Huddersfield.

Bob is particularly concerned that, although analytical chemistry forms a major part of the UK chemical industry's efforts, it is still not considered by many to be a subject worthy of special consideration. Consequently, experimental design is often not employed when it should be and safeguards to ensure accuracy and precision of analytical measurements are often lacking. He would argue that although the terms *accuracy* and *precision* can be defined by rote, their meanings, when applied to analytical measurements, are not appreciated by many members of the scientific community.

He is therefore pleased to be associated with this series, which he hopes will help to address this problem.

# Chapter 1
# Introduction

**Learning Objectives**

- To understand the need to interface liquid chromatography and mass spectrometry.
- To understand the requirements of an interface between liquid chromatography and mass spectrometry and the performance of the combined system.

The combination of chromatography and mass spectrometry (MS) is a subject that has attracted much interest over the last forty years or so. The combination of gas chromatography (GC) with mass spectrometry (GC–MS) was first reported in 1958 and made available commercially in 1967. Since then, it has become increasingly utilized and is probably the most widely used 'hyphenated' or 'tandem' technique, as such combinations are often known. The acceptance of GC–MS as a routine technique has in no small part been due to the fact that interfaces have been available for both packed and capillary columns which allow the vast majority of compounds amenable to separation by gas chromatography to be transferred efficiently to the mass spectrometer. Compounds amenable to analysis by GC need to be both volatile, at the temperatures used to achieve separation, and thermally stable, i.e. the same requirements needed to produce mass spectra from an analyte using either electron (EI) or chemical ionization (CI) (see Chapter 3). In simple terms, therefore, virtually all compounds that pass through a GC column can be ionized and the full analytical capabilities of the mass spectrometer utilized.

This is not the case when high performance liquid chromatography (HPLC) and MS are considered where, due to the incompatibilities of the two techniques, they cannot be linked directly and an interface must be used, with its prime purpose being the removal of the chromatographic mobile phase. Unfortunately, no single interface exists which possesses similar capabilities to those available for GC–MS, i.e. one that will allow mass spectra to be obtained from any compound that elutes from an HPLC column, and thus LC–MS has not been guaranteed to provide the required analytical information. In addition, the complexity of the mass spectrometer has meant that the majority of chromatographers have not had direct access to the instrumentation and have had to rely on a service facility to provide results. They were therefore unable to react rapidly to the results of an analysis and consequently found it a particularly inconvenient detector to contemplate using. The different interfaces that have been made available commercially, and the applications to which they have been put, are the subjects of the following chapters.

Before discussing these in detail, it is appropriate to consider a number of general questions, namely:

(1) What are the advantages of linking HPLC with mass spectrometry?

(2) What capabilities are required of such a combination?

(3) What problems, if any, have to be addressed to allow the combination to function, *and function effectively*?

## 1.1 What are the Advantages of Linking High Performance Liquid Chromatography with Mass Spectrometry?

In order to answer the first question, the limitations of the individual techniques must be considered and whether the combination will allow all or some of these to be overcome. Before doing this, however, the analytical tasks to which the combination will be applied must be defined.

In many analyses, the compound(s) of interest are found as part of a complex mixture and the role of the chromatographic technique is to provide separation of the components of that mixture to allow their identification or quantitative determination. From a qualitative perspective, the main limitation of chromatography in isolation is its inability to provide an unequivocal identification of the components of a mixture even if they can be completely separated from each other. Identification is based on the comparison of the retention characteristics, simplistically the retention time, of an unknown with those of reference materials determined under identical experimental conditions. There are, however, so many compounds in existence that even if the retention characteristics of an unknown and a reference material are, within the limits of experimental error, identical, the analyst cannot say with absolute certainty that the two compounds are the same. Despite a range of chromatographic conditions being available to the analyst, it is not always possible to effect complete separation of all of the components of a mixture and this may prevent the precise and accurate quantitative determination of the analyte(s) of interest.

The power of mass spectrometry lies in the fact that the mass spectra of many compounds are sufficiently specific to allow their identification with a high degree of confidence, if not with complete certainty. If the analyte of interest is encountered as part of a mixture, however, the mass spectrum obtained will contain ions from all of the compounds present and, particularly if the analyte of interest is a minor component of that mixture, identification with any degree of certainty is made much more difficult, if not impossible. The combination of the separation capability of chromatography to allow 'pure' compounds to be introduced into the mass spectrometer with the identification capability of the mass spectrometer is clearly therefore advantageous, particularly as many compounds with similar or identical retention characteristics have quite different mass spectra and can therefore be differentiated. This extra specificity allows quantitation to be carried out which, with chromatography alone, would not be possible.

The combination of HPLC with mass spectrometry therefore allows more definitive identification and the quantitative determination of compounds that are not fully resolved chromatographically.

## 1.2 What Capabilities are Required of the Combination?

Ideally, the capabilities of both instruments should be unaffected by their being linked. These include the following (adapted from Snyder and Kirkland [1]):

- The interface should cause no reduction in chromatographic performance. This is particularly important for the analysis of complex multi-component mixtures (although the specificity of the mass spectrometer may, in certain circumstances, compensate for some loss of performance – see Chapter 3).
- No uncontrolled chemical modification of the analyte should occur during its passage through the interface or during its introduction into the mass spectrometer.
- There should be high sample transfer to the mass spectrometer or, if this takes place in the interface, ionization efficiency. This is of particular importance when trace-level components are of interest or when polar and/or labile analytes are involved.
- The interface should give low chemical background, thus minimizing possible interference with the analytes.
- The interface should be reliable and easy to use.
- The interface should be simple and inexpensive (a subjective assessment).
- Operation of the interface should be compatible with all chromatographic conditions which are likely to be encountered, including flow rates from around 20 nl $min^{-1}$ to around 2 ml $min^{-1}$, solvent systems from 100% organic phase to 100% aqueous phase, gradient elution, which is of particular importance in

the biological field in which mixtures covering a wide range of polarities are often encountered, and buffers, both volatile and involatile.

- Operation of the interface should not compromise the vacuum requirements of the mass spectrometer and should allow all capabilities of the mass spectrometer to be utilized, i.e. ionization modes, high resolution, etc.
- The mass spectrum produced should provide unambiguous molecular weight information from the wide range of compounds amenable to analysis by HPLC, including biomolecules with molecular weights in excess of 1000 Da. The study of these types of molecule by mass spectrometry may be subject to limitations associated with their ionization and detection and the mass range of the instrument being used.
- The mass spectrometer should provide structural information that should be reproducible, interpretable and amenable to library matching. Ideally, an electron ionization (EI) (see Chapter 3) spectrum should be generated. An interface that fulfils both this requirement and/or the production of molecular weight information, immediately lends itself to use as a more convenient alternative to the conventional solid-sample insertion probe of the mass spectrometer and some of the interfaces which have been developed have been used in this way.
- The interface should provide quantitative information with a reproducibility better than 10% with low limits of detection and have a linear response over a wide range of sample sizes (low picograms to micrograms).

## 1.3 What Problems, if Any, Have to be Addressed to Allow the LC–MS Combination to Function, and Function Effectively?

It is possible to carry out a chromatographic separation, collect all, or selected, fractions and then, after removal of the majority of the volatile solvent, transfer the analyte to the mass spectrometer by using the conventional inlet (probe) for solid analytes. The direct coupling of the two techniques is advantageous in many respects, including the speed of analysis, the convenience, particularly for the analysis of multi-component mixtures, the reduced possibility of sample loss, the ability to carry out accurate quantitation using isotopically labelled internal standards, and the ability to carry out certain tasks, such as the evaluation of peak purity, which would not otherwise be possible.

There are two major incompatibilities between HPLC and MS. The first is that the HPLC mobile phase is a liquid, often containing a significant proportion of water, which is pumped through the stationary phase (column) at a flow rate of typically 1 ml min$^{-1}$, while the mass spectrometer operates at a pressure of around $10^{-6}$ torr ($1.333\,22 \times 10^{-4}$ Pa). It is therefore not possible simply to pump the eluate from an HPLC column directly into the source of a mass spectrometer and an important function of any interface is the removal of all, or a significant portion, of the mobile phase. The second is that the majority of analytes that are likely to be separated by HPLC are relatively involatile and/or thermally labile and therefore not amenable to ionization by using either EI or CI. Alternative ionization methods have therefore to be developed.

In the following chapters, the basic principles of HPLC and MS, in as far as they relate to the LC–MS combination, will be discussed and seven of the most important types of interface which have been made available commercially will be considered. Particular attention will be paid to the electrospray and atmospheric-pressure chemical ionization interfaces as these are the ones most widely used today. The use of LC–MS for identification and quantitation will be described and appropriate applications will be discussed.

# Summary

In this chapter, the reader has been introduced to the analytical advantages to be gained by linking high performance liquid chromatography to mass spectrometry with particular regard to the limitations of the two techniques when they are used independently.

The characteristics of an ideal liquid chromatography–mass spectrometry interface have been discussed, with emphasis having been placed upon the major incompatibilities of the two component techniques that need to be overcome to allow the combination to function effectively.

# References

1. Snyder, L. R. and Kirkland, J. J., *Introduction to Modern Liquid Chromatography*, Wiley, New York, 1974.

Chapter 2

# Liquid Chromatography

**Learning Objectives**

- To understand those aspects of high performance liquid chromatography which are essential to the application of LC–MS.

## 2.1 Introduction

The International Union of Pure and Applied Chemistry (IUPAC) defines chromatography as follows [1]:

> 'Chromatography is a physical method of separation in which the components to be separated are distributed between two phases, one of which is stationary (the stationary phase), while the other (the mobile phase) moves in a definite direction. A mobile phase is described as "a fluid which percolates through or along the stationary bed in a definite direction". It may be a liquid, a gas or a supercritical fluid, while the stationary phase may be a solid, a gel or a liquid. If a liquid, it may be distributed on a solid, which may or may not contribute to the separation process.'

A chromatographic system may be considered to consist of four component parts, as follows:

- a device for sample introduction
- a mobile phase
- a stationary phase
- a detector

A number of different chromatographic techniques are in use and these differ in the form of these four components and their relative importance. For example, in *gas chromatography* the injector used for sample introduction is of paramount importance and must be chosen in light of the properties of the analytes under investigation (their stability and volatility) and the amounts of the analytes present. An incorrect choice could prevent a successful analysis. In high performance liquid chromatography (HPLC) the injector is simply required to allow introduction of the analytes into a flowing liquid stream without introducing any discrimination effects and a single type, the loop injector, is used almost exclusively.

The two components which are associated with the separation that occurs in a chromatographic system are the mobile and stationary phases.

In HPLC, the mobile phase is a liquid delivered under high pressure (up to 400 bar ($4 \times 10^7$ Pa)) to ensure a constant flow rate, and thus reproducible chromatography, while the stationary phase is packed into a column capable of withstanding the high pressures which are necessary.

A chromatographic separation occurs if the components of a mixture interact to different extents with the mobile and/or stationary phases and therefore take different times to move from the position of sample introduction to the position at which they are detected. There are two extremes, as follows:

(i) All analytes have total affinity for the mobile phase and do not interact with the stationary phase – all analytes move at the same rate as the mobile phase, they reach the detector very quickly and are not separated.

(ii) All analytes have total affinity for the stationary phase and do not interact with the mobile phase – all analytes are retained on the column and do not reach the detector.

The role of the chromatographer is therefore, based on a knowledge of the analytes under investigation, to manipulate the properties of the stationary and/or mobile phases to move from these extremes and effect the desired separation.

A number of detectors may be used in conjunction with HPLC (see Section 2.2.5 below), with the type chosen being determined by the type of analysis, i.e. qualitative or quantitative, being undertaken. The requirements for each of these are often quite different, as described in the following:

- **Qualitative** (identification) applications depend upon the comparison of the retention characteristics of the unknown with those of reference materials. In the case of gas chromatography, this characteristic is known as the retention index and, although collections of data on 'popular' stationary phases exist, it is unlikely that any compound has a unique retention index and unequivocal identification can be effected. In liquid chromatography, the situation is more complex because there is a much larger number of combinations of stationary and mobile phases in use, and large collections of retention characteristics on any single 'system' do not exist. In addition, HPLC is a less efficient separation technique than GC and this results in wider 'peaks' and more imprecision in retention time measurements, and thus identification.

- **Quantitative** accuracy and precision (see Section 2.5 below) often depend upon the selectivity of the detector because of the presence of background and/or co-eluted materials. The most widely used detector for HPLC, the UV detector, does not have such selectivity as it normally gives rise to relatively broad signals, and if more than one component is present, these overlap and deconvolution is difficult. The related technique of fluorescence has more selectivity, since both absorption and emission wavelengths are utilized, but is only applicable to a limited number of analytes, even when derivatization procedures are used.

**DQ 2.1**

What is meant by the 'selectivity' of a detector? Define the 'limit of detection' of a detector.

*Answer*

*The **selectivity** of a detector is its ability to determine an analyte of interest without interference from other materials present in the analytical system, i.e. the sample matrix, solvents used, etc.*

*The **limit of detection** is the smallest amount of an analyte that is required for reliable determination, identification or quantitation. More mathematically, it may be defined as that amount of analyte which produces a signal greater than the standard deviation of the background noise by a defined factor. Strictly for quantitative purposes, this should be referred to as the 'limit of determination'. The factor used depends upon the task being carried out and for quantitative purposes a higher value is used than for identification. Typical values are 3 for identification and 5 or 10 for quantitation.*

*The selectivity of a detector is often related to its limit of detection, i.e. the more selective it is, then the lower the background noise is likely to be, and consequently the lower the limit of detection.*

*The term 'sensitivity' is often used in place of the 'limit of detection'. The **sensitivity** actually refers to the degree of response obtained from a detector, i.e. the increase in output signal obtained from an increasing amount or concentration of analyte reaching the detector. Care must therefore be taken when these terms are being used or when they are encountered to ensure that their meanings are unambiguous.*

*The terms defined above are all important in the consideration of the overall performance of an analytical method. The greatest 'sensitivity' (response) does not necessarily imply the lowest 'limit of detection/determination' as a more intense signal may also be observed from any interferences present. An inherently less sensitive but more selective detector may provide a 'better' analysis with lower 'limits of detection/determination'.*

*The performance of a detector is therefore intimately linked to the samples being analysed.*

Mass spectrometry (see Chapter 3) is capable of providing molecular weight and structural information from picogram amounts of material and to provide selectivity by allowing the monitoring of ions or ion decompositions characteristic of a single analyte of interest. These are the ideal characteristics of both a qualitative and a quantitative detector.

## 2.2 High Performance Liquid Chromatography

There are a number of specialist texts in which high performance liquid chromatography (HPLC) is described in varying amounts of detail (Lindsay [2]; Robards *et al.* [3]; Meyer [4]). It is not, therefore, the intention of this author to provide a comprehensive description of the technique but merely to discuss those aspects which are essential to the successful application of the LC–MS combination.

A block diagram of an HPLC system, illustrating its major components, is shown in Figure 2.1. These components are discussed in detail below.

### 2.2.1 Pump

The pump must provide stable flow rates from between 10 $\mu l\,min^{-1}$ and 2 $ml\,min^{-1}$ with the LC–MS requirement dependent upon the interface being used and the diameter of the HPLC column. For example, the electrospray interface, when used with a microbore HPLC column, operates at the bottom end of this range, while with a conventional 4.6 mm column such an interface usually operates towards the top end of the range, as does the atmospheric-pressure chemical ionization (APCI) interface. The flow rate requirements of the different interfaces are discussed in the appropriate section of Chapter 4.

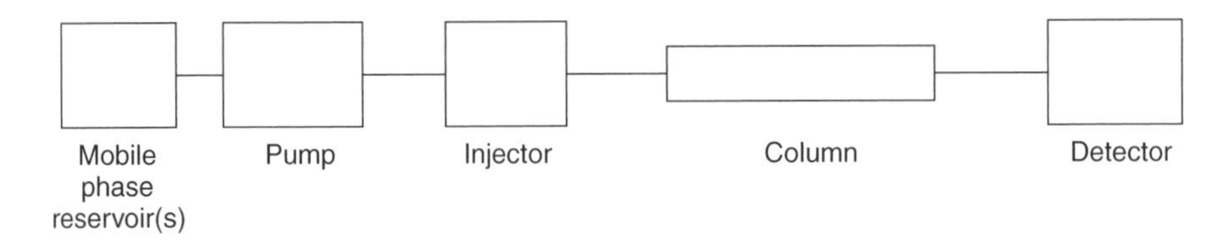


**Figure 2.1** Block diagram of a typical HPLC system.

A number of different types of pump are available and these are described elsewhere [2, 3], but the most popular pump used today is the *reciprocating* pump.

From a mass spectrometry perspective, the pump must be pulse free, i.e. it must deliver the mobile phase at a constant flow rate. Pulsing of the flow causes the total-ion-current (TIC) trace (see Chapter 3) – the primary piece of information used for spectral analysis – to show increases in signal intensity when analytes are not being eluted and this makes interpretation more difficult.

### 2.2.2 Sample Introduction (Injector)

In contrast to gas chromatography, in which a number of different types of injector are available and the selection of which is often crucial to the success (or otherwise) of the analysis, a single type of injector is used almost exclusively in HPLC.

The *loop injector* (sometimes known as the valve injector) is, as mentioned previously, merely a convenient way of introducing a liquid sample into a flowing liquid stream and consists of a loop of a nominal volume into which sample is introduced by using a conventional syringe. While the loop is being filled, mobile phase is pumped, at the desired flow rate, through the valve to the column to keep the column in equilibrium with the mobile phase and maintain chromatographic performance. When 'injection' is required, a rotating switch is moved and the flow is diverted through the loop, thus flushing its contents onto the top of the column.

From a quantitative perspective, the way in which the injector functions is crucial to the precision and accuracy which may be obtained and therefore these two parameters are of paramount importance.

*Quantitative precision* will be dependent upon, among other things, the extent to which the loop may be filled repeatably. It is usual to fill the loop completely by having a greater volume in the conventional syringe than the loop capacity (excess goes to waste) and it is important to ensure, as much as is possible, that air bubbles are not introduced in place of the sample. To obtain the best precision and *accuracy* during quantitative measurements, an internal standard should be used (this will be discussed further in Section 2.5 below), and if insufficient sample is available to allow complete filling of the loop, i.e. it is only partially filled, an internal standard **must** be used if meaningful quantitative results are to be obtained.

Loops are not calibrated accurately and a loop of nominally 20 μl is unlikely to have this exact volume. This will not affect either the precision of measurement and, as long as the same loop is used for obtaining the quantitative calibration and for determining the 'unknowns', the accuracy of measurement.

From a mass spectrometry perspective, the injector is of little concern other than the fact that any bubbles introduced into the injector may interrupt the liquid flow, so resulting in an unstable response from the mass spectrometer.

### 2.2.3 Mobile Phase

Unlike gas chromatography, in which the mobile phase, i.e. the carrier gas, plays no part in the separation mechanism, in HPLC it is the relative interaction of an analyte with both the mobile and stationary phases that determines its retention characteristics. Hence, it is the varying degrees of interaction of different analytes with the mobile and stationary phases which determines whether or not they will be separated by a particular HPLC system.

A number of different retention mechanisms operate in HPLC and interested readers may find further details elsewhere [2–4]. It is sufficient to say here that the interaction may be considered in terms of the relative polarities of the species involved. As indicated in Section 2.1 above, there are two extremes of interaction, neither of which is desirable if separation is to be achieved.

HPLC requires a mobile phase in which the analytes are soluble. The majority of HPLC separations which are carried out utilize reversed-phase chromatography, i.e. the mobile phase is more polar then the stationary phase. In these systems, the more polar analytes elute more rapidly than the less polar ones.

It is not always possible to achieve an adequate separation by using a mobile phase containing a single solvent and often mixtures of solvents are used. A wide range of mobile phases are therefore available and yet, despite this, a particular problem exists when the mixture under investigation contains analytes of widely differing polarities. A mobile phase that gives adequate separation of highly polar analytes will lead to excessively long retention times for non-polar analytes, and vice versa. Under these circumstances, separation is often achieved only by varying the composition of the mobile phase in a controlled way, during the analysis.

A separation involving a mobile phase of constant composition (irrespective of the number of components it contains) is termed *isocratic elution*, while that in which the composition of the mobile phase is changed is termed *gradient elution*. In the latter, a mobile phase is chosen which provides adequate separation of the early eluting analytes and a solvent which is known to elute the longer-retained compounds is added over a period of time. The rate at which the composition is changed may be determined by 'trial and error', or more formal optimization techniques may be used [5–7].

Buffers are used in HPLC to control the degree of ionization of the analyte and thus the tailing of responses and the reproducibility of retention. A range of buffers is available but those most widely used are inorganic, and thus involatile, materials, such as potassium or sodium phosphate.

One of the functions of an LC–MS interface is to remove the mobile phase and this results in buffer molecules being deposited in the interface and/or the source of the mass spectrometer with a consequent reduction in detector performance. Methods involving the use of volatile buffers, such as ammonium acetate, are therefore preferred.

The effect of the mobile-phase composition on the operation of the different interfaces is an important consideration which will be discussed in the appropriate chapter of this book but mobile-phase parameters which affect the operation of the interface include its boiling point, surface tension and conductivity. The importance of degassing solvents to prevent the formation of bubbles within the LC–MS interface must be stressed.

Some LC–MS interfaces have been designed such that mobile phase is not pumped directly into the source of the mass spectrometer, thus minimizing contamination and increasing the time over which the interface operates at optimum performance. One such is the ‘Z-spray’ interface from Micromass, with a comparison of the spray trajectories of an in-line and a Z-spray interface being shown in Figure 2.2. In the Z-spray interface, the HPLC mobile phase is sprayed across (orthogonal to) a sampling cone to which is applied a voltage that attracts appropriately charged ions with a velocity which causes them to pass through this cone into the mass spectrometer. Solvent and buffer molecules pass by this arrangement and are pumped directly to waste, thus reducing contamination and prolonging the performance of the system.

The effect of the quality of the mobile phase on the operation of the detector being employed is of importance whatever that detector may be.

The mobile phase is pumped through the column at a flow rate of, typically, 1 ml $min^{-1}$. If we assume an impurity is present at a level of 0.000 001%, this is equivalent to such a compound being **continually** introduced into the mass spectrometer at a rate of ca. 1 ng $s^{-1}$.

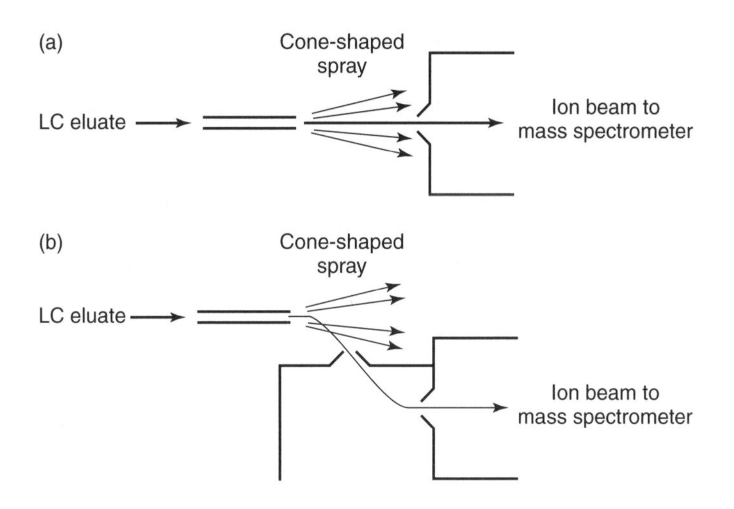


**Figure 2.2** Schematics of (a) in-line and (b) Z-spray electrospray interfaces. From applications literature published by Micromass UK Ltd, Manchester, UK, and reproduced with permission.

A full-scan mass spectrum can easily be obtained from this amount of material and it should be clear, therefore, that even high-purity (and usually expensive!) solvents can give rise to a significant mass spectral background, hence rendering the interpretation of both qualitative and quantitative data difficult.

Fortunately, this background is often less of a problem than might be anticipated from the above. The majority of ionization techniques employed in LC–MS are 'soft' ionization techniques which provide primarily molecular ions that occur at relatively high values of mass-to-charge ratio ($m/z$), rather than fragment ions which occur at relatively low $m/z$ values. In the majority of cases, the molecular weight of the analyte is higher than those of the solvent impurities and the effect of these may therefore be minimized.

The primary piece of LC–MS data considered by the analyst is the total-ion-current (TIC) trace which shows the sum of the intensity of each of the ions observed in each of the mass spectra that have been acquired during the chromatographic separation. As with other detectors, a 'peak' signifies the elution of a component from the column followed by its ionization. If solvent impurities are continually being ionized, a high background TIC is observed and the elution of an analyte may cause a minimal increase in this, i.e. 'peaks' may not be readily apparent. This situation may be improved by either (a) modifying the scan range of the mass spectrometer to exclude the ions from the background, e.g. scan only from $m/z$ 150 upwards, or (b) acquiring data over the complete $m/z$ range but then use computer manipulation of these data to construct a TIC trace from only those ions that do not arise from the background.

The advantage of the latter approach is that all data are stored during acquisition and if any ions of analytical significance are subsequently found below $m/z$ 150, they may be examined. If the mass spectrometer has only been scanned above $m/z$ 150, then this is not possible.

This methodology will be discussed further in Chapter 3 but is illustrated here in Figure 2.3. In this, Figure 2.3(a) shows the TIC trace from an LC–MS analysis in which data over the $m/z$ range from 35 to 400 have been acquired. A number of responses may be observed but the trace is dominated by a constant background amounting to around 70% of the maximum TIC value. Figure 2.3(b) shows the TIC from the same analysis, constructed by using the intensity of ions with $m/z$ only in the range of 200 to 400. In this case, the constant background amounts to less than 5% of the maximum of the TIC value and the presence of components may be much more readily observed.

### *2.2.4 Stationary Phase*

As has previously been stated, the majority of HPLC analyses which are carried out employ reversed-phase systems.

The most widely used columns contain a chemically modified silica stationary phase, with the chemical modification determining the polarity of the column. A

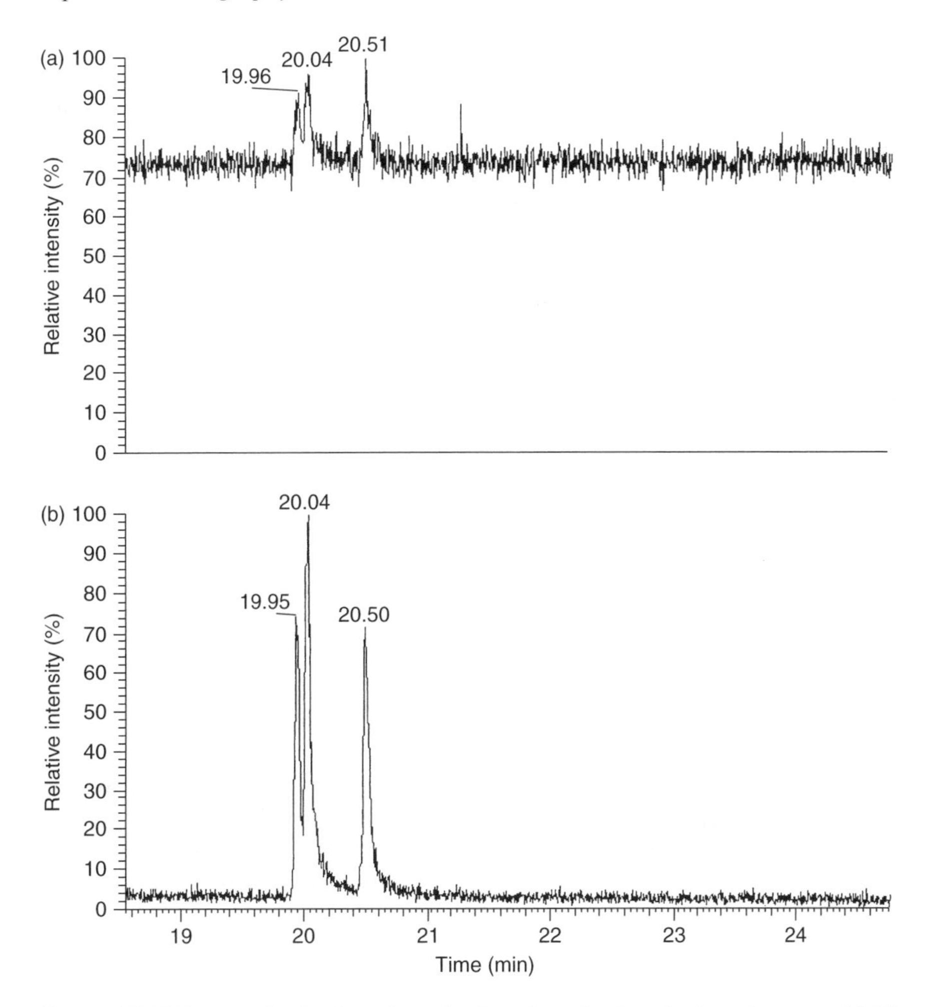


**Figure 2.3** TIC traces, having been brought about by using ions in the $m/z$ ranges (a) 35 to 400, and (b) 200 to 400, showing the improvement in signal-to-noise ratio obtained by excluding background ions.

very popular stationary phase is one in which a $C_{18}$ alkyl group is bonded to the silica surface.

In contrast to GC, in which, particularly at high temperatures, the stationary phase may give rise to a continuous background at the detector, this is not normally observed in HPLC unless the pH of the mobile phase is such that degradation of the stationary phase occurs. Under these circumstances, both an increased background and a reduction in chromatographic performance may be observed.

### *2.2.5 Detectors*

The choice of detector is often crucial to the success of a particular HPLC method. A number are in routine use, including the UV, fluorescence, electrochemical, conductivity and refractive index detectors, and each has particular advantages and disadvantages, details of which can be found elsewhere [2–4].

A more general discussion of their attributes will, hopefully, provide an insight into some of the ways in which the mass spectrometer may be used to advantage as a detector.

Detectors may be classified in a number of ways, including their use as the following:

- solute- or solvent-property detectors
- selective or general (universal) detectors
- mass- or concentration-sensitive detectors

#### *2.2.5.1 Solute- or Solvent-Property Detectors*

This classification is concerned with whether the detector monitors a property of the solute (analyte), e.g. the UV detector, or a change in some property of the solvent (mobile phase) caused by the presence of an analyte, e.g. the refractive index detector.

#### *2.2.5.2 Selective or General Detectors*

This classification is concerned with whether the detector responds to a specific feature of the analyte of interest or whether it will respond to a large number of analytes, irrespective of their structural properties. In terms of the previous classification, it may be considered that solute detectors are also usually *selective* detectors, while solvent detectors are *general* detectors.

The most widely used HPLC detector methodology is, arguably, UV absorption, and this has capabilities as both a specific or general detector, depending upon the way it is used.

If the wavelength of maximum absorption of the analyte ($\lambda_{max}$) is known, it can be monitored and the detector may be considered to be selective for that analyte(s). Since UV absorptions are, however, generally broad, this form of detection is rarely sufficiently selective. If a diode-array instrument is available, more than one wavelength may be monitored and the ratio of absorbances measured. Agreement of the ratio measured from the 'unknown' with that measured in a reference sample provides greater confidence that the analyte of interest is being measured, although it still does not provide absolute certainty.

Many organic molecules absorb UV radiation, to some extent, at 254 nm and if this wavelength is used it may be considered to be a general detection system.

It must be remembered, however, that not all compounds absorb UV radiation. In these circumstances, the use of indirect UV detection, in which a UV-active compound is added to the mobile phase, may be employed. This gives a constant (hopefully) background signal which is reduced when a compound that does not absorb UV radiation elutes from the HPLC column. Care must be taken if the mass spectrometer is used in series with indirect UV detection that the UV-active compound added to the mobile phase does not produce an unacceptably high background signal which hinders interpretation of either the TIC trace or the resulting mass spectra.

A widely used general detector is the refractive index detector which monitors changes in the refractive index of the mobile phase as an analyte elutes from the column. If gradient elution is being used, the refractive index of the mobile phase also changes as its composition changes, thus giving a continually varying detector baseline. The determination of both the position and intensity of a low-intensity analytical signal on a varying baseline is less precise and less accurate than the same measurement on a constant baseline with zero background signal.

It is usually recognized that general detectors are less sensitive than specific detectors, have a lower dynamic range (see below) and do not give the best results when gradient elution is used.

Like the UV detector, the mass spectrometer may be employed as either a general detector, when full-scan mass spectra are acquired, or as a specific detector, when selected-ion monitoring (see Section 3.5.2.1) or tandem mass spectrometry (MS–MS) (see Section 3.4.2) are being used.

#### *2.2.5.3 Mass- or Concentration-Sensitive Detectors*

The final classification concerns whether the intensity of detector response is proportional to the concentration of the solute or the absolute amount of solute reaching it. This classification is particularly important for *quantitative* applications. If the mobile phase flow rate is increased, the concentration of analyte reaching the detector remains the same, but the amount of analyte increases. Under these circumstances, the signal intensity from a concentration-sensitive detector will remain constant, although the peak width will decrease, i.e. the area of the response will decrease. A change in flow rate will also reduce the width of the response from a mass-sensitive detector, while, in contrast to a concentration-sensitive detector, the signal intensity will increase as the absolute amount of analyte reaching the detector has increased. Since the overall response increases, this may be used to improve the quality of the signal obtained.

Under many experimental conditions, the mass spectrometer functions as a mass-sensitive detector, while in others, with LC–MS using electrospray ionization being a good example, it can behave as a concentration-sensitive detector. The reasons for this behaviour are beyond the scope of this present book (interested readers should consult the text by Cole [8]) but reinforce the need to ensure that adequate calibration and standardization procedures are incorporated into any quantitative methodology to ensure the validity of any results obtained.

An advantage of the mass spectrometer as a detector is that it may allow differentiation of compounds with similar retention characteristics or may allow the identification and/or quantitative determination of components that are only partially resolved chromatographically, or even those that are totally unresolved. This may reduce the time required for method development and is discussed in more detail in Chapter 3.

**SAQ 2.1**

You require to develop an HPLC method for the determination of a high-molecular-weight aliphatic alcohol that has no UV absorption. Unfortunately, you only have a UV detector available. How would you attempt the analysis?

## 2.3 Chromatographic Properties

In carrying out a chromatographic separation, an analyst is concerned with whether the components of a mixture can be separated sufficiently for the analytes of interest, and this is not always all of them, to be identified and/or for the amounts present to be determined. Our ability to carry out these tasks successfully will depend upon the 'performance' of the chromatographic system as a whole.

The *performance* may be described in terms of a number of theoretical parameters, although the 'performance' required for a particular analysis will depend upon the separation that is required. This, in turn, depends upon the similarity in the behaviour in the chromatographic system of the analyte(s) of interest to each other and to other compounds present in the mixture.

The time taken for an analyte to elute from a chromatographic column with a particular mobile phase is termed its *retention time*, $t_{an}$. Since this will vary with column length and mobile phase flow rate, it is more useful to use the *capacity factor*, $k'$. This relates the retention time of an analyte to the time taken by an unretained compound, i.e. one which passes through the column without interacting with the stationary phase, to elute from the column under identical conditions ($t_0$). This is represented mathematically by the following equation:

$$k' = \frac{t_{an} - t_0}{t_0} \tag{2.1}$$

To give adequate resolution in a reasonable analysis time, $k'$ values of between 1 and 10 are desirable.

The separation of two components, e.g. A and B, is termed the *selectivity* or *separation factor* ($\alpha$) and is the ratio of their capacity factors (by convention, $t_B > t_A$ and $\alpha \geq 1$), as shown by the following equation:

$$\alpha = \frac{k'_B}{k'_A} = \frac{t_B - t_0}{t_A - t_0} \tag{2.2}$$

The separation of two components is of particular importance when one is being determined in the presence of the other and this is defined as the resolution ($R$), given as follows:

$$R = \frac{t_B - t_A}{0.5(w_A + w_B)} \tag{2.3}$$

where $w_A$ and $w_B$ are the peak widths of the detector responses from the two components (it does not matter which units these are measured in, e.g. time, volume or distance, on the chart recorder, as long as the same units are used for both measurements). These parameters are represented graphically in Figure 2.4.

The column performance (efficiency) is measured either in terms of the *plate height* ($H$), the efficiency of the column per unit length, or the *plate number* ($N$), i.e. the number of plates for the column. This number depends upon the column length ($L$), whereas the plate height does not. The mathematical relationships between the number of plates, the retention time of the analyte and the width of the response is shown in the following equations:

$$N = 16\left(\frac{t_A}{w_A}\right)^2 = 5.54\left(\frac{t_A}{w_{0.5}}\right)^2 \tag{2.4}$$

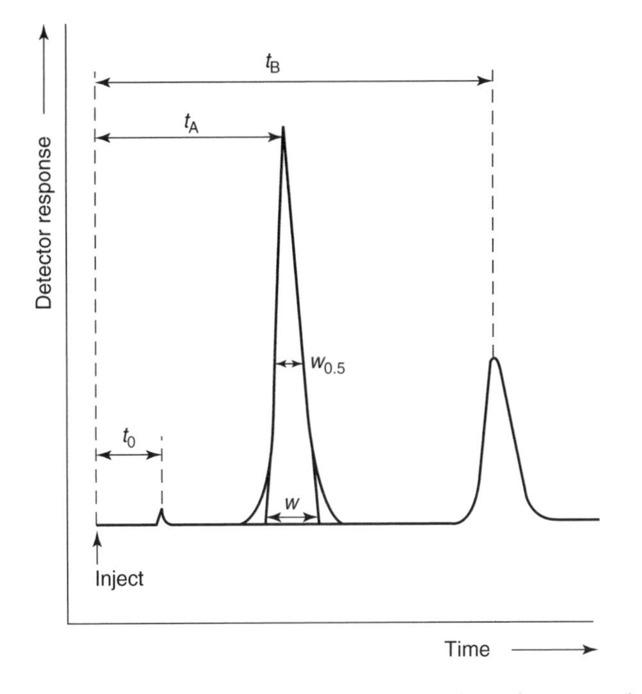


**Figure 2.4** Illustration of HPLC parameters: $t_0$, retention time of a non-retained component; $t_A$ and $t_B$, retention times of analytes A and B; $w$, width of peak at base; $w_{0.5}$, width at half-height.

$$H = \frac{L}{N} \tag{2.5}$$

The expression for resolution may now be written as follows [9]:

$$R = 0.25\left(\frac{\alpha - 1}{\alpha}\right)\left(\frac{k'_{\mathrm{B}}}{1 + k'_{\mathrm{B}}}\right)\sqrt{N} \tag{2.6}$$

where $k'_{\mathrm{B}}$ is the capacity factor of the second of the two components and $N$ is the number of theoretical plates measured for that component. The effect that each of the three terms in equation (2.6) has on the resolution that may be achieved is discussed elsewhere [9].

**DQ 2.2**

Why is liquid chromatography a less efficient separation technique, as measured by the number of theoretical plates per column, than gas chromatography?

*Answer*

*In all forms of column chromatography, analytes are deposited as a narrow band on the top of the separation column. As they move through the column, a number of mechanisms cause this band to broaden and these reduce the efficiency of the separation being carried out. The Van Deemter equation may be used to assess efficiency and this contains three terms which account for the major causes of peak broadening. These relate to eddy diffusion (the A-term), longitudinal diffusion (the B-term) and the resistance to mass transfer in both the mobile and stationary phases (the C-term). The relative importance of these three terms is different for each of the chromatographic techniques, e.g. the A-term is of importance in HPLC but not in capillary GC, while the C-term is of importance in both techniques. A more detailed consideration of the relative importance of each of these mechanisms can be obtained from the chromatography texts indicated at the end of this chapter but in general terms band broadening is greater in HPLC than in capillary GC, and thus the efficiency is reduced to a greater extent.*

How then does the performance of the chromatographic system affect the quality of the analytical information that may be obtained?

In order to answer this question, we should not consider the chromatographic resolution in isolation but in conjunction with the selectivity of the detector. If the detector is not selective, i.e. we cannot isolate the signal resulting from the analyte from those representing the other compounds present, we must rely on the chromatographic resolution to provide a signal which is measurable with sufficient precision and accuracy. If, however, the detector has sufficient selectivity for the response from an analyte not to be affected by the presence of other compounds, the chromatographic separation, or more importantly, the lack of chromatographic separation, will not affect the ability to determine the analyte precisely and accurately.

The great advantage of the mass spectrometer is its ability to use mass, more accurately the mass-to-charge ratio, as a discriminating feature. In contrast to, for example, the UV detector, which gives rise to broad signals with little selectivity, the ions in the mass spectrum of a particular analyte are often characteristic of that analyte. Under these conditions, discrete signals, which may be measured accurately and precisely, may be obtained from each analyte when they are only partially resolved or even completely unresolved from the other compounds present.

## 2.4 Identification Using High Performance Liquid Chromatography

As in other forms of chromatography, the identification of analytes is effected by the comparison of the retention characteristic of an unknown with those of reference materials determined under identical experimental conditions.

Often, the retention time is used but, as discussed above in Section 2.3, this absolute parameter changes with column length and flow rate and this precludes the use of reference data obtained in other laboratories. To make use of these reference data, the capacity factor ($k'$), which removes such variability, must be employed.

It should be clear from a simple consideration of the large number of organic compounds amenable to analysis by HPLC, the peak widths obtained and from the desirability to obtain $k'$ values of between 1 and 10, that it is likely that a number of compounds will have closely similar $k'$ values. Identification using this parameter alone will not therefore be possible.

> **Note** – *Unequivocal identification of a total unknown using a single HPLC retention characteristic (or indeed a single retention characteristic from any form of chromatography) should* ***not*** *be attempted.*

A general approach to the problem of identification, should more definitive detectors not be available, is to change the chromatographic 'system', which in the case of HPLC is usually the mobile phase, and redetermine the retention parameter. The change obtained is often more characteristic of a single analyte than is the capacity factor with either of the mobile phases.

The $k'$ values of the barbiturates, secbutobarbitone and vinbarbitone, determined by using an octadecyl silyl (ODS) stationary phase and mobile phases of

**Table 2.1** HPLC capacity factors for secbutobarbitone and vinbarbitone with an octadecyl silyl stationary phase and mobile phases of methanol/0.1 M sodium dihydrogen phosphate (40:60) at (a) pH 3.5, and (b) pH 8.5. From Moffat, A. C. (Ed.), *Clarke's Isolation and Identification of Drugs*, 2nd Edn, The Pharmaceutical Press, London, 1986. Reproduced by permission of The Royal Pharmaceutical Society

| Compound | $k'$(a) | $k'$(b) |
|---|---|---|
| Secbutobarbitone | 4.89 | 3.32 |
| Vinbarbitone | 4.83 | 2.32 |

methanol/0.1 M sodium dihydrogen phosphate (40:60) at (a) pH 3.5, and (b) pH 8.5, are shown in Table 2.1 [10].

**SAQ 2.2**

Calculate the resolution of secbutobarbitone and vinbarbitone on systems (a) and (b) (see Table 2.1) when using a column of 3500 theoretical plates.

In mobile phase (a), the two compounds have virtually identical $k'$ values and if a single response were to be measured with a $k'$ value of 4.86 it would not be possible to say, unequivocally, which, if either, of these analytes was present. In mobile phase (b), the $k'$ of vinbarbitone has changed to a significantly greater extent than that of secbutobarbitone. This change would allow these two compounds to be differentiated, although an unequivocal identification on these limited data would still be dangerous.

If the identity of the analyte is genuinely unknown, a further problem is encountered. In contrast to GC, there are no HPLC systems, combinations of mobile and stationary phases, that are routinely used for general analyses. Therefore, there are no large collections of $k'$ values that can be used. For this reason, retention characteristics are often used for identification after the number of possible compounds to be considered has been greatly reduced in some way, e.g. the class of compound involved has been determined by colour tests or UV spectroscopy.

A more definitive identification may be obtained by combining retention characteristics with more specific information from an appropriate detector. Arguably, the most 'information-rich' HPLC detectors for the general identification problem are the diode-array UV detector, which allows a complete UV spectrum of an analyte to be obtained as it elutes from a column, and the mass spectrometer. The UV spectrum often allows the class of compound to be determined but the spectra of analogues are usually very similar and the $k'$ value is essential for definitive identification. The mass spectrometer provides both molecular weight and structural information for, when equipped with the right interface(s), a wider range of analytes than any other single detector.

## 2.5 Quantitation Using High Performance Liquid Chromatography

Quantitation in high performance liquid chromatography, as with other analytical techniques, involves the comparison of the intensity of response from an analyte ('peak' height or area) in the sample under investigation with the intensity of response from known amounts of the analyte in standards measured under identical experimental conditions.

**SAQ 2.3**

When making quantitative measurements, should peak height or area be used?

There are a number of properties of a detector that determine whether they may be used for a particular analysis, with the most important being (a) the noise obtained during the analysis, (b) its limit of detection, (c) its linear range, and (d) its dynamic range. The last three are directly associated with the analyte being determined.

*Noise* is defined as the change in detector response over a period of time in the absence of analyte.

An ideal detector response in the absence of an analyte is shown in Figure 2.5(a).

In practice, the absence of some form of noise on a detector trace is unusual, particularly when high-sensitivity detection is employed. There are two components of noise, namely the short-term random variation in signal intensity, the 'noise level', shown in Figure 2.5(b), and the 'drift', i.e. the increase or decrease in the average noise level over a period of time.

The short-term noise shown in Figure 2.5(b) arises primarily from the electronic components of the system and stray signals in the environment. Drift may also arise from electronic components of the system, particularly just after an instrument has been turned on and while it is stabilizing.

Another important form of noise is 'chemical noise'. This may be defined as signals from species other than the analyte present in the system or sample which cannot be resolved from that of the analyte. Chemical noise may originate from the actual chromatographic system being used or from discrete chromatographic components in the sample matrix. As the mobile phase composition changes during gradient elution, for example, any resulting noise will appear as

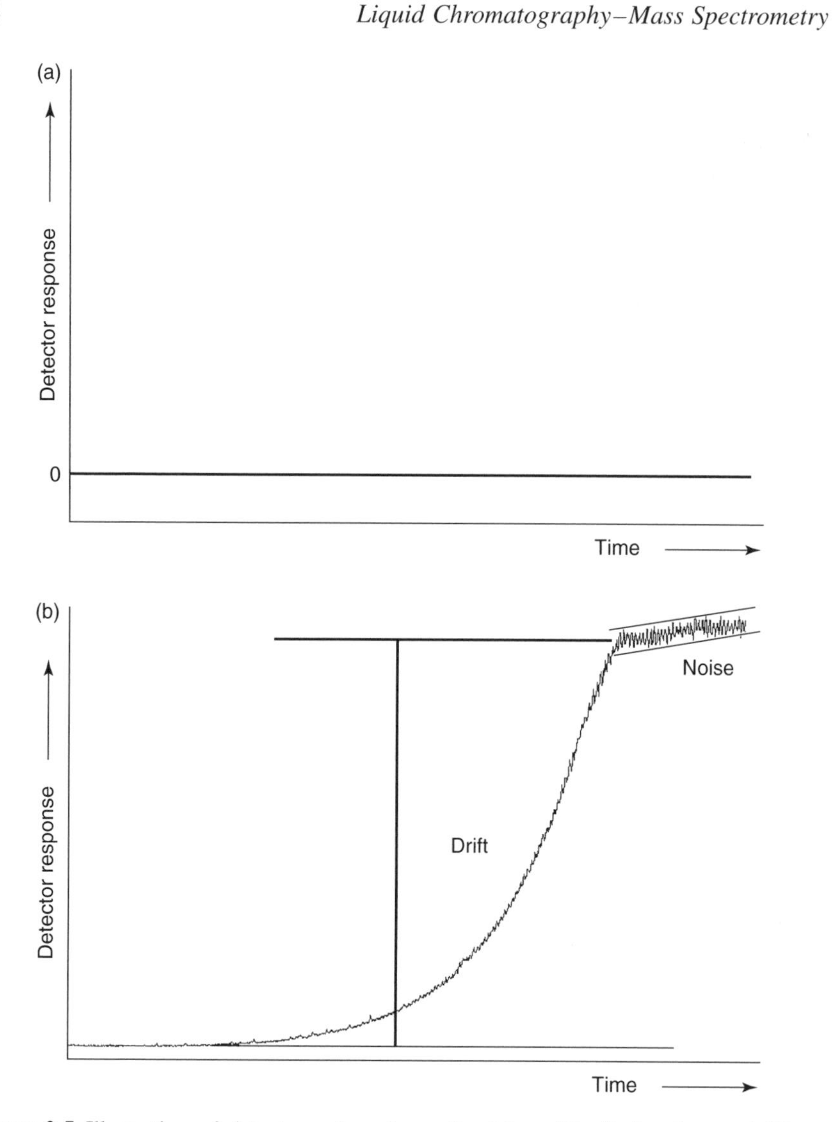


**Figure 2.5** Illustration of detector noise, shown for (a) an 'ideal' situation, and (b) an example of a 'real' situation.

a regular increase or decrease in signal (or drift). In the latter case, a separate chromatographic 'peak' is obtained.

The intensity of signal from the analyte compared to that from the noise is termed the signal-to-noise ratio (S/N). This is used by the analyst to determine, in the first instance, whether a detector signal can be said to be 'real', and therefore whether an analyte is present, and then to calculate the accuracy and precision with which that analyte can be quantified.

The *limit of detection* (LOD) (see Figure 2.6) is defined as the smallest quantity of an analyte that can be 'reliably' detected. This is a subjective definition and to introduce some objectivity it is considered to be that amount of analyte which produces a signal that exceeds the noise by a certain factor. The factor used, usually between 2 and 10 [11], depends upon the analysis being carried out. Higher values are used for quantitative measurements in which the analyst is concerned with the ability to determine the analyte accurately and precisely.

The **linear range** (see Figure 2.6) is defined as that range for which the analytical signal is directly proportional to the amount of analyte present.

When the linear range is exceeded, the introduction of more analyte continues to produce an increase in response but no longer is this directly proportional to the amount of analyte present. This is referred to as the **dynamic range** of the detector (see Figure 2.6). At the limit of the dynamic range, the detector is said to be *saturated* and the introduction of further analyte produces no further increase in response.

It is important for obtaining precise results that the signals from the samples to be determined should lie on the linear part of the calibration graph as elsewhere within the dynamic range a small change in detector response corresponds to a relatively large range of concentrations.

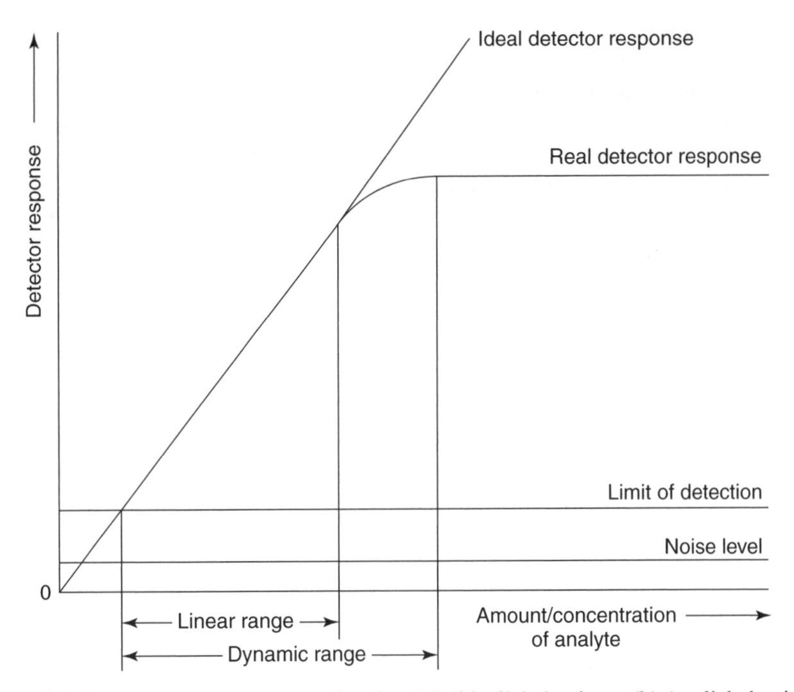


**Figure 2.6** Detector response curve showing (a) 'ideal' behaviour, (b) 'real' behaviour, (c) its linear range, (d) its dynamic range, (e) the noise level, and (f) the limit of detection at three times the noise level.

The precision and accuracy of a quantitative measurement are of great importance.

Precision is assessed by carrying out a determination a number of times, with the result being given in the form $1.48 \pm 0.01\ \mu g\,ml^{-1}$. The '±' value gives an indication of the precision of the measurement and without an indication of this the result is rendered meaningless.

**SAQ 2.4**

Prosecution of a company occurs if they are found to discharge more than $1.00\ mg\,ml^{-1}$ of a certain chemical into a local river. The analysis of river water at the outflow from the factory showed levels of pollutant to be (a) $1.48 \pm 1.00\ mg\,ml^{-1}$, and (b) $1.48 \pm 0.10\ mg\,ml^{-1}$. In each case, indicate whether you would recommend prosecution and justify your decision.

The accuracy of a method can only be determined if the true 'answer' is known and, of course, for the majority of analyses it is not. The accuracy of a method is determined during its validation procedure by the analysis of samples containing known amounts of analyte. In order to ensure that the method accuracy is maintained during routine use, samples containing known amounts of analyte are analysed among the unknowns as part of a quality control regime [12, 13].

To reiterate, a quantitative result is obtained by comparing the intensity of analytical signal obtained from the 'unknown' with those obtained from samples containing known amounts/concentrations of the analyte (standards).

There are three commonly encountered methods of employing these standards, namely the use of external standards, the use of standard additions and the use of internal standards.

The use of an **external standard** procedure is probably the simplest methodology that may be employed. In this situation, a number of samples containing known amounts of the analyte of interest are made up and analysed. The intensity of the analytical signal from these 'standards' is then plotted against the known concentration of analyte present and a calibration graph of the form shown above in Figure 2.6 is obtained. It is important that the range of concentrations covered by the standards includes the concentrations encountered in the 'unknowns' – interpolation of the result is required, rather than extrapolation. Care should also be taken to fit the correct form of 'curve' to the calibration data, i.e. to ensure that the data genuinely obey a linear relationship, not a gentle curve, before using linear regression [14] to define the relationship between signal intensity and concentration.

Although widely employed, the use of external standardization takes no account of matrix effects, i.e. the effect on the analytical signal caused by the interaction of analyte with the matrix in which it is found, or losses of analyte from the 'unknowns' during sampling, storage and work-up.

The use of **standard additions** addresses the influence of matrix effects. In this method, the standard is again the analyte itself. An analytical measurement is made on the 'unknown' and the signal intensity noted. A known amount of the analyte is then added to the 'unknown' and a second analytical measurement made. From the increase in analytical signal, a response factor, i.e. the signal per unit concentration, can be calculated. The concentration of analyte in the original sample may then be obtained by dividing the signal from the original sample by the response factor. A result obtained from a single determination of the response factor is liable to a greater imprecision than had it been obtained from multiple additions and it is more normal to add further known amounts of analyte and determine the analyte signal after each addition. A graph may then be drawn, as shown in Figure 2.7. The concentration of analyte in the original sample may then be obtained by extrapolation (using the equation of the linear regression straight line) of this graph to intercept the $x$-axis.

It should be noted that this method assumes that the matrix has the same effect on added analyte as it had on the analyte in the unknown, but this is not always the case.

Neither of the two methods described above take into account the possibility of loss of analyte between sampling and analysis. They may, therefore, provide a precise measurement but the result obtained may not give an accurate indication of the amount of analyte present in the original sample.

The use of an **internal standard** is designed to overcome this major source of inaccuracy and also to improve precision. An internal standard is a suitable compound added to the sample as early in the analytical procedure as is possible, ideally at the sampling stage. Analytical signals from both the analyte and internal standard are measured during each determination of both standards and 'unknowns', and it is the ratio of these two signal intensities that are used to generate the calibration graph and to determine the amount of analyte present in each of the 'unknowns'. In order for this methodology to have validity, losses of the analyte must be exactly and/or reproducibly mirrored by losses in the

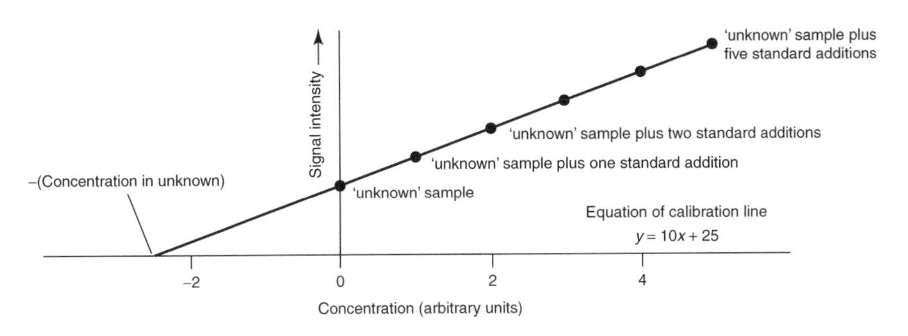


**Figure 2.7** A typical calibration curve for the method of standard additions.

internal standard and for this reason the choice of internal standard is crucial to its success.

An hypothetical set of analytical results is shown in Table 2.2, in which, for simplicity, both the analyte and internal standard give one unit of detector response per mg $ml^{-1}$ of concentration.

Let us consider the analysis of an 'unknown' containing 4 mg $ml^{-1}$ of analyte (A) but with a 50% loss during sampling, storage and work-up, i.e. the sample actually presented for analysis contains the equivalent of 2 mg $ml^{-1}$. An analytical signal intensity of 2 units will be obtained from this sample which, using the data in Table 2.2, corresponds to a concentration of 2 mg $ml^{-1}$ in the unknown (in this example, the precision of the measurements has deliberately been ignored).

We now consider the same 'unknown', i.e. containing 4 mg $ml^{-1}$ of analyte, but to which has been added, at the sampling stage, 3 mg $ml^{-1}$ of an internal standard (IS). If we assume the same 50% loss of both analyte and internal standard, the sample presented for analysis will contain the equivalent of 2 mg $ml^{-1}$ of the analyte (as before) but also the equivalent of 1.5 mg $ml^{-1}$ of internal standard. As before, an analytical signal of 2 units will be obtained from the analyte and, in this case, an additional signal of 1.5 units from the internal standard will be observed. The ratio of these two signals, A/IS, is 1.33 which, from Table 2.2, indicates a concentration of 4 mg $ml^{-1}$ in the 'unknown', i.e. the 'correct' answer.

How can we try to ensure that the analyte and internal standard are 'lost' equally?

Often, the choice of an appropriate internal standard is very difficult but the following criteria should be considered:

(i) The internal standard should not be present in the samples to be analysed or otherwise the ratios will be disturbed from those obtained from the standards and accuracy will be reduced.
(ii) The internal standard should be chemically and physically similar to the analyte so that the losses occurring during the work-up procedure are mirrored.

**Table 2.2** Hypothetical experimental results from the quantitative determination of a sample containing an analyte (A) and an internal standard (IS)

| [A] (mg $ml^{-1}$) | [IS] (mg $ml^{-1}$) | Signal intensity from A | Signal intensity from IS | Signal ratio, A/IS |
|---|---|---|---|---|
| 0 | 3 | 0 | 3 | 0 |
| 1 | 3 | 1 | 3 | 0.33 |
| 2 | 3 | 2 | 3 | 0.67 |
| 3 | 3 | 3 | 3 | 1.00 |
| 4 | 3 | 4 | 3 | 1.33 |
| 5 | 3 | 5 | 3 | 1.67 |
| 10 | 3 | 10 | 3 | 3.33 |

For example, if the internal standard is much more volatile than the analyte it is likely that more will be lost during storage, if it is much more polar it may be extracted either much more or less efficiently than the analyte during sample work-up, or if the internal standard is different chemically then derivatization procedures may be more or less efficient (it may be argued that as long as this differential behaviour is reproducible, it can be allowed for in the calculation procedures but significant extra work would be required to confirm such reproducibility).

(iii) The internal standard should elute in a similar time to the analyte in order that peak widths, and therefore the precision of measuring signal intensity, will be similar but the analytical signals must be capable of resolution from each other so that both may be measured with similar accuracy and precision.

(iv) The signal from the internal standard should be of similar intensity to that from the analyte so that they may be determined with similar precision, thus maximizing the precision of the overall result.

**DQ 2.3**

A soil sample was taken from a field, transported back to the laboratory by road and stored for three weeks prior to analysis. The analytical procedure consisted of drying the soil in an oven at 100°C for 24 h before the analyte was extracted using 200 $cm^3$ of dichloromethane. This extract was reduced in volume to 200 μl and a 20 μl aliquot then analysed by HPLC. A calibration was set up by measuring the response from a number of solutions containing known concentrations of the analyte. The result obtained from the 'unknown', after suitable mathematical manipulation, indicated the original soil sample contained $20 \pm 0.05$ mg $kg^{-1}$ of the analyte. Comment on the accuracy of this result.

*Answer*

*It is impossible to comment in this case upon the accuracy which relates to the closeness of the experimentally determined value to the true value. What has been determined is the amount of analyte present in the sample introduced into the chromatograph and the results from replicate determinations will give an indication of the precision of the methodology. At each stage of the procedure outlined above, there is the possibility of loss of sample and no attempt has been made to assess the magnitude of any of these losses.*

The advantage of using a mass spectrometer as the detector is associated with cases (ii) and (iii) above. In particular, because mass may be used as a discriminating feature, it is possible to use an isotopically labelled analyte as an internal standard. These have virtually identical physical and chemical properties to the unlabelled analogue, and are therefore likely to experience similar losses during any work-up procedures, and have, to all intents and purposes, identical chromatographic properties. This precludes the use of many detectors, e.g. the UV detector, which do not possess sufficient selectivity to differentiate between the analyte and its labelled analogue, but not the mass spectrometer which uses mass to provide the extra selectivity.

An isotope that is used extensively is deuterium (heavy hydrogen), often in the form of a deuteromethyl ($-CD_3$) group. The molecular weight of this compound is thus 3 Da higher than the unlabelled precursor and this is often sufficient to ensure that the ions in the molecular ion region of the unlabelled compound do not occur at the same $m/z$ ratios as those from the labelled molecule.

The two compounds will be ionized with equal efficiency and therefore provide similar signal intensities.

## 2.6 The Need for High Performance Liquid Chromatography–Mass Spectrometry

HPLC was a widely used analytical technique before the development of LC–MS and continues to be used in many laboratories with other forms of detector. What are the advantages of the mass spectrometer when used as a detector?

- With the LC–MS interfaces now available, a wide range of analytes, from low-molecular-weight drugs and metabolites (<1000 Da) to high-molecular-weight biopolymers (>100 000 Da), may be studied.
- The mass spectrometer provides the most definitive identification of all of the HPLC detectors. It allows the molecular weight of the analyte to be determined – this is the single most discriminating piece of information that may be obtained – which, together with the structural information that may be generated, often allows an unequivocal identification to be made.
- The high selectivity of the mass spectrometer often provides this identification capability on chromatographically unresolved or partially resolved components.
- This selectivity allows the use of isotopically labelled analytes as internal standards and this, coupled with high sensitivity, allows very accurate and precise quantitative determinations to be carried out.

These capabilities will be discussed in greater detail in Chapter 3.

## Summary

In this chapter, an HPLC system has been described in terms of its component parts and the effect of each of these on the use of a mass spectrometer as a detector has been discussed. The performance characteristics of a detector for qualitative and quantitative analysis have been discussed and the classification of detectors in terms of their response to varying amounts of analyte has been introduced. The ways in which standards may be used to improve the precision and accuracy of quantitative determinations have been outlined and various criteria for the choice of an appropriate internal standard put forward.

# References

1. *Recommendations on Nomenclature for Chromatography, Pure Appl Chem.*, **65**, 819–872 (1993).
2. Lindsay, S., *High Performance Liquid Chromatography*, ACOL Series, Wiley, Chichester, UK, 1992.
3. Robards, K., Haddad, P. R. and Jackson, P. E., *Principles and Practice of Modern Chromatographic Methods*, Academic Press, London, 1994.
4. Meyer, V. R., *Practical High Performance Liquid Chromatography*, Wiley, Chichester, UK, 1994.
5. Dolan, J. W., Snyder, L. R., Djordjevic, N. M., Hill, D. W. and Waeghe, T. J., *J. Chromatogr., A*, **857**, 1–20 (1999).
6. Dolan, J. W., Snyder, L. R., Djordjevic, N. M., Hill, D. W. and Waeghe, T. J., *J. Chromatogr., A*, **857**, 21–39 (1999).
7. Dolan, J. W., Snyder, L. R., Wolcott, R. G., Haber, P., Baczer, T., Kaliszan, R. and Sander, L. C., *J. Chromatogr., A*, **857**, 41–68 (1999).
8. Cole R. B. (Ed.), *Electrospray Ionization Mass Spectrometry – Fundamentals, Instrumentation and Applications*, Wiley, New York, 1997.
9. Sewell, P. A. and Clarke, B., *Chromatographic Separations*, ACOL Series, Wiley, Chichester, UK, 1987, p. 123.
10. Moffat, A. C. (Ed.), *Clarke's Isolation and Identification of Drugs*, 2nd Edn, The Pharmaceutical Press, London, 1986.
11. Harris, D. C., *Quantitative Chemical Analysis*, 4th Edn, W. H. Freeman, New York, 1995.
12. Prichard, E. (Co-ordinating Author), *Quality in the Analytical Laboratory*, ACOL Series, Wiley, Chichester, UK, 1995.
13. Currell, G., *Analytical Instrumentation: Performance Characteristics and Quality*, AnTS Series, Wiley, Chichester, UK, 2000.
14. Miller, J. N. and Miller, J. C., *Statistics and Chemometrics for Analytical Chemistry*, 4th Edn, Prentice Hall, Harlow, UK, 2000.

# Chapter 3
# Mass Spectrometry

**Learning Objectives**

- To understand those aspects of mass spectrometry which are essential to the application of LC–MS.
- To appreciate the types of analytical information that may be obtained from each of the different types of mass spectrometer likely to be encountered when carrying out LC–MS.
- To appreciate the ways in which mass spectral data may be processed to utilize fully the selectivity and sensitivity of the mass spectrometer as a detector for HPLC.

## 3.1 Introduction

A number of different types of HPLC detector have been discussed in the previous chapter. In comparison to these, a mass spectrometer is a relatively expensive detector and there need to be considerable advantages associated with its use to make the significant financial investment worthwhile. What are these advantages? In order to answer this question, we must first consider what it is we are trying to achieve when using chromatography.

Simplistically, chromatography can be regarded as the separation of the components of a mixture to allow the identification and/or quantitation of some or all of them. Identification is initially carried out on the basis of the chromatographic retention characteristic. This is not sufficient to allow unequivocal identification because of the possibility of more than one analyte having virtually identical retention characteristics. Further information is usually required from an auxiliary technique – often some form of spectroscopy.

The most widely used LC detector, and the one which, other than the mass spectrometer, gives the most insight into the identity of an analyte, is probably the UV detector, although a UV spectrum very rarely allows an unequivocal identification to be made. It may allow the class of compound to be identified and this, together with the retention characteristics of the analyte, can provide the analyst with a better indication of the identity of the analyte. In the vast majority of cases, however, identification with complete certainty cannot be achieved.

How then can this problem be addressed? From a chromatographic standpoint, the usual method is to change either the stationary phase or, more usually in the case of HPLC, the mobile phase, and look for a change in the retention characteristics. The change observed is usually more characteristic of a single analyte than is the actual retention on any individual chromatographic single system. Even this, however, does not always provide an unequivocal identification.

The advantage of the mass spectrometer is that in many cases it can provide that absolute identification. It provides not only structural information from the molecule under investigation but it may also provide the molecular weight of the analyte. This is, in most cases, the single most important and discriminating piece of information available to the analyst which, when determined, immediately reduces, dramatically, the number of possible structures for the analyte.

For example, the 1998 version of the NIST/EPA/NIH library [1], compiled in the USA under the auspices of the Standard Reference Data Program by the National Institute of Standards and Technology and commercially available to be used with many mass spectrometers, contains 107 886 reference spectra. For example, if we were able to determine the molecular weight of our analyte to be 283, then the number of entries with this molecular weight, and thus the number of potential identities for the molecule, is 260, an overall reduction of 99.75%. Similarly, if the molecular weight were determined to be 134, the corresponding reduction is 99.7%, while if the molecular weight were 328, it is 99.7%. These molecular weights have been chosen at random and the reduction obtained will be different for each value, although these figures clearly indicate the discriminating power of this parameter.

Another advantage of mass spectrometry is its sensitivity – a full-scan spectrum, and potentially an identification, can be obtained from picogram (pg) amounts of analyte. In addition, it may be used to provide quantitative information, usually to low levels, with high accuracy and precision.

There are a number of specific texts devoted to the various aspects of mass spectrometry [2–7]. In this chapter, a brief overview of the technique will be provided, with particular attention being paid to those aspects necessary for the application of LC–MS. In addition, a number of manufacturers provide educational material on their websites (for further details, see the *Bibliography* section at the end of this text).

A mass spectrometer can be considered to comprise four component parts, as follows:

- a method of sample introduction
- a method of ion production
- a method of ion separation
- facilities for ion detection and data manipulation

Before considering these in detail, it is necessary to revisit the inherent incompatibilities between mass spectrometry and liquid chromatography. These are, as discussed previously, that HPLC utilizes a liquid mobile phase, often containing significant amounts of water, flowing typically at 1 $ml\,min^{-1}$, while the mass spectrometer must be maintained under conditions of high vacuum, i.e. around $10^{-6}$ torr ($1.333\,22 \times 10^{-4}$ Pa).

**DQ 3.1**

Why is the mass spectrometer operated under conditions of high vacuum?

*Answer*

*Ions formed in the source of the mass spectrometer must reach the detector for them to be of any value. The average distance that an ion travels between collisions – the mean free path – at atmospheric pressure is around $10^{-8}$ m, and it is therefore unlikely that it will reach the detector under these conditions. Since the mean free path is inversely proportional to the pressure, reducing this to $10^{-6}$ torr will increase the mean free path to around 10 m, and thus allow ions to reach the detector of the mass spectrometer.*

The mass spectrometer inlet system for liquid chromatography, often termed the 'interface' between the two component techniques, must therefore remove as much of the unwanted mobile phase as possible while still passing the maximum amount of analyte into the mass spectrometer. This must be done in such a way that the mass spectrometer is still able to generate all of the analytical information of which it is capable.

**SAQ 3.1**

Calculate the volume of vapour produced when 1 g of (a) methanol, (b) acetonitrile, and (c) water, is vaporized at STP. Calculate the volume if this were to occur at $10^{-6}$ torr, the operating pressure of the mass spectrometer.

## 3.2 Ionization Methods

Ionization methods that may be utilized in LC–MS include electron ionization (EI), chemical ionization (CI), fast-atom bombardment (FAB), thermospray (TSP), electrospray (ESI) and atmospheric-pressure chemical ionization (APCI).

EI may be used with the moving-belt and particle-beam interfaces, CI with the moving-belt, particle-beam and direct-liquid-introduction interfaces, and FAB with the continuous-flow FAB interface. A brief description of these ionization methods will be provided here but for further details the book by Ashcroft [8] is recommended.

TSP, ESI and APCI effect ionization from solution and in these cases it is not possible to separate a description of the processes involved in the ionization of an analyte from a description of the interface. These ionization techniques will therefore be described in detail in Chapter 4.

### *3.2.1 Electron Ionization*

In electron ionization (EI), the analyte of interest, in the vapour phase, is bombarded with high-energy electrons (usually 70 eV) (1 ev = $1.602\,177\,33 \times 10^{-19}$ J). Analyte molecules absorb some of this energy (typically around 20 eV) and this causes a number of processes to occur. The simplest of these is where the analyte is ionized by the removal of a single electron. This yields a radical cation, termed the molecular ion ($M^{+\bullet}$), the $m/z$ of which corresponds to the molecular weight of the analyte. This process typically requires some 10 eV of energy and the ion so formed is therefore likely to possess around 10 eV of excess energy which may bring about its fragmentation – bond energies in organic molecules are typically around 4–5 eV. Two types of process may occur, i.e. simple scission of bonds and, when certain spatial arrangements of atoms occur within the molecule, fragmentation after rearrangement of the molecular structure. The latter process produces ions which would not immediately be expected from a simple examination of the structure of the analyte molecule involved. The presence of rearrangement ions within a mass spectrum is usually highly significant in terms of deriving the structure of the analyte concerned. The processes occurring in electron ionization are summarized in Figure 3.1.

Interpretation of an EI spectrum involves a consideration of the chemical significance of the ions observed in the mass spectrum and then using this information to derive an unequivocal structure. For a detailed consideration of the interpretation of EI mass spectra, the text by McLafferty and Turecek [7] is recommended.

### *3.2.2 Chemical Ionization*

One of the major limitations of EI is that the excess energy imparted to the analyte molecule during electron bombardment may bring about such rapid fragmentation that the molecular ion is not observed in the mass spectrum. Under

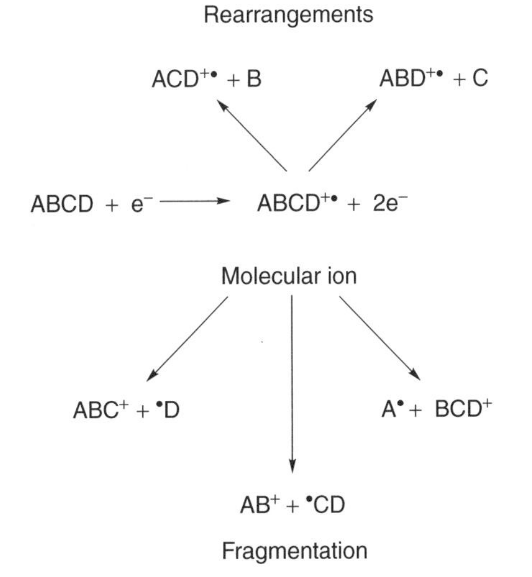


**Figure 3.1** Processes occurring in the production of a mass spectrum by electron ionization.

these circumstances, one of the most important pieces of analytical information is lost and the value of mass spectrometry is much reduced.

Chemical ionization (CI) is a technique that has been developed specifically to enhance the production of molecular species, i.e. to reduce the fragmentation associated with ionization. A number of such techniques exist and these are known collectively as 'soft ionization techniques'.

In this approach, analyte molecules, in the vapour phase (as with EI), are introduced into a mass spectrometer source containing a reagent gas. This mixture is then bombarded with electrons, as described above for EI, and ionization occurs. Since the reagent gas is present in vast excess when compared to the analyte molecules (typically $> 1000:1$), it is the reagent gas, almost exclusively, which is ionized. Ion–molecule reactions then take place between the reagent gas ions and the neutral analyte molecules in the high-pressure regime of the mass spectrometer source. The specific reactions which take place depend upon the thermodynamics of the processes that are possible but typically lead to the formation of adducts of reagent ions with analyte molecules in relatively low-energy processes which lead to little fragmentation.

The most commonly used reagent gases are methane, isobutane and ammonia, with the processes involved when methane is used being summarized in Figure 3.2. When interpreting spectra generated in this way it must be remembered that the $m/z$ of the ion observed in the molecular ion region does not give the molecular weight directly as it arises from the combination of the analyte with an adduct. The mass of that adduct, 1 in the case of methane and

$$CH_4 + e^- \longrightarrow CH_4^{+\bullet} + 2e^-$$

$$CH_4^{+\bullet} \longrightarrow CH_3^+ + H^\bullet$$

$$CH_4^{+\bullet} + CH_4 \longrightarrow CH_5^+ + CH_3^\bullet$$

$$CH_3^+ + CH_4 \longrightarrow C_2H_5^+ + H_2$$

$$CH_5^+ + M \longrightarrow MH^+ + CH_4$$

**Figure 3.2** Processes occurring in chemical ionization mass spectrometry using methane as the reagent gas.

isobutane, and 18 in the case of ammonia, must be subtracted from the $m/z$ value observed.

**SAQ 3.2**

The positive-ion CI spectrum, using ammonia as a reagent gas, of an analyte containing carbon, hydrogen, oxygen and two nitrogen atoms, has a molecular species at $m/z$ 222. What is the molecular weight of the compound involved?

CI is not the only ionization technique where this aspect of interpretation must be considered carefully; fast-atom bombardment, thermospray, electrospray and atmospheric-pressure chemical ionization, described below in Sections 3.2.3, 4.6, 4.7 and 4.8, respectively, all produce adducts in the molecular ion region of their spectra.

### *3.2.3 Fast-Atom Bombardment*

Both EI and CI require the analyte of interest to be in the vapour phase before ionization can take place and this precludes the study of a significant number of polar, involatile and thermally labile analytes.

Fast-atom bombardment (FAB) is one of a number of ionization techniques which utilize a matrix material, in which the analyte is dissolved, to transfer sufficient energy to the analyte to facilitate ionization. In FAB, the matrix material is a liquid, such as glycerol, and the energy for ionization is provided by a high-energy atom (usually xenon) or, more recently, an ion ($Cs^+$) beam. In conventional FAB, the solution of analyte in the matrix material is applied to the end of a probe which is placed in the source of the mass spectrometer where it is bombarded with the atom/ion beam.

When FAB is utilized for LC–MS, often known as continuous-flow FAB, a matrix material is added to the HPLC eluent, either pre- or post-column, and this mixture continuously flows to the tip of a probe inserted into the source of the mass spectrometer where it is bombarded by the atom beam (Figure 3.3).

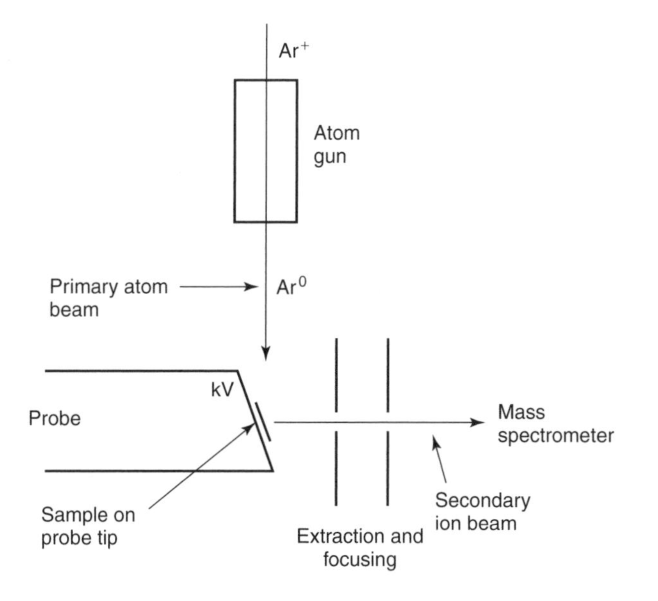


**Figure 3.3** Schematic of fast-atom bombardment ionization.

In contrast to conventional FAB where the analyte is dissolved in the matrix material, it has been found that FAB performance can be obtained when the mobile phase contains as little as 5% of the matrix material, thus reducing the chemical background associated with the technique. It should be noted that if the matrix material is added before the column it may have an effect on the separation achieved.

**SAQ 3.3**

Why might the addition of the FAB matrix to the HPLC mobile phase have an effect on the separation obtained?

### *3.2.4 Matrix-Assisted Laser Desorption Ionization*

Matrix-assisted laser desorption ionization (MALDI) is not yet a technique that has been used extensively for LC–MS applications. It is included here because it often provides analytical information complementary to that obtained from LC–MS with electrospray ionization, as illustrated later in Chapter 5.

MALDI shares many features with FAB in that it employs a matrix which transfers energy to an analyte molecule to facilitate the ionization of polar and thermally labile, high-molecular-weight molecules. In this case, however, the energy is provided by a pulsed laser at a wavelength which may be absorbed by a matrix material such as nicotinic or sinapinic acids. The ability to obtain mass spectra from biomolecules using FAB depends as much on the performance of the mass spectrometer, in particular the range of $m/z$ ratios that it can separate, as it does on the FAB system, the matrix and bombarding species used. The molecular weight limit for FAB is usually taken to be around 10 000 Da. MALDI extends the mass range of compounds that can be successfully ionized and detected to around 500 000 Da.

A major difference between MALDI and FAB is that a solid rather than a liquid matrix is used and a mixture of the analyte and matrix is dried on the laser target. For this reason, the effective combination of HPLC with MALDI is not as readily achieved although, since MALDI is largely free of the suppression effects experienced with FAB, it is able to provide useful analytical data directly from mixtures.

When a pulsed laser is used, ions are only produced for the duration of the pulse, i.e. they are not produced continuously and the mass spectrometer used must be capable of producing a mass spectrum from these 'pulses' of ions. As discussed below in Section 3.3.4, the time-of-flight (ToF) mass analyser is the most appropriate for this purpose and has the added advantage of being able to measure very high $m/z$ ratios. Indeed, the recent dramatic developments in the performance of the ToF mass analyser have largely been occasioned by the requirement to produce useful spectra from MALDI.

### *3.2.5 Negative Ionization*

For many years, electron ionization, then more usually known as electron impact, was the only ionization method used in analytical mass spectrometry and the spectra encountered showed exclusively the positively charged species produced during this process. Electron ionization also produces negatively charged ions although these are not usually of interest as they have almost no structural significance. Other ionization techniques, such as CI, FAB, thermospray, electrospray and APCI, however, can be made to yield negative ions which are of analytical utility.

These arise either by an analogous process to that described above for CI, i.e. the 'adduction' of a negatively charged species such as $Cl^-$, and the abstraction of a proton to generate an $(M - H)^-$ ion, or by electron attachment to generate an $M^-$ ion. The ions observed in the mass spectrum are dependent on the species generated by the reagent gas and the relative reactivities of these with each other and with the analyte molecule.

Negative-ion CI tends to be a very selective method of ionization which generates reduced levels of background, thus allowing sensitive analyses to be developed for appropriate analytes.

## 3.3 Ion Separation

A mass spectrum may be considered to be a plot of the number of ions of each $m/z$ ratio produced by an analyte upon ionization. Having produced the ions by using an appropriate ionization method, it is therefore necessary to separate the ions of different $m/z$ ratios, determine these $m/z$ values and then measure the relative intensities of each group of ions.

There are a number of different mass separation devices – analysers – used in mass spectrometry and each has its own advantages and disadvantages. Those most likely to be encountered by users of LC–MS are described briefly below, while more detailed descriptions may be found elsewhere [2–4]. One property that is important in defining the analytical capabilities of a mass analyser is the resolution which it may achieve.

*Resolution* is a term encountered in many areas of analytical science and refers to the ability to differentiate between closely related signals. In mass spectrometry, these 'signals' are the $m/z$ ratios of the ions, with the resolution being defined mathematically as follows:

$$R = m/\Delta m \tag{3.1}$$

where $R$ is the resolution, $m$ is the '$m/z$' to be measured and $\Delta m$ is the difference (in Da) between this and the ion from which it is to be separated.

Low-resolution devices are those that can separate and measure $m/z$ ratios to the nearest integer value and have a numerical resolution of up to around 1000. As such, they can separate (resolve), for example, ions at $m/z$ 28 and 29, i.e. they allow the analyst to differentiate between $CO^+$ and $CHO^+$, or $C_2H_4{}^+$ and $C_2H_5{}^+$. Using these types of instrument, we need only consider the masses of the isotopes as integers, e.g. $^{12}C = 12$ Da, $^{1}H = 1$ Da, $^{14}N = 14$ Da and $^{16}O = 16$ Da.

In fact, the masses of the isotopes are not exactly integer, e.g. $^{12}C$ has a mass of 12.0000 Da, $^{1}H$ 1.0078 Da, $^{14}N$ 14.0031 Da and $^{16}O$ 15.9949 Da [7], while the masses of $CO^+$, $C_2H_4{}^+$ and $N_2{}^+$, all nominally 28 Da, are actually 27.9949, 28.0312 and 28.0062 Da, respectively. If it were possible to resolve these ions and then determine their $m/z$ ratios accurately, the specific atomic composition of the ion could be assigned. A resolution of around 2500 is required to separate $CO^+$ from $N_2{}^+$.

### 3.3.1 The Quadrupole Mass Analyser

This, as the name suggests, consists of four rods arranged as shown in Figure 3.4. The opposite pairs are connected electrically and a voltage, consisting of both radiofrequency (RF) and direct-current (DC) components, is applied, with the RF components on the two pairs of rods being 180° out-of-phase. At a specific value of these voltages, ions of a particular $m/z$ follow a stable trajectory through the rods and reach the detector. A mass spectrum is therefore produced by varying the RF and DC voltages in a systematic way to bring ions of increasing or decreasing $m/z$ ratios to the detector.

The quadrupole analyser is an ideal detector for chromatography as it is capable of fast scanning and uses low voltages which make it tolerant of relatively high operating pressures, such as those encountered in LC–MS.

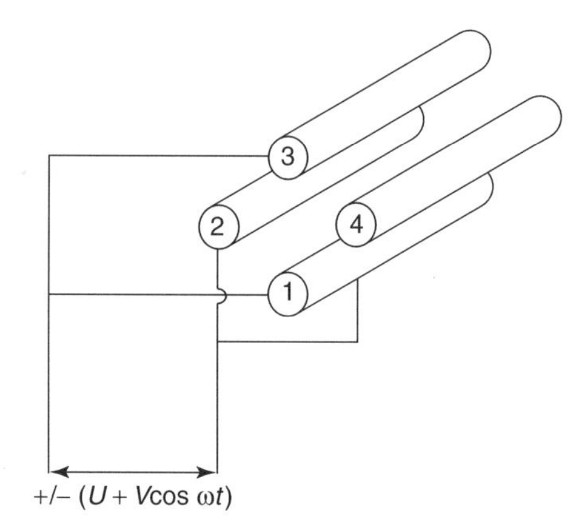


**Figure 3.4** Schematic of a quadrupole mass analyser.

The quadrupole is classified as a low-resolution device, i.e. it is capable of measuring the $m/z$ ratio of an ion to the nearest integer value, and thus is unable to provide the elemental composition of an ion.

### 3.3.2 The (Quadrupole) Ion-Trap Mass Analyser

The quadrupole ion-trap, usually referred to simply as the *ion-trap*, is a three-dimensional quadrupole. This type of analyser is shown schematically in Figure 3.5. It consists of a ring electrode with further electrodes, the end-cap electrodes, above and below this. In contrast to the quadrupole, described above, ions, after introduction into the ion-trap, follow a stable (but complex) trajectory, i.e. are trapped, until an RF voltage is applied to the ring electrode. Ions of a particular $m/z$ then become unstable and are directed toward the detector. By varying the RF voltage in a systematic way, a complete mass spectrum may be obtained.

Again, this is a low-resolution device, capable of fast scanning and tolerant of relatively high operating pressures.

### 3.3.3 The Double-Focusing and Tri-Sector Mass Analysers

The description of mass spectrometry in many texts begins with a consideration of the magnetic-sector mass analyser as this was the first one to be developed. Very few instruments utilizing only a magnetic sector are now used for LC–MS. The magnetic sector is, however, encountered in the double-focusing and tri-sector mass spectrometers – these are termed high-resolution instruments, which enable the $m/z$ of an ion to be determined with greater accuracy and thus provide an indication of the atomic composition of the ion.

One of the major reasons why ions with similar $m/z$ ratios cannot be resolved is that they are not all formed at exactly the same place in the source of the

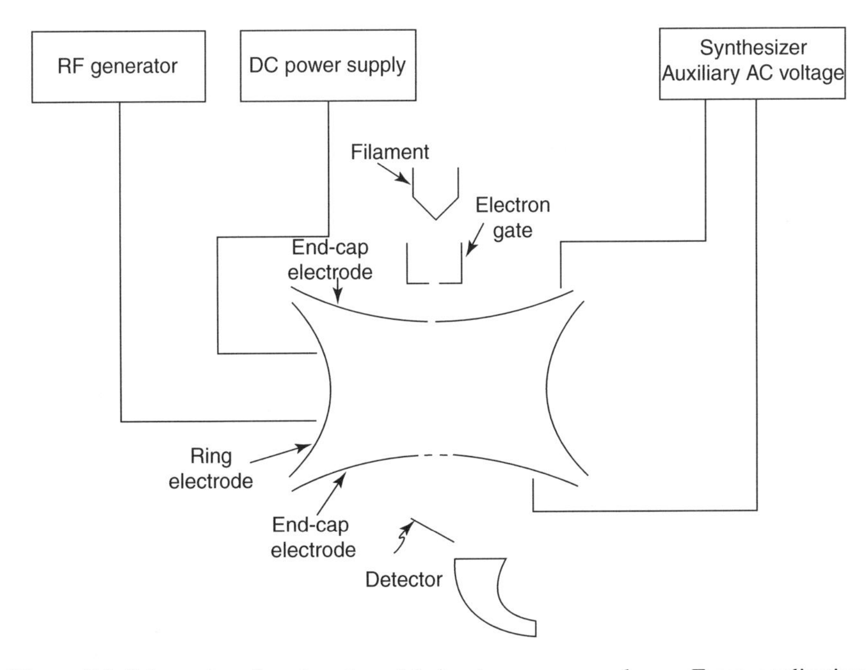


**Figure 3.5** Schematic of a (quadrupole) ion-trap mass analyser. From applications literature published by Thermofinnigan, Hemel Hempstead, UK, and reproduced with permission.

mass spectrometer and therefore ions enter the mass analyser with a small, but finite, range of energies. As they pass through the mass separation device, beams from ions with similar $m/z$ ratios overlap and cannot be completely separated by low-resolution devices such as the magnetic-sector or quadrupole systems.

The double-focusing mass spectrometer consists of both magnetic sector and electrostatic analysers (ESAs), the latter being a device which focuses ions with the same $m/z$ values but differing energies. The extent to which the beams of ions of closely similar $m/z$ ratios overlap is thus reduced so that in many cases they may be separated. This then allows their $m/z$ ratios to be determined with more accuracy and precision and the atomic composition of the ion to be determined.

Three configurations of electrostatic and magnetic analysers have been used commercially and all are capable of making accurate mass measurements. In the forward-geometry instrument, the electrostatic analyser precedes the magnetic analyser (Figure 3.6(a)), in the reverse-geometry instrument it is found after the magnet (Figure 3.6(b)), and in the tri-sector instrument an electrostatic analyser is found both before and after the magnet (Figure 3.6(c)). The configuration has no effect on the ability to make accurate mass measurements; however, the advantages and disadvantages of each of these are beyond the scope of this present text.

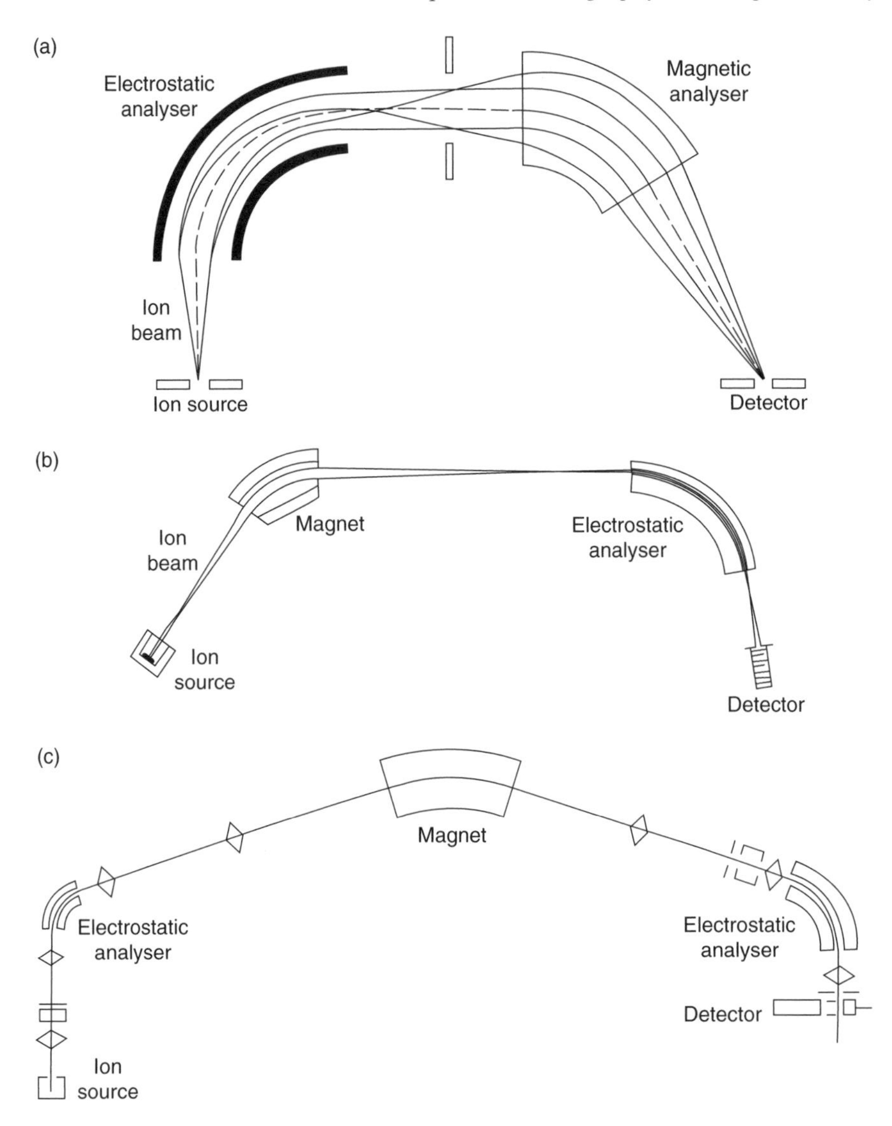


**Figure 3.6** Schematics of three configurations of mass spectrometer capable of accurate mass measurement: (a) forward-geometry; (b) reverse-geometry; (c) tri-sector. From applications literature published by Micromass UK Ltd, Manchester, UK, and reproduced with permission.

## 3.3.4 *The Time-of-Flight Mass Analyser*

In some respects, the time-of-flight (ToF) analyser is the simplest of the mass separation devices. This system relies on the fact that if all of the ions produced

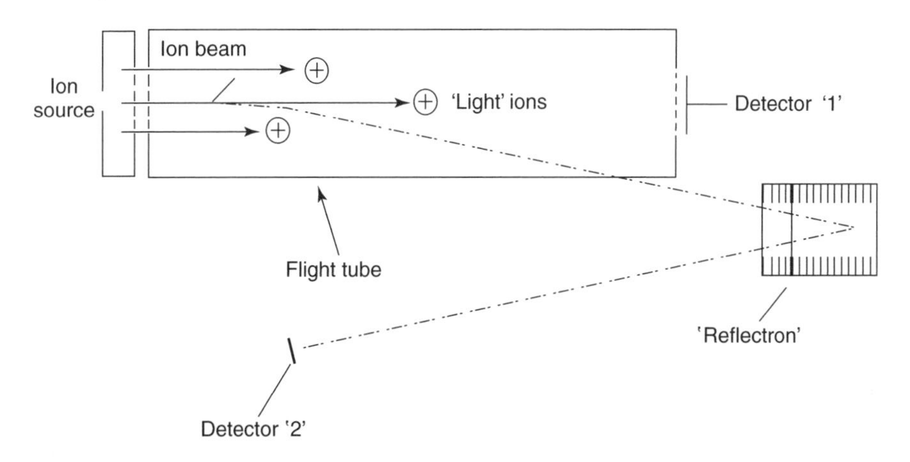


**Figure 3.7** Schematic of a time-of-flight mass analyser, involving the use of a 'reflectron'.

in the source of a mass spectrometer, by whatever technique, are given the same kinetic energy then the velocity of each will be inversely proportional to the square root of its mass. As a consequence, the time taken for them to traverse a field-free region (the flight tube of the mass spectrometer) will be related in the same way to the $m/z$ of the ion. A complete mass spectrum is obtained simply by allowing sufficient time for all of the ions of interest to reach the detector. A schematic of the time-of-flight instrument is shown in Figure 3.7.

The operation of this type of device is fundamentally different to those described previously in which ions of one $m/z$ ratio at a time enter the mass analyser. By varying the conditions in the mass analyser, e.g. magnetic field, quadrupole field, etc., ions of different $m/z$ values are brought to the detector and a corresponding mass spectrum obtained.

In the time-of-flight instrument, it is essential that ions of all $m/z$ ratios present in the source are transferred, simultaneously and instantaneously, into the mass analyser at a known time so that their times of flight, and thus their $m/z$ ratios, may be determined accurately. Were ions to be introduced continuously it would be impossible to determine exactly when each began its passage through the flight tube and therefore to calculate its $m/z$ ratio. A complete mass spectrum at a specific time is therefore obtained and when this has been recorded, a matter of milliseconds later, a further set of ions can be transferred from the source. This is sometimes referred to as a 'pulsed' source. Fast scanning, only limited by the time it takes the heaviest ion to travel from the source to the detector, is possible and any distortion of ion intensity brought about by changes in analyte concentration during the scanning process is removed.

In the first generation of ToF instruments, ions passed directly from the source to detector '1' (see Figure 3.7), and only low-resolution spectra were obtained. With the increasing interest in the production of ions with high $m/z$ ratios (in excess of 50 000), an increase in resolution is required to obtain spectra from which useful information may be obtained.

The resolution of the ToF analyser is dependent upon the ability to measure the very small differences in time required for ions of a similar $m/z$ to reach the detector. Increasing the distance that the ions travel between source and detector, i.e. increasing the length of the flight tube, would accentuate any such small time-differences. The implication of such an increase is that the instrument would be physically larger and this goes against the current trend towards the miniaturization of all analytical equipment.

A neat solution is to use one or more ion mirrors, known as *reflectrons*. These reflect the ion beam, in the case of an instrument with a single reflectron, towards a detector (detector '2' in Figure 3.7) located at the same end of the instrument as the source. The distance that the ion travels is thus doubled without an equivalent increase in the length of the flight tube. Some instruments now incorporate more than one reflectron, so increasing the distance travelled by the ions even more and in this way the resolution of current ToF instruments has been increased to more than 10 000. The implications of high mass spectrometer resolution on the quality of data produced from high-molecular-weight compounds and in tandem mass spectrometry (MS–MS) experiments are discussed further in Sections 4.7.3 and 3.4.1.4, respectively.

The advantage of the ToF instrument, in addition to its simplicity, is its fast scanning capability and for this reason it is increasingly being encountered in LC–MS instrumentation, particularly when fast analysis or high chromatographic resolution is involved.

## 3.4 Tandem Mass Spectrometry (MS–MS)

The great strength of mass spectrometry as a technique is that it can provide both the molecular weight of an analyte (the single most discriminating piece of information in structure elucidation) and information concerning the structure of the molecule involved.

The ionization techniques most widely used for LC–MS, however, are termed 'soft ionization' in that they produce primarily molecular species with little fragmentation. It is unlikely that the molecular weight alone will allow a structural assignment to be made and it is therefore desirable to be able to generate structural information from such techniques. There are two ways in which this may be done, one of which, the so-called 'cone-voltage' or 'in-source' fragmentation, is associated specifically with the ionization techniques of electrospray and APCI and is discussed later in Section 4.7.4. The other, termed mass spectrometry–mass spectrometry (MS–MS) or tandem mass spectrometry, is applicable to all forms of ionization, provided that appropriate hardware is available, and is described here.

Tandem mass spectrometry (MS–MS) is a term which covers a number of techniques in which one stage of mass spectrometry, not necessarily the first, is used to isolate an ion of interest and a second stage is then used to probe the relationship of this ion with others from which it may have been generated or which it may generate on decomposition. The two stages of mass spectrometry are related in specific ways in order to provide the desired analytical information. There are a large number of different MS–MS experiments that can be carried out [9, 10] but the four most widely used are (i) the product-ion scan, (ii) the precursor-ion scan, (iii) the constant-neutral-loss scan, and (iv) selected-decomposition monitoring.

Before considering these four scan modes in detail, it is worthwhile considering the types of instrument that have MS–MS capability because, as two stages of mass spectrometry are involved, not all systems will provide this facility.

## 3.4.1 Instrumentation

### 3.4.1.1 The Triple Quadrupole

This is probably the most widely used MS–MS instrument. The hardware, as the name suggests, consists of three sets of quadrupole rods in series (Figure 3.8). The second set of rods is not used as a mass separation device but as a collision cell, where fragmentation of ions transmitted by the first set of quadrupole rods is carried out, and as a device for focusing any product ions into the third set of quadrupole rods. Both sets of rods may be controlled to allow the transmission of ions of a single $m/z$ ratio or a range of $m/z$ values to give the desired analytical information.

### 3.4.1.2 The Hybrid Mass Spectrometer

When the first quadrupole of a triple quadrupole is replaced by a double-focusing mass spectrometer, the instrument is termed a *hybrid* (i.e. a hybrid of magnetic sector and quadrupole technologies). Figure 3.9 shows the $MS_1$ unit as a forward-geometry instrument although there is no reason why this could not be of reversed- or even tri-sector geometry.

The advantage of this configuration is that the $MS_1$ instrument can be used under high-resolution conditions to select the ion of interest. Despite the efficiency of modern chromatography, complete resolution of components, especially

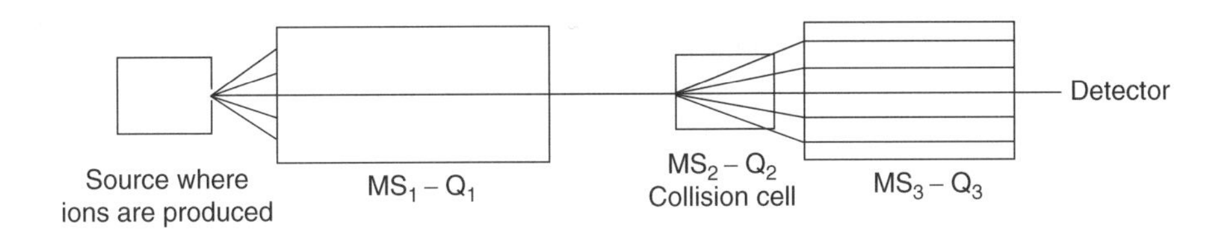


**Figure 3.8** Schematic of a triple quadrupole mass spectrometer.

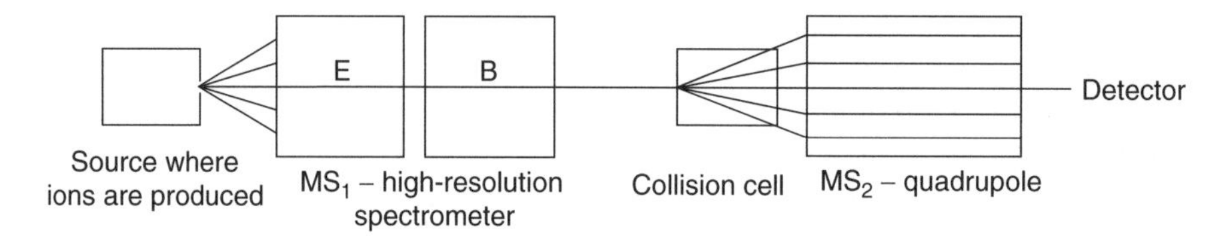


**Figure 3.9** Schematic of a hybrid MS–MS instrument.

if complex mixtures are involved, is not always possible and background signals may be observed. Low-resolution mass spectrometers allow us to differentiate between components with similar retention properties but whose mass spectra contain ions at different $m/z$ ratios, while high-resolution mass spectrometers allow us to do this but also to differentiate between components whose mass spectra contain ions with the same nominal mass but which have different atomic compositions (see Section 3.3 above). In this way, further specificity is conferred on the analysis.

### *3.4.1.3 Tandem Mass Spectrometry on the Time-of-Flight Analyser*

Ions are produced in the source of the mass spectrometer and are accelerated into the flight tube for mass analysis. If an ion fragments during its passage through the flight tube, and there are a number of mechanisms which may cause this, product ions are formed – this is analogous to the formation of fragment ions in a collision cell in other forms of MS–MS, which have the same velocity as their precursor but a reduced kinetic energy. The kinetic energy of the product ions is directly related to the ratio of their mass to that of the precursor, i.e. a product ion with half the mass of the precursor ion will have half its energy.

In instruments without a reflectron (see Figure 3.7 above), both the precursor and product ions reach the detector at the same time and are not separated. The reflectron, however, is an energy analyser and product ions with different energies, after passage through the reflectron, will have different flight times to the detector and may be separated and their $m/z$ ratios determined. This is known as *post-source decay* (PSD) [11].

### *3.4.1.4 The Quadrupole–Time-of-Flight Instrument*

In this instrument, the final stage of the triple quadrupole is replaced by an orthogonal time-of-flight (ToF) mass analyser, as shown in Figure 3.10. The configuration is typical of the latest generation of ToF instruments in which a number of reflectrons, in this case two, are used to increase the flight path of the ions and thus increase the resolution that may be achieved.

The difference between this and other MS–MS instruments is the way in which the $MS_2$ unit operates, as discussed in Section 3.3.4 above. To reiterate, in contrast to other mass analysers which are scanned sequentially through the

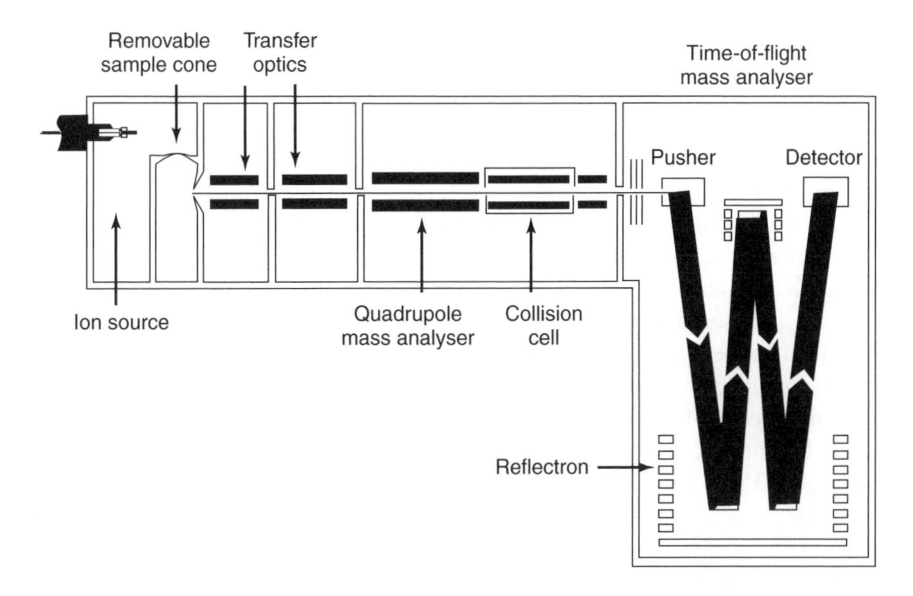


**Figure 3.10** Schematic of a quadrupole–time-of-flight mass spectrometer. From applications literature published by Micromass UK Ltd, Manchester, UK, and reproduced with permission.

$m/z$ range of interest and provide MS–MS spectra of user-selected masses, the ToF analyser detects **all** of the ions that enter it at a specific time. It is therefore possible, particularly in view of the high-scan-speed capability of this instrument, to provide, continuously, a full MS–MS product-ion spectrum (see below) of each ion produced in the source of the mass spectrometer. The disadvantage of this mode of operation is that it renders the Q–ToF system unable to carry out precursor and constant-neutral-loss scans, as described later in Sections 3.4.2.2 and 3.4.2.3, respectively.

#### *3.4.1.5 Tandem Mass Spectrometry on the Ion-Trap*

More recently, certain MS–MS scans have been made available on the ion-trap instrument. This type of system differs from those described previously in that the MS–MS capability is associated only with the way in which the ion-trap is operated, i.e. it is software controlled, and does not require the addition of a collision cell and a further analyser. This is because ion selection, decomposition and the subsequent analysis of the product ions are all carried out in the same part of the instrument, with these processes being separated solely in time, rather than time and space as is the case for the instruments described previously.

As with the Q–ToF instrument, only two types of MS–MS experiment are available with the ion-trap, i.e. the product-ion scan and selected-decomposition monitoring, as described in Sections 3.4.2.1 and 3.4.2.4, respectively.

*3.4.1.6 Other Tandem Mass Spectrometry Instrumentation*

Although it is likely that the majority of those involved in LC–MS and LC–MS–MS will have one of the instruments described above, it is also worthwhile mentioning two other types of instrument that have MS–MS capabilities, with these being instruments containing both a magnetic sector and an electrostatic analyser (see Section 3.3.3 above), and four-sector systems.

Tri-sector, and reversed-geometry double-focusing instruments, are able to produce mass-analysed ion kinetic energy spectrometry (MIKES) spectra. This technique is based on the fact that when a precursor ion fragments, its energy is distributed between the product ions in the ratio of their masses – this same principle is involved in post-source decay (see Section 3.4.1.3 above). This means that if the precursor ion has a mass of $M$ and an energy of $E$ and this fragments to give two product ions with masses of $0.33M$ and $0.67M$, these ions will have energies of $0.33E$ and $0.67E$, respectively.

As discussed above in Section 3.3.3, the electrostatic analyser (ESA) is an energy-focusing device. When a high-resolution mass spectrum is obtained, the ESA is operated at a fixed voltage as all ions produced in the source have nominally the same energy and energy focusing is required. To obtain a MIKES spectrum, the ESA voltage is 'scanned' from its initial value, $V$. As the voltage is decreased, ions with reduced energy pass through the device, e.g. at a voltage of $0.67V$, ions with energy $0.67E$ (and consequently mass $0.67M$) reach the detector. Measuring the ESA voltage at which an ion reaches the detector enables its mass to be determined. Scan rates for this technique, however, are relatively slow and largely incompatible with the peak widths encountered in chromatography.

Forward-geometry double-focusing instruments, including tri-sectors, are also capable of providing MS–MS data through a number of techniques known as *linked-scanning*. These are again based on the same principles as MIKES, i.e. that the energy of a product ion is dependent upon the initial energy of the precursor and its mass relative to that of the precursor. When a conventional mass spectrum is obtained in a double-focusing instrument, the ESA voltage remains constant and the magnetic field is scanned to focus ions of differing $m/z$ values at the detector. During linked-scanning, both the ESA voltage ($E$) and the magnetic field ($B$) are varied but in such a way that precise mathematical relationships are satisfied. A product-ion scan is obtained when the magnetic field of the mass spectrometer is scanned (with $E$ constant) until the precursor ion mass is reached. From this point, both $B$ and $E$ are scanned with the ratio $B/E$ being kept constant, the precursor-ion scan is obtained when the ratio $B^2/E$ is kept constant, and a constant-neutral-loss scan is obtained when the ratio $B/E\sqrt{1-E}$ is kept constant. As with MIKES, scan speeds are relatively slow, thus making the use of linked-scanning with chromatography problematic and therefore the alternative instrumental configurations described earlier are more widely used.

The four-sector mass spectrometer is the ultimate in MS–MS instrumentation and consists of two high-resolution mass spectrometers in series. The strength of these instruments is in terms of their high-mass and high-resolution capabilities for both precursor-ion selection and product-ion analysis. Their cost, however, precludes their primary use for LC–MS and therefore they will not be considered any further here [12].

## 3.4.2 Techniques

### 3.4.2.1 The Product-Ion Scan

In this scan, the first stage of mass spectrometry ($MS_1$ in Figure 3.8 above) is used to isolate an ion of interest – in LC–MS, this is often the molecular species from the analyte. Fragmentation of the ion is then effected; the means by which this is achieved is dependent on the type of instrument being used but is often by collision with gas molecules in a collision cell, i.e. $MS_2$ in the triple quadrupole (see Figure 3.8). The second-stage mass spectrometer ($MS_3$) is scanned (as a conventional instrument) to provide a mass spectrum of the ions formed in the collision cell, i.e. the product (fragment) ions. Interpretation of this spectrum is carried out in a similar way to the interpretation of an electron-ionization spectrum, i.e. a consideration of the structural significance of the ions observed, although it must be remembered that the mechanisms occurring in MS–MS are not identical to those occurring in EI.

In the ion-trap, ionization of the sample is carried out as in conventional operation and ions of all $m/z$ ratios take up stable trajectories within the trap. In the production of a conventional full-scan mass spectrum, ions of different $m/z$ values are then sequentially made unstable and ejected from the trap to the detector. In MS–MS operation, ions of all $m/z$ ratios, except that required for further study, are made unstable and ejected from the trap. The ions remaining in the trap, only those of the selected $m/z$ ratio, are now ‘excited’ to bring about their dissociation. The resulting product ions are then sequentially made unstable and sent to the detector to generate the product-ion spectrum.

A feature of MS–MS using the ion-trap is that having generated product ions from a selected precursor ion, any one of these ‘product’ ions may be isolated, dissociated and a further product-ion spectrum obtained. This is termed MS–MS–MS, or $MS^3$. This process can then be repeated to obtain further stages of MS–MS, or $MS^n$ [13].

**SAQ 3.4**

What is the difference between a product-ion MS–MS scan on a molecular ion and a conventional EI mass spectrum?

#### *3.4.2.2 The Precursor-Ion Scan*

In this scan, the second stage of mass spectrometry ($MS_3$ in Figure 3.8 above) is set to transmit a single $m/z$ ratio, namely that of the product (fragment) ion of interest, while the first stage ($MS_1$) is set to scan through the mass range of interest, with the fragmentation of ions passing through $MS_1$ being again carried out in $MS_2$, the collision cell. A signal is seen at the detector only when ions are being transmitted by both $MS_1$ and $MS_3$, i.e. when an ion being transmitted by $MS_1$ fragments to give the desired ion.

The ion-trap and Q–ToF instruments are, because of the way that they operate, unable to carry out precursor-ion scans. Computer manipulation of data generated during product-ion scans of the Q–ToF system, however, can yield equivalent data to that produced directly by precursor-ion scans on other instruments and an evaluation of this software-based approach has been carried out [14].

#### *3.4.2.3 The Constant-Neutral-Loss Scan*

In addition to simple fragmentation reactions it has been pointed out (see Section 3.2.1 above) that molecules containing certain structural features also undergo rearrangement reactions in the source of the mass spectrometer. Probably the best known of these is the McLafferty rearrangement [7], in which a hydrogen atom migrates to an unsaturated group (often a carbonyl group) with elimination of a neutral molecule (Figure 3.11).

These rearrangement reactions may also occur in MS–MS instruments and the constant-neutral-loss scan enables the analyst to observe **all** of the ions in the mass spectrum that fragment with a particular mass loss and therefore contain a specific structural feature. This knowledge can be of great value when attempting to interpret the mass spectrum of an unknown material.

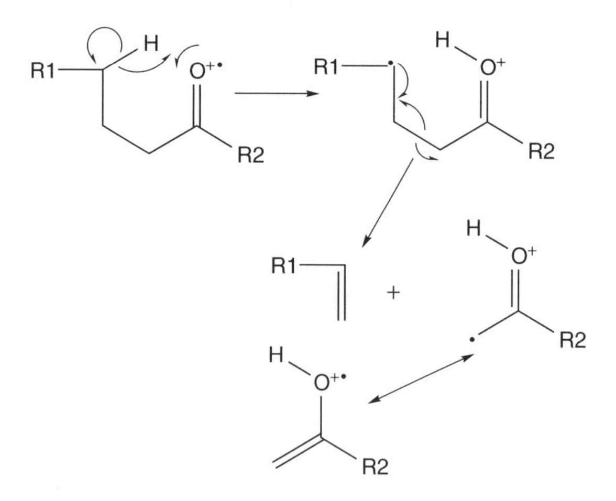


**Figure 3.11** Schematic of the McLafferty rearrangement [7].

The constant-neutral-loss scan is carried out by scanning both of the stages of mass spectrometry with a constant, specific, mass difference between them, e.g. if the constant neutral loss of interest is 42 Da, scanning of $MS_1$ (see Figure 3.8) might start at $m/z$ 100 at which time $MS_3$ would be set to transmit $m/z$ 58. As $MS_1$ then moves to $m/z$ 101, $MS_3$ would move in conjunction to $m/z$ 59, and when $MS_1$ moves to $m/z$ 102, $MS_3$ would move to $m/z$ 60, etc. A signal is only obtained at the detector when both $MS_1$ and $MS_3$ are transmitting ions, i.e. only when the ion being transmitted by $MS_1$ fragments with loss of the mass of interest. If it fragments by any other loss, the resulting product ion is not transmitted by $MS_3$. Again, ion-traps and Q–ToF instruments are not capable of carrying out this type of scan.

#### *3.4.2.4 Selected-Decomposition Monitoring*

In conventional mass spectrometry, quantitative determinations are often carried out by using selected-ion monitoring (SIM), i.e. by monitoring the intensities of a small number of ions characteristic of the analyte of interest (see Section 3.5.2.1 below).

The MS–MS equivalent of this technique is known as *selected-decomposition monitoring* (SDM) or *selected-reaction monitoring* (SRM), in which the fragmentation of a selected precursor ion to a selected product ion is monitored. This is carried out by setting each of the stages of mass spectrometry to transmit a single ion, i.e. the precursor ion by $MS_1$ and the product ion by $MS_3$ (see Figure 3.8 above).

In the ion-trap, the precursor is selected by the methodology described in Section 3.4.2.1 above for the product-ion scan. In SDM, however, the product ion 'scan' is confined to only one $m/z$ ratio, namely that of the product ion of interest.

## 3.5 Data Acquisition

Having separated the ions of differing $m/z$ values, a complete mass spectrum is obtained by determining the $m/z$ ratios and relative numbers of each of the ions present. A description of these processes is beyond the scope of this present text and may be found elsewhere [2, 4].

There are two ways in which we use mass spectral data, namely (a) for identification purposes, and (b) for determining the amount of an analyte(s) present in a sample (quantitation), with the ways in which we acquire data for these two purposes being usually quite different.

### *3.5.1 Identification*

For identification purposes, a mass spectrum covering all of the $m/z$ ratios likely to be generated by the analyte is required.

Very rarely, however, will a single mass spectrum provide us with complete analytical information for a sample, particularly if mass spectral data from a chromatographic separation, taking perhaps up to an hour, is being acquired. The mass spectrometer is therefore set up to scan, repetitively, over a selected $m/z$ range for an appropriate period of time. At the end of each scan, the mass spectrum obtained is stored for subsequent manipulation before a further spectrum is acquired.

## 3.5.2 Quantitation

Quantitation using mass spectrometry is no different to quantitation using other techniques and, as discussed above in Section 2.5, involves the comparison of the intensity of a signal generated by an analyte in a sample to be determined with that obtained from standards containing known amounts/concentrations of that analyte.

### 3.5.2.1 Selected-Ion Monitoring

A mass spectrum typically contains a large number of ions of differing $m/z$ ratios and all of these could be used for quantitation purposes and will, theoretically, provide the same result. Why then acquire data for each $m/z$ value if the majority provide no benefit to the analysis being carried out? As long as the identity of the analyte is assured, i.e. the analyst is certain that the signal being measured is derived from the desired analyte and not from an interfering compound, there are significant sensitivity advantages associated with not obtaining a full-scan spectrum but obtaining data from only a small number of $m/z$ ratios. This technique is known variously as *selected-ion monitoring* (SIM), *selected-ion recording* (SIR) and *multiple-ion detection* (MID).

Consider a full scan from $m/z$ 45 to $m/z$ 545, i.e. 500 $m/z$ ratios, carried out in 1 s. Simplistically, an individual $m/z$ ratio is being monitored for 1/500 (0.002) s or, considering this from another perspective, for 499/500 (0.998) s ions of a particular $m/z$ ratio are not reaching the detector and play no part in the analysis. If instead of monitoring 500 ions – effectively the process that is carried out during full scanning – only 4 ions are selected, then 1/4 (0.25) s are spent monitoring each. The amount of time spent monitoring each $m/z$ ratio is therefore increased by a factor of over 100 and many more ions of each $m/z$ value are used in the analysis, hence resulting in a proportional increase in sensitivity.

If compounds other than the target analyte are present, but these do not produce ions at the same $m/z$ value as the analyte of interest, they play no part in the analysis. The selectivity of the detector is therefore enhanced.

How many ions should be monitored and how are appropriate ions for SIM selected? There are a number of general guidelines and these should be applied equally to any internal standard(s) that may be employed, as described in the following.

(a) A single ion is usually inadequate as it is, in isolation, unlikely to be unique to the analyte of interest. If chromatography is involved, the combination of retention time and a single ion may be considered adequate, especially if it is a molecular ion generated by a soft-ionization technique, although the use of more than one ion is advisable. The relative intensity of the chosen ions is determined from a standard and then a check is made that this ratio is reproduced in each analytical sample. Deviation from the standard value indicates interference at one or both of the ions and requires the analyst to consider the quantitative result further.

The number of ions chosen has an effect on the sensitivity of the analysis. The advantage of SIM is achieved through spending more time monitoring the ions of interest – the more ions being monitored, then the less time will be spent on each of them and the lower the increase in sensitivity of SIM over full scanning.

If SIM is being carried out in conjunction with chromatography, a further consideration is that an adequate number of cycles of measurement must be made to define the shape and intensity of the chromatographic response exactly, or otherwise inaccurate and imprecise measurements will be made (Figure 3.12).

Too many cycles is as undesirable as too few, since, as in the case of monitoring too many ions, the amount of time spent on each ion of interest, and consequently the sensitivity of the analysis, is reduced.

The number of cycles that is attainable is also a function of the chromatographic peak width – the narrower the peak, then the faster the cycle rate required to define that peak accurately. The peak widths encountered in HPLC, which are relatively wide compared to GC, are such that a compromise between scan speed and sensitivity is less likely to be required.

As shown in Figure 3.12, around 10 cycles of the masses is required to define the peak shape accurately and allow precise analytical measurements to be made.

(b) The ions should be characteristic of the analyte under investigation. If the analyte contains an element with a characteristic isotopic distribution, such as

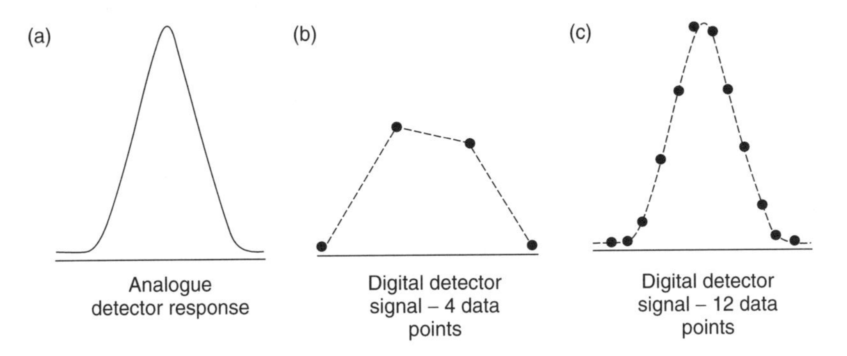


**Figure 3.12** Comparison of the chromatographic peak shapes obtained from (a) analogue and (b,c) digital detectors, and the effect on peak shape of the number of data points defining the signal.

chlorine, the isotope ion may be used as the second SIM mass, but it must be remembered that this is not wholly specific as any compound containing the same number of atoms of these elements will give the same intensity at the second mass, e.g. all species containing one chlorine atom will show two ions 2 Da apart, with relative intensities of 100 and 33%. It is better, therefore, if at all possible, to use a fragment-ion rather than that from an isotope.

(c) The ions should be intense to confer sensitivity on the analysis. Of particular importance is the intensity of the ion being used to derive the quantitative result as this will provide a good limit of detection. It is desirable that the second ion signal is also intense but this is of slightly less importance as it is being used primarily as a specificity check.

(d) The ions should be at high $m/z$ values – this reduces the possibility of background interference which tends to be at relatively low $m/z$ ratios. In addition, the higher the $m/z$ value chosen, then the more likely that it will be characteristic of the analyte of interest.

The final choice will have to be made during method development and/or analysis of the 'real' samples, e.g. one of the ions selected may provide superb data from standard solutions but show a high matrix background on all or, perversely, on only a small number of samples, which will preclude its/their use.

**DQ 3.2**

Below are three mass spectra from compounds for which you wish to develop an SIM assay. Which ions would you choose to monitor and why?

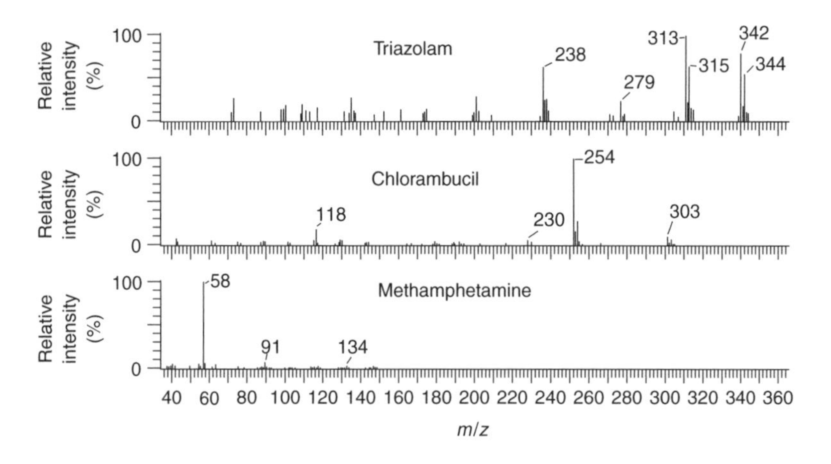


*Answer*

***Triazolam*** *– this is fairly obvious, i.e. the ions at* m/z *342 and* m/z *313, which are intense and at high mass. A loss of 29 Da is more significant* *that using (M + 2) as all chlorine-containing species will give the same pattern here. If sensitivity is not a problem, we could monitor the ions at either* m/z *344 or* m/z *315 to give further confirmation of the presence of chlorine (the larger ion for quantitation, with the smaller ion as confirmation). If excessive chemical background is observed at either* m/z *342 or* m/z *313, we are lucky to have another reasonably intense ion at* m/z *238.*

***Chlorambucil** – there is no problem with the quantitation ion (at* m/z *254), although the second ion proves to be a little difficult. While the ion at* m/z *303 is the obvious choice, this is not very intense and therefore for samples containing small amounts of analyte the precision of measurement of this ion will be reduced and it may not be detectable at all levels at which the quantitation ion is observed. We could possibly consider the (M + 2) ion, as the combination of* m/z *254 (high mass, and therefore reasonable specificity), the presence of one chlorine, and the chromatographic retention time could be considered sufficient for definitive identification in those cases in which the intensity of* m/z *303 is insufficient.*

***Methamphetamine** – this is a more difficult problem. The most intense ion is observed at* m/z *58 and may be used for quantitation, although this is a common ion in the mass spectra of amines. The use of a second ion to provide specificity is therefore crucial. The next most intense ion is, however,* m/z *91 (of 6% relative intensity), although this ion equally occurs in the spectra of a number of compounds containing the benzyl group and cannot, in this author's opinion, be considered to be specific enough to provide any confirmation of identity. Another ion present in the spectrum that might provide more specificity, i.e. at* m/z *134, is, however, only of 1% relative intensity and use of this is likely to increase the detection limit significantly. All of the ions available are at relatively low mass and therefore susceptible to background interference. Although the use of retention times may improve the situation, it is still not clear-cut. We should, therefore, possibly consider the use of a soft-ionization technique to provide the extra specificity of the molecular ion at higher* m/z *values.*

If SDM (see Section 3.4.2.4 above) is to be used, product-ion scans on ions associated with the analyte must be carried out to determine whether fragmentation(s) occur that are appropriate to monitor.

**DQ 3.3**

What are the potential advantages of selected-decomposition monitoring (SDM) over selected-ion monitoring (SIM)?

*Answer*

*The great advantage of SDM over SIM is that it usually confers even greater selectivity onto the analysis and reduces the amount of chemical noise observed. Selected-ion monitoring is concerned with particular m/z values and while an ion may be chosen because it occurs in the mass spectrum of the analyte of interest, it is also likely to occur, not necessarily at any great abundance, in the mass spectra of many other compounds. Hence, the need to monitor another ion to provide an intensity check. When SDM is used, not only must the precursor ion be present but it must also fragment to give a defined product ion. Since the fragmentation is related, to some extent, to the structure of the precursor ion, the likelihood of an interfering compound fragmenting in the same way as the analyte of interest is significantly less and the chances of obtaining an interfering response is reduced. Although the absolute intensities of MS–MS dissociations may be less than some of the fragmentation processes in electron ionization, the associated reduction in noise often leads to an increase in the signal-to-noise ratio compared to SIM and a consequent reduction in detection limits.*

## 3.6 Processing of Mass Spectral Data

While it is true that in many cases the quality of data acquired during analysis is directly proportional to the quality of the result that may be obtained, it is also true that in many cases the power of modern computer systems attached to analytical equipment of all sorts can be used to provide 'better' results than might be thought possible from a cursory examination of the raw data. Even when chromatography is used to separate the components of a mixture and simplify the job of the analyst, the computer may still allow information hidden in the vast amount of data generated to be extracted.

A full description of a mass spectrometry data system is beyond the scope of this present text, but its major roles may be considered to be the following:

(a) control of the mass spectrometer (ionization mode and scan mode);

(b) the acquisition of mass spectral data and its subsequent storage;

(c) the processing of the stored data to allow an analytical result to be obtained.

### *3.6.1 The Total-Ion-Current Trace*

During a chromatographic separation, the mass spectrometer is set to scan, repetitively, over an appropriate mass range, with the upper mass limit being defined by the highest molecular weight of the analyte(s) thought/likely to be present, and the lower limit by the background ions – in the case of LC–MS, usually from the mobile phase. At the end of each scan, a spectrum is stored and the sum of the intensities of the ions present, i.e. the total-ion current, is computed and stored. After an appropriate time, determined by the retention time of the

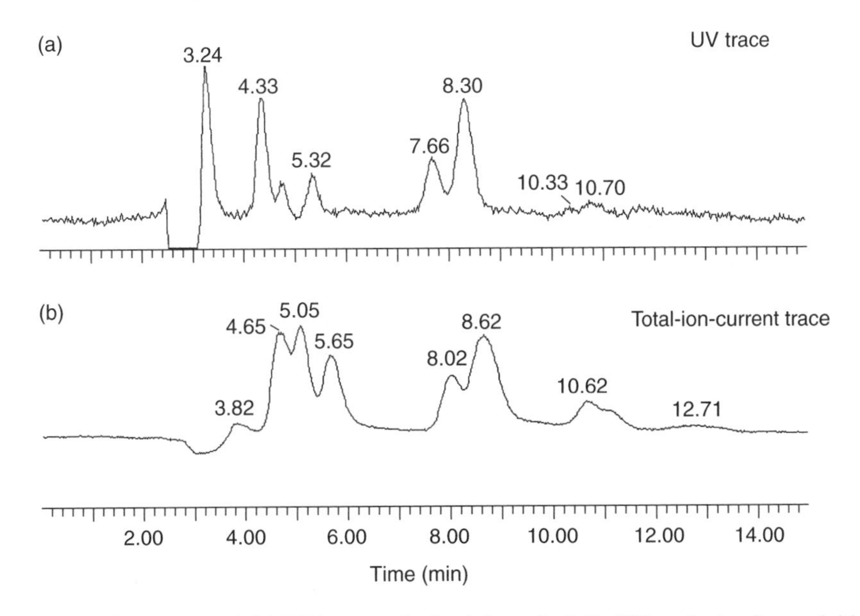


**Figure 3.13** (a) UV and (b) TIC traces obtained from the LC–MS analysis of a pesticide mixture. From applications literature published by Micromass UK Ltd, Manchester, UK, and reproduced with permission.

longest retained component, scanning of the mass spectrometer is halted and the data are available for processing.

The fundamental piece of information on which the subsequent spectral analysis is based is the total-ion-current (TIC) trace. Such a trace, obtained from the LC–MS analysis of a pesticide mixture, is shown in Figure 3.13, together with the UV trace recorded simultaneously. For the purposes of this discussion, the HPLC and MS conditions used to generate the data, other than the fact that electrospray ionization was used, are irrelevant.

There are a number of features worthy of note in this figure. For example, there is a difference in retention times, determined by the two detectors, of ca. 0.32 min, and this reflects the fact that they are used in series, i.e. the column effluent passes through the UV detector on its way to the mass spectrometer.

Secondly, the intensity of response for a certain compound from one type of detector is not necessarily the same as that obtained from the other detector. This should not be unexpected, since the two detectors are measuring quite different properties of the analyte, in this case UV absorption at a particular wavelength and how readily it is ionized and fragmented under the conditions employed. These properties are unrelated.

The third feature is that of peak widths. One of the properties of an ideal detector, as described earlier in Chapter 1, is that it should not degrade the chromatographic performance. In this particular example, there is an increase in peak width – the response at 8.30 min in the UV trace is approximately 18 s wide, while that at 8.62 min in the TIC trace is 35 s wide. Peak broadening occurs for a number of reasons [15], including, in this case, the presence of the UV detector (a disadvantage of having a second in-line detector) and the LC–MS interface. Before trying to eliminate the peak broadening, often a time-consuming process unless the cause is readily apparent, e.g. a leaking tubing joint, its effect on the analysis being carried out should be assessed. Is the analyst able to obtain good quality mass spectra of individual analytes from which identification may be attempted? Can quantitative responses of adequate signal-to-noise ratios be obtained to allow the amount of analyte to be determined with adequate accuracy and precision? If the answer to whichever of these questions is relevant is 'yes', spending time searching for the causes of peak broadening may well not improve the quality of the result obtained and therefore will not be time well spent.

### 3.6.2 Qualitative Analysis

As previously mentioned, the passage of a continuous flow of mobile phase into the mass spectrometer may give rise to a significant amount of background, as can be seen from the TIC trace shown in Figure 3.13(b). A background mass spectrum, obtained after ca. 0.8 min, is shown in Figure 3.14.

Ions at $m/z$ 55, 60, 214 and 236 are observed but do some or all of these arise from the background and are present throughout the analysis, or are they present in only a few scans, i.e. are they from a component with insufficient overall intensity to appear as a discrete 'peak' in the TIC trace? An examination of reconstructed ion chromatograms (RICs) from these ions generated by the data system may enable the analyst to resolve this dilemma. The TIC shows the variation, with time, of the total number of ions being detected by the mass spectrometer, while an RIC shows the variation, with time, of a single ion with a chosen $m/z$ value. The RICs for the four ions noted above are shown in Figure 3.15. These ions have similar profiles and show a reduction in intensity as analytes elute from the column. The reduction in intensity is a suppression effect,

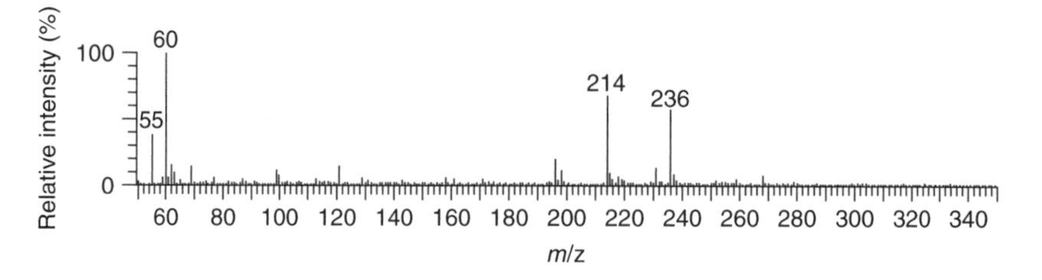


**Figure 3.14** Background mass spectrum obtained from the LC–MS analysis of a pesticide mixture. From applications literature published by Micromass UK Ltd, Manchester, UK, and reproduced with permission.

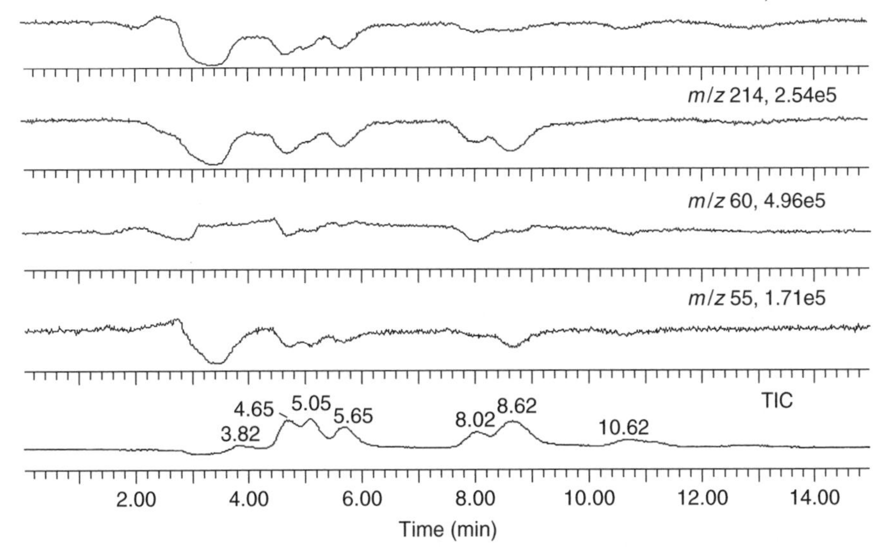


**Figure 3.15** Reconstructed ion chromatograms obtained for ions in the background mass spectrum from the LC–MS analysis of a pesticide mixture. From applications literature published by Micromass UK Ltd, Manchester, UK, and reproduced with permission.

one which is often encountered with electrospray ionization if more than one species is being ionized at the same time (see Section 4.7.2 below). The otherwise relatively constant level of ion intensity is characteristic of background ions.

Another ion profile often encountered from background ions is one in which the intensity increases or falls at a regular rate throughout the analysis. This often occurs in LC–MS during gradient elution when the ion is associated with only one component of the mobile phase.

In Figure 3.15, the additional information given next to the $m/z$ values, e.g. 1.71e5 by $m/z$ 55, gives an indication of the maximum absolute intensity measured for that ion. Comparison can then be made of the relative intensities of these ions and also with ions generated by genuine components of the mixture.

What effect will the background have on the analysis? While the background, in isolation, may look to be significant, its real importance can only be determined by examination of the mass spectrum of a compound as it elutes from the HPLC column. The mass spectrum recorded at the TIC maximum after 4.65 min is shown in Figure 3.16.

The intensities of the major background ions of $m/z$ 214 and 236 are 3.5 and 2.8%, respectively, of the ion of $m/z$ 229. It is known from the RICs of $m/z$ 214 and 236 (see Figure 3.15) that these ions do not increase in intensity as the analyte elutes (in fact, their intensity decreases). They may, therefore, be considered to

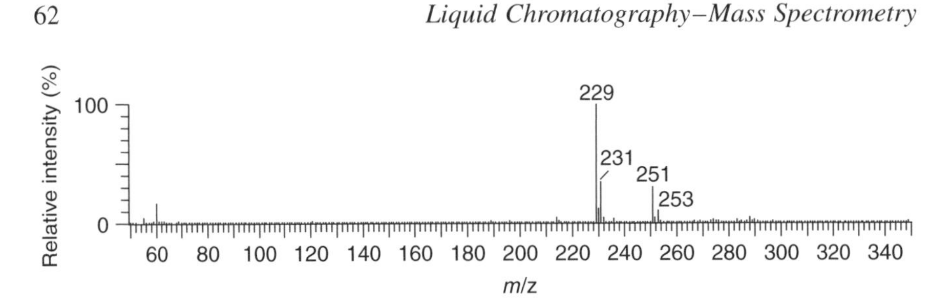


**Figure 3.16** Mass spectrum obtained from the component eluting after 4.65 min in the LC–MS analysis of a pesticide mixture. From applications literature published by Micromass UK Ltd, Manchester, UK, and reproduced with permission.

be insignificant and it is legitimate to proceed with the interpretation and simply ignore the background ions. The ions at $m/z$ 55 and 60, while of greater intensity, are at low $m/z$ values and are unlikely to correspond to molecular ions produced by a soft ionization technique such as electrospray. These too, can justifiably be ignored.

If preferred, the data system may be used to subtract the background spectrum (Figure 3.14) from that of the analyte (Figure 3.16). This manipulation yields the spectrum shown in Figure 3.17, in which the ions from the background are now totally absent. Care must be taken when adopting this procedure to ensure that any contribution from the analyte is not removed when ions at the same $m/z$ value arise from both the background and the analyte of interest.

'Background' may also occur when two components are not completely resolved and a significant contribution from the mass spectrum of the early eluting analyte is found in the mass spectrum of that eluting subsequently. For interpretation purposes, it is equally important that this type of 'background' is removed. The spectrum recorded at the TIC maximum after 5.05 min is shown in Figure 3.18.

This spectrum shows ions at $m/z$ 163, 195, 214, 217, 229, 236 and 251. The spectrum with the background subtracted, obtained as explained above, is shown in Figure 3.19. As before, the ions at $m/z$ 214 and 236 have been removed.

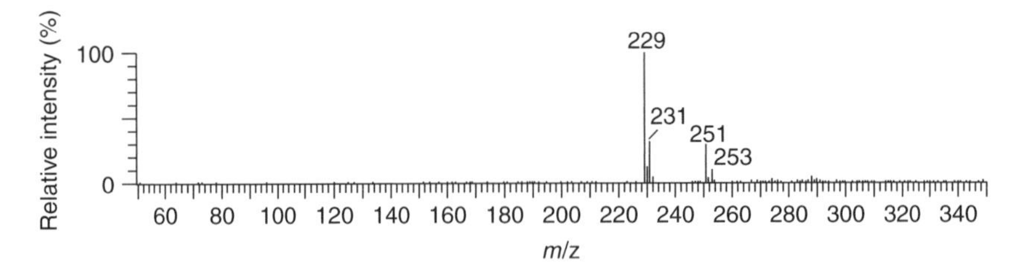


**Figure 3.17** Background-subtracted mass spectrum obtained from the component eluting after 4.65 min in the LC–MS analysis of a pesticide mixture. From applications literature published by Micromass UK Ltd, Manchester, UK, and reproduced with permission.

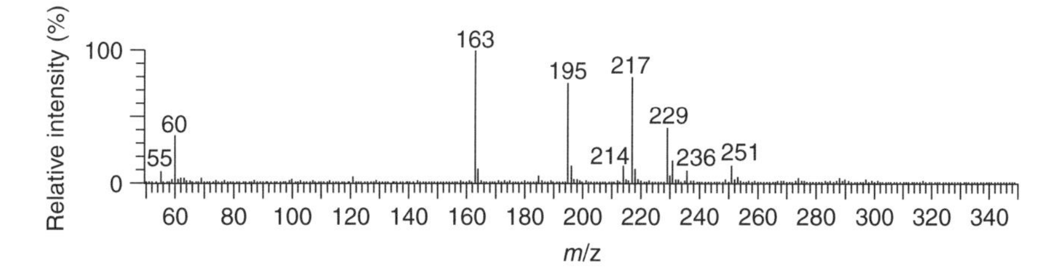


**Figure 3.18** Mass spectrum obtained from the component eluting after 5.05 min in the LC–MS analysis of a pesticide mixture. From applications literature published by Micromass UK Ltd, Manchester, UK, and reproduced with permission.

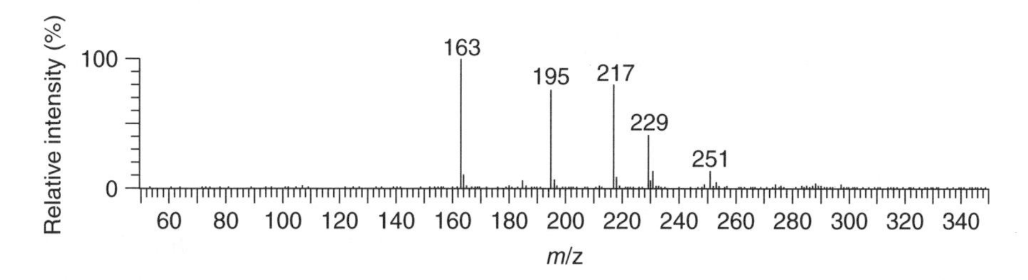


**Figure 3.19** Background-subtracted mass spectrum obtained from the component eluting after 5.05 min in the LC–MS analysis of a pesticide mixture. From applications literature published by Micromass UK Ltd, Manchester, UK, and reproduced with permission.

While $m/z$ 229 and 251 were observed in the spectrum of the previously eluting component (see Figure 3.16), it is necessary to confirm that their presence in this spectrum is solely from that source and not from another component whose mass spectrum also contains these ions. If this can be done, the spectrum of the second component can be obtained by background subtraction.

The RICs for the five ions are shown in Figure 3.20. It is clear that the ions of $m/z$ 229 and 251 do not 'maximize' at the same time as the ions of $m/z$ 163, 195 and 217, and from this it may be concluded that the ions at $m/z$ 229 and 251 are not associated with the second component.

A closer examination of the RICs presented in Figure 3.20, however, shows that, because of the difference in absolute intensity of the ions involved, a mass spectrum at the second peak maximum, after 5.10 min, will show significant contributions from the compound eluting after 4.67 min. Under these circumstances, the relationship between particular ions can be investigated in two other ways, i.e. (a) the relative intensities of the ions may be examined, and (b) the chemical significance of these ions may be considered.

Compare the intensities of the ions of $m/z$ 163, 195, 217, 229 and 251 presented in Figures 3.19 and 3.21, which show the spectra obtained after 5.05 and 5.34 min, respectively. The relative intensities of $m/z$ 163, 195 and 217 remain

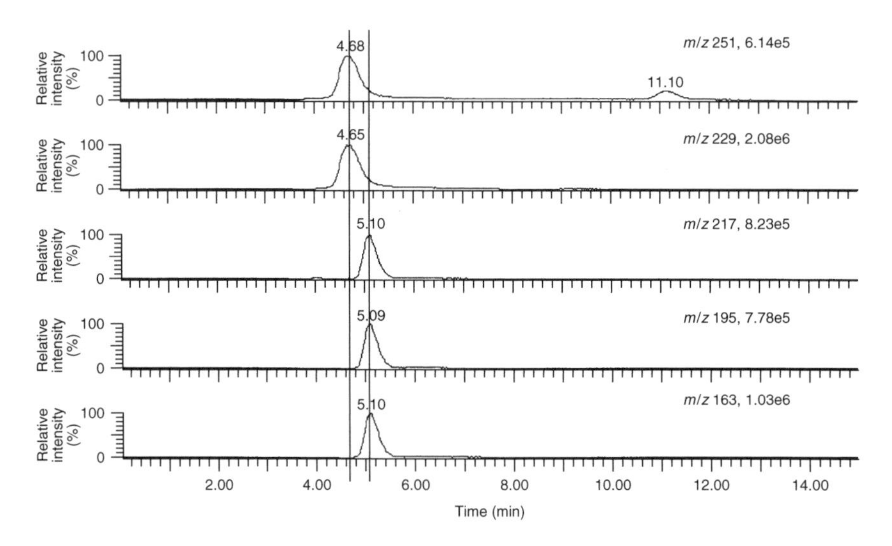


**Figure 3.20** Reconstructed ion chromatograms for the ions observed in the background-subtracted mass spectrum obtained from the component eluting after 5.05 min in the LC–MS analysis of a pesticide mixture. From applications literature published by Micromass UK Ltd, Manchester, UK, and reproduced with permission.

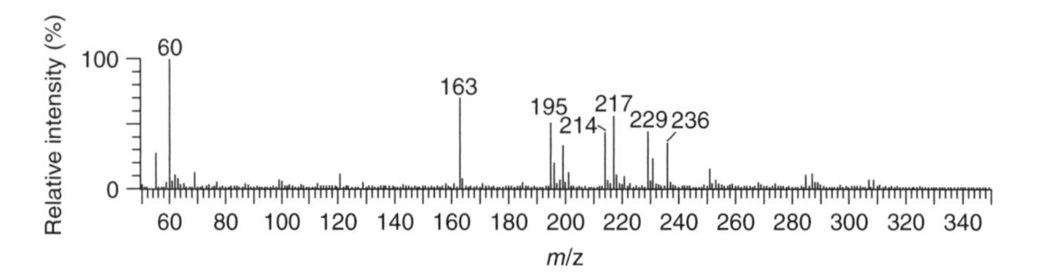


**Figure 3.21** Mass spectrum obtained from the component eluting after 5.34 min in the LC–MS analysis of a pesticide mixture. From applications literature published by Micromass UK Ltd, Manchester, UK, and reproduced with permission.

constant, as does that of $m/z$ 229 and 251, although the relative intensities of the two groups of ions have changed. This behaviour would support the existence of two sets of ions with no cross-contribution from one to the other; should any of these ions occur in the mass spectra of both components, the relative intensities of the ions within each of these groups would not remain constant.

**SAQ 3.5**

Rationalize the presence of the ions at $m/z$ 201 and $m/z$ 221 in the mass spectrum shown in Figure 3.21

A full interpretation of the spectra is not required but a consideration of the observed $m/z$ values can give some indications of the relationship between the ions. The ions at $m/z$ 163, 195 and 217 have tentatively been associated with a single compound. For this to be true, it must be possible to interpret the losses in a meaningful way. These spectra have been obtained by electrospray ionization which is a soft ionization technique and would be expected to provide, primarily, molecular species. Why then do we see three ions? There are three possibilities, i.e. (a) the molecular ion region is around $m/z$ 217 and the other two ions observed, $m/z$ 195 and $m/z$ 163, are both fragments from the dissociation of this ion, (b) the molecular ion region is at $m/z$ 195 and the ion at $m/z$ 217 is an adduct, while that at $m/z$ 163 is a fragment, or (c) the molecular ion is at $m/z$ 163 and the ions with higher $m/z$ values are both adducts.

If the first possibility were to be the case, a change in the cone voltage (see Section 4.7.4 below) is likely to influence the extent of fragmentation and the relative intensities of the molecular species when compared to the fragments would vary. It is also possible that, depending upon the fragmentation mechanisms which are operative, the relative intensity of the two fragments would change.

If the molecular species were at $m/z$ 195 (case (b) above) the ion of $m/z$ 163 is generated by a loss of 32 Da. A similar loss from the adduct ion would not be unusual, in this case an ion at $m/z$ (217 – 32), i.e. $m/z$ 185, would be expected. No significant ion of this $m/z$ value is observed and while this is not conclusive it would suggest that this is not the explanation.

The third possibility is that the molecular species is $m/z$ 163 and that $m/z$ 195 and $m/z$ 217 are both adducts. If this is the case, it must be possible to explain the differences of 32 Da and 54 Da easily. Can this be done? Commonly occurring adducts in electrospray involve the mobile phase and either sodium (relative molecular mass (RMM) 23) or potassium (RMM 39). The simplest interpretation of this spectrum is that the molecular weight of the analyte is 162, the ion at $m/z$ 163 is the protonated species, that at $m/z$ 195 is $(M + H + CH_3OH)^+$ and that at $m/z$ 217 is $(M + Na + CH_3OH)^+$. Since the HPLC mobile phase contained methanol (molecular weight of 32), this is not an unreasonable conclusion.

The second series consists of only two ions, $m/z$ 229 and $m/z$ 251, and therefore either $m/z$ 251 is an adduct of the molecular species of $m/z$ 229, or $m/z$ 229 is a fragment arising from the dissociation of the ion at $m/z$ 251. The mass difference is 22 Da, which is most easily explained in terms of $m/z$ 251 arising from a sodium adduct, with the ion at m/z 229 corresponding to the $(M + H)^+$ species.

The RICs shown in Figure 3.20 show a small contribution from ions at $m/z$ 251 and 229 to the mass spectrum of the component eluting after 5.10 min. If these two ions were genuinely associated with those of $m/z$ 163, 195 and 217, it must be possible to explain them as adducts of the molecular species postulated above, i.e. at $m/z$ 163. No adducts readily spring to mind to account for the mass differences of 66 and 88 Da.

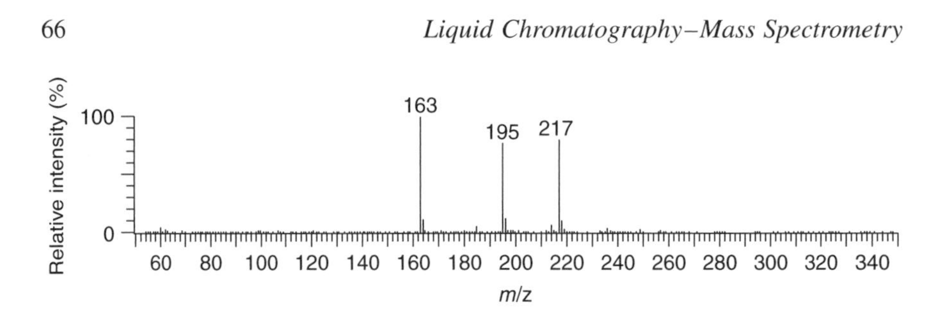


**Figure 3.22** Background-subtracted mass spectrum obtained from the component eluting after 5.34 min in the LC–MS analysis of a pesticide mixture. From applications literature published by Micromass UK Ltd, Manchester, UK, and reproduced with permission.

Further experimental work involving cone-voltage studies may provide further confirmatory evidence but the most likely explanation is that the mass spectrum of the component with retention time 4.65 min is that shown in Figure 3.17, while the mass spectrum of the second component is that obtained by background subtraction, and is shown in Figure 3.22.

Further examination of Figure 3.20 reveals another interesting feature. The peak widths of the RICs for the ions of $m/z$ 229 and $m/z$ 251 are significantly greater than those of the other three ions. Since the two components have very similar retention times, this cannot be explained in terms of the chromatographic process. A possible explanation is that a further unresolved low-intensity component (or components) might be present and in these circumstances further insight into what is giving rise to this apparent peak broadening might be obtained by examination of a mass spectrum on the trailing edge of the RIC trace, i.e. at around 4.8 min. The high-mass region of this spectrum is shown in Figure 3.23.

The RICs for the annotated ions are shown in Figure 3.24. The ions at $m/z$ 267 and $m/z$ 283 have similar profiles to those of $m/z$ 229 and $m/z$ 251 (see Figure 3.20 above), while the ions at $m/z$ 274, $m/z$ 288 and $m/z$ 297 have broader profiles, with the suggestion of unresolved components being present. Should it be necessary, further examination of mass spectra and RICs, using similar methodology to that

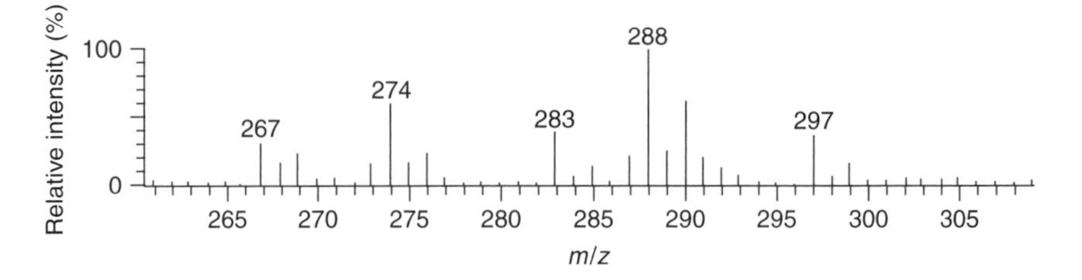


**Figure 3.23** Mass spectrum obtained after 4.8 min in the LC–MS analysis of a pesticide mixture. From applications literature published by Micromass UK Ltd, Manchester, UK, and reproduced with permission.

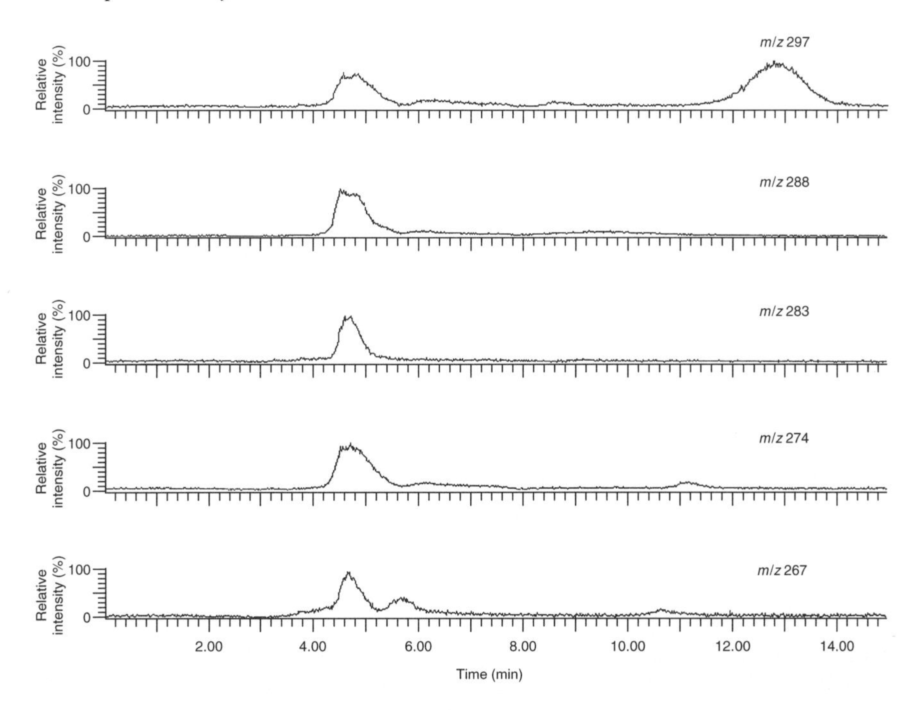


**Figure 3.24** Reconstructed ion chromatograms of ions in the high-mass portion of the mass spectrum obtained after 4.8 min in the LC–MS analysis of a pesticide mixture. From applications literature published by Micromass UK Ltd, Manchester, UK, and reproduced with permission.

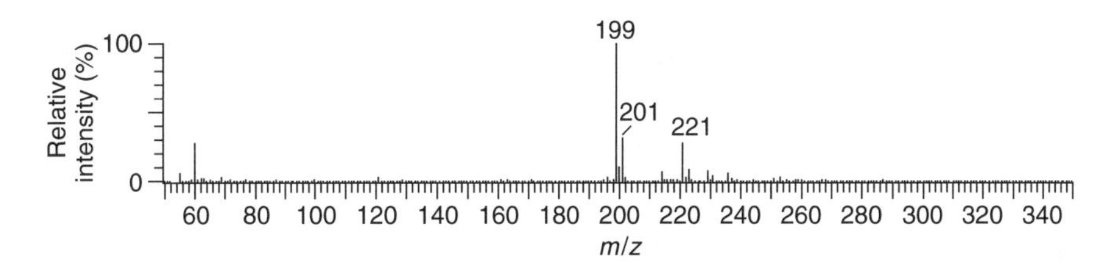


**Figure 3.25** Mass spectrum obtained from the component eluting after 5.65 min in the LC–MS analysis of a pesticide mixture. From applications literature published by Micromass UK Ltd, Manchester, UK, and reproduced with permission.

described above, or further experimentation utilizing MS–MS, could be undertaken in an attempt to identify the unresolved components.

A similar exercise involving the region between the second and third 'peaks' is even more revealing. The mass spectrum of the component eluting after 5.65 min is shown in Figure 3.25, and displays a species at $m/z$ 199. Examination of the

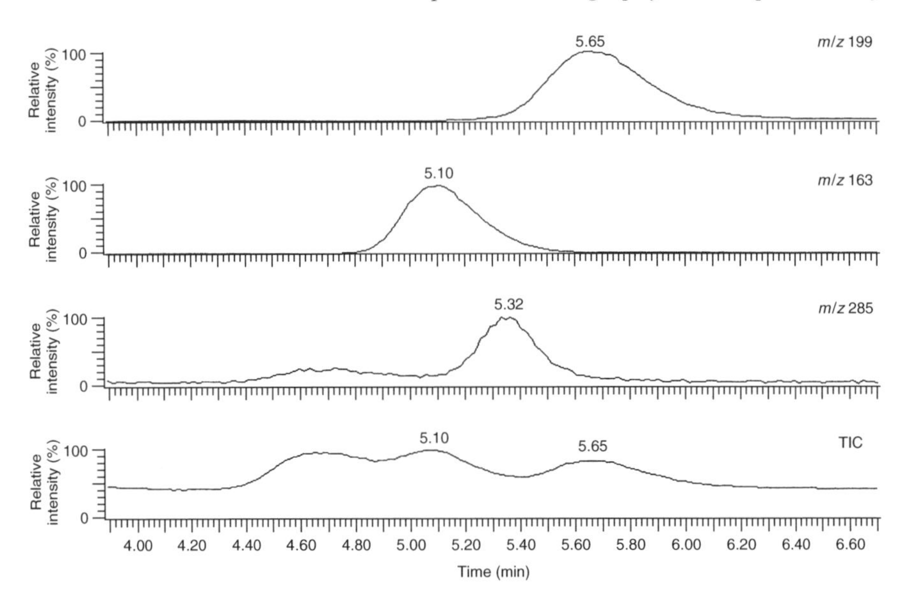


**Figure 3.26** Reconstructed ion chromatograms obtained from the ions of $m/z$ 199, 163 and 285 for retention times between 3.95 and 6.65 min in the LC–MS analysis of a pesticide mixture. From applications literature published by Micromass UK Ltd, Manchester, UK, and reproduced with permission.

spectra between 5.10 and 5.65 min shows a number of ions, including that of $m/z$ 285, which do not appear in the mass spectra recorded after 5.10 and 5.65 min. The RICs for $m/z$ 163, from the component eluting after 5.10 min, for $m/z$ 199, from the compound eluting after 5.65 min, and for $m/z$ 285, are shown in Figure 3.26. A component, not readily observed in either the TIC or UV traces, eluting after 5.32 min, is clearly present and may be investigated further.

### *3.6.3 Quantitative Analysis*

To enable quantitative measurements to be made, the analyst requires the ability to determine the areas or heights of the detector responses of analyte(s) and any internal standard that may be present and then, from these figures, to derive the amount(s) of analyte(s) present in the unknown sample. The software provided with the mass spectrometer allows this to be done with a high degree of automation if the analyst so desires.

The result of using such a procedure on the TIC trace shown in Figure 3.13(b) above is illustrated in Figure 3.27. Here, the annotation for each ‘peak’ consists of three figures, where the first, e.g. 3.84, is the retention time of the peak, the second, e.g. 422 449, is the peak height, and the third, e.g. 785 853, is the peak area.

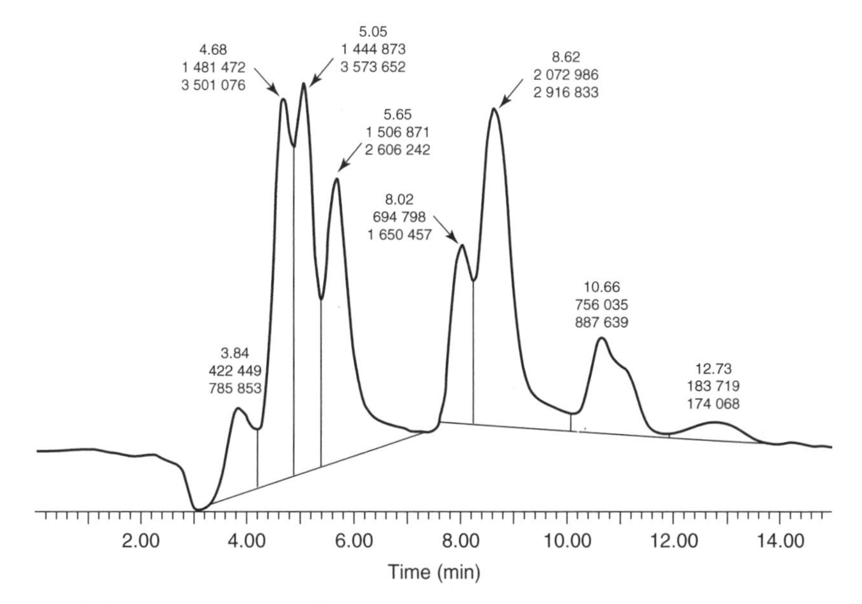


**Figure 3.27** TIC trace obtained from the LC–MS analysis of a pesticide mixture, showing integrated peak areas and heights. From applications literature published by Micromass UK Ltd, Manchester, UK, and reproduced with permission.

From a quantitative perspective, each 'peak' is defined by two parameters, i.e. the position of its baseline and the retention time boundaries, with those derived by the computer system being shown in Figure 3.27. It is not the intention of this present author to discuss how these have been determined but simply to point out that their positions may have a significant effect on the accuracy and precision of any quantitative measurements, especially, as in Figure 3.27, when the baseline is not horizontal and the signals from each of the components are not fully resolved. It is usual for the software to allow the analyst to override the parameters chosen by the computer to provide what they consider to be more appropriate peak limits and/or baseline positions.

As stated previously, the advantage of the mass spectrometer is that mass can be used as a discriminating feature and this may allow quantitative measurements to be made on unresolved components.

Figure 3.26 shows an example of the use of RICs to locate a compound which is not chromatographically resolved from the major components of the mixture and which is not obviously present from an examination of the TIC trace. In this case, integration of these RICs, as shown in Figure 3.28, would allow quantitative measurements to be made on all three components, although with different degrees of precision.

The data considered above have been derived from a TIC trace, i.e. have been acquired from full scanning, but the same methodology is used for analysing

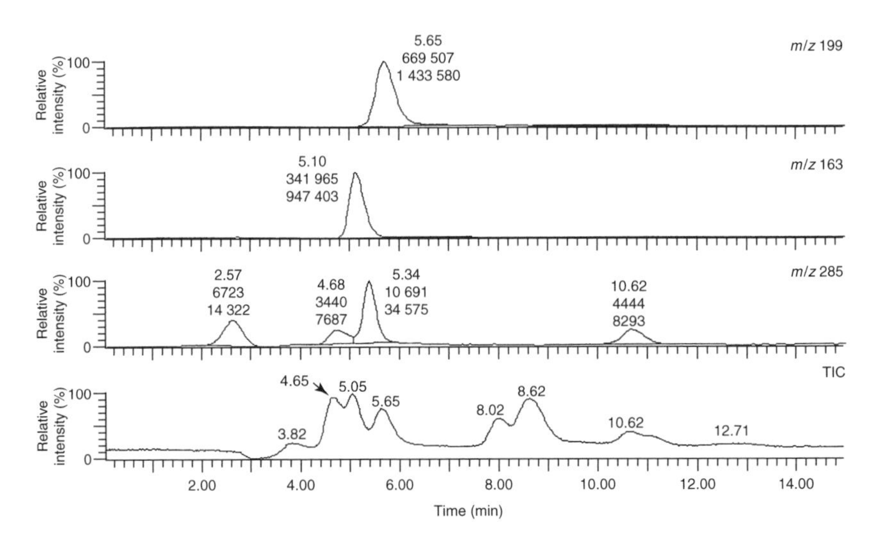


**Figure 3.28** Representative reconstructed ion chromatograms obtained from the LC–MS analysis of a pesticide mixture, showing integrated peak area and height measurements. From applications literature published by Micromass UK Ltd, Manchester, UK, and reproduced with permission.

data generated by SIM (see Section 3.5.2.1 above). It should be noted, however, that the selectivity obtained using RICs is the same as would be obtained by using SIM, whereas the sensitivity is not.

### DQ 3.4

Why is the sensitivity obtained when using reconstructed ion chromatograms (RICs) for quantitation less than that achieved when employing selected-ion monitoring (SIM) to monitor the same ions?

*Answer*

*To answer this question, we must consider the ways in which the data are acquired. An RIC is generated, post-acquisition, from consecutive full scans in which a small amount of time is spent monitoring each ion, as discussed above. The data produced in an SIM experiment are generated by monitoring only a small number of ions, thus taking advantage of the increased time spent monitoring each ion.*

It should not be concluded that the above examples of the evaluation of qualitative and quantitative data comprise an exhaustive analysis of this particular set of LC–MS data. They have been included primarily for those not used to the analysis of mass spectral data, to show the principles involved, and to demonstrate how powerful the mass spectrometer can be as a chromatographic detector.

It can also be readily seen that the analysis of data may well take a considerable amount of time, often significantly more time than the acquisition of the data itself, and that the use of profiles of specific masses may allow information not readily obvious from other detectors, and even the TIC trace, to be obtained.

### *3.6.4 The Use of Tandem Mass Spectrometry*

This will be discussed in greater detail in Chapter 5, but at this point it is worthwhile to consider a brief example of the analytical capabilities of MS–MS.

A reported method for the screening for transformation products of a number of pesticides [16] provides an elegant example of the complementary nature of the product-ion, precursor-ion and constant-neutral-loss scans (see Section 3.4.2 above).

The product-ion spectra of the $(M + H)^+$ ions of atrazine, the structure of which is shown in Figure 3.29, and three of its transformation products showed that if the isopropyl side-chain was present in the structure a constant neutral

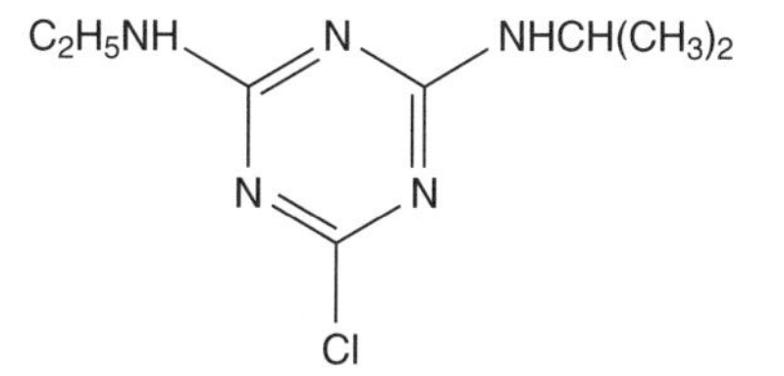


**Figure 3.29** Structure of atrazine. Reprinted from *J. Chromatogr., A*, **915**, Steen, R. J. C. A., Bobeldijk, I. and Brinkman, U. A. Th., 'Screening for transformation products of pesticides using tandem mass spectrometric scan modes', 129–137, Copyright (2001), with permission from Elsevier Science.

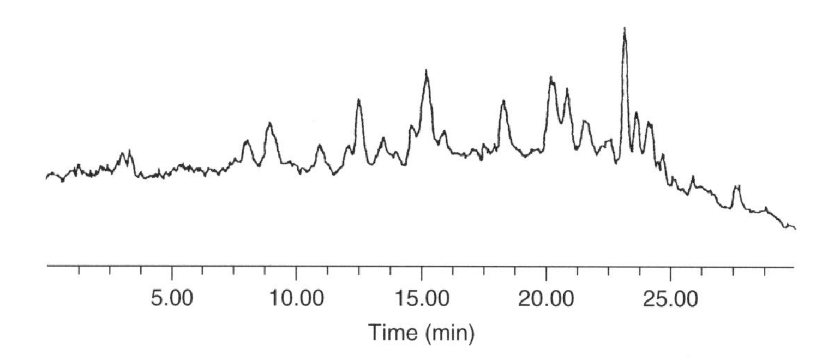


**Figure 3.30** TIC trace obtained from the LC–MS analysis of atrazine and its degradation products. Reprinted from *J. Chromatogr., A*, **915**, Steen, R. J. C. A., Bobeldijk, I. and Brinkman, U. A. Th., 'Screening for transformation products of pesticides using tandem mass spectrometric scan modes', 129–137, Copyright (2001), with permission from Elsevier Science.

loss of 42 Da was detected, while if the chlorine atom remained attached to the ring an intense product ion at $m/z$ 68, attributed to $[N{=}C{-}NH{-}C{=}N{=}H]^+$, was observed.

The TIC trace from the LC–MS analysis of an extracted river water sample, spiked with 3 $\mu g\,dm^{-3}$ of atrazine and three of its degradation products, is shown in Figure 3.30. The presence of significant levels of background makes confirmation of the presence of any materials related to atrazine very difficult. The TIC traces from the constant-neutral-loss scan for 42 Da and the precursor-ion scan for $m/z$ 68 are shown in Figure 3.31 and allow the signals from the target compounds to be located readily.

Further information on each of these components can then be obtained by examining the mass spectra at the positions of the TIC maxima in the traces. The spectra from the two components marked OH and DEA in the constant-neutral-loss (CNL) TIC trace (Figure 3.31(a)) are shown in Figure 3.32. The molecular weights of the

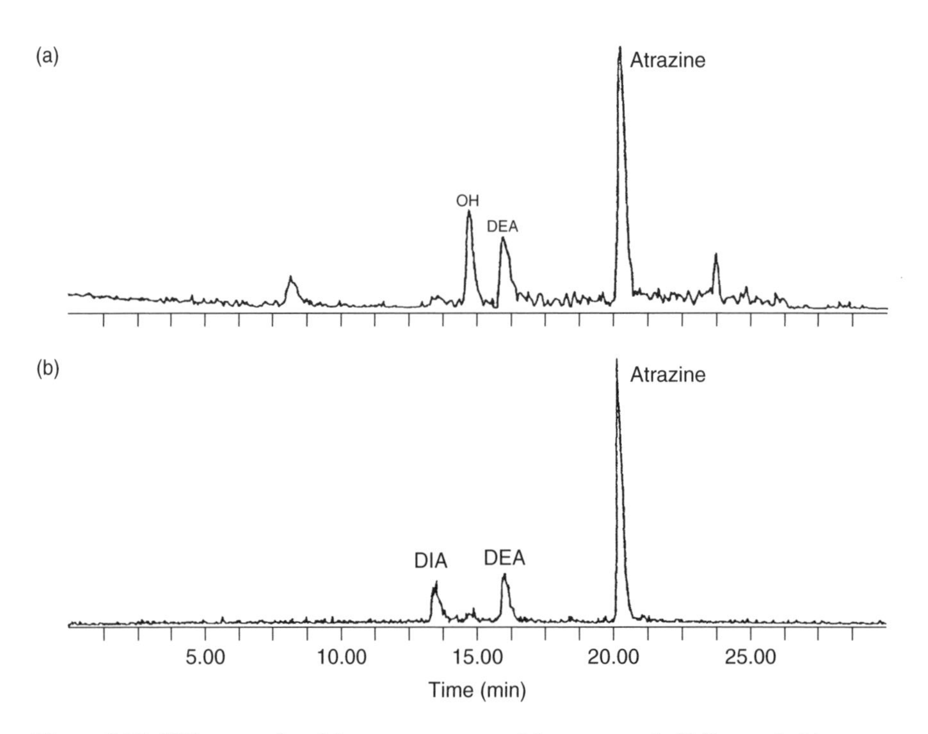


**Figure 3.31** TIC traces for (a) a constant-neutral-loss scan of 42 Da, and (b) a precursor-ion $m/z$ 68 scan, obtained from the LC–MS analysis of a mixture of atrazine and its degradation products. Reprinted from *J. Chromatogr., A*, **915**, Steen, R. J. C. A., Bobeldijk, I. and Brinkman, U. A. Th., 'Screening for transformation products of pesticides using tandem mass spectrometric scan modes', 129–137, Copyright (2001), with permission from Elsevier Science.

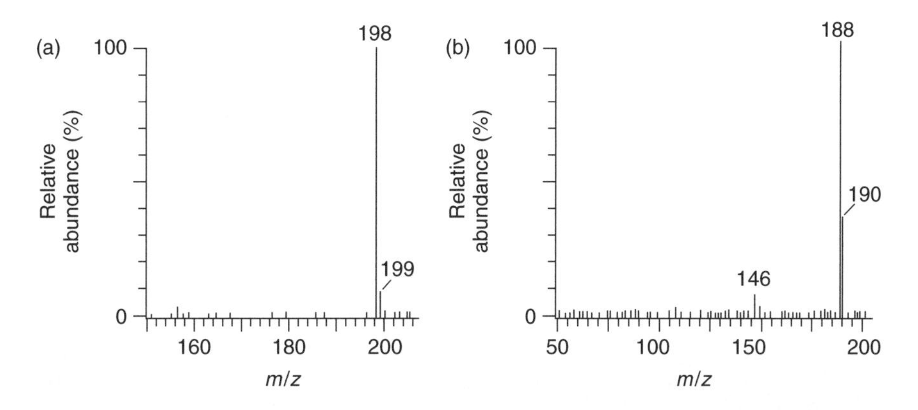


**Figure 3.32** Full-mass spectra at peak maxima of the constant-neutral-loss TIC trace shown in Figure 3.31(a), obtained after (a) 13.25 and (b) 15.9 min, from the LC–MS analysis of a mixture of atrazine and its degradation products. Reprinted from *J. Chromatogr., A*, **915**, Steen, R. J. C. A., Bobeldijk, I. and Brinkman, U. A. Th., 'Screening for transformation products of pesticides using tandem mass spectrometric scan modes', 129–137, Copyright (2001), with permission from Elsevier Science.

two products can be seen to be 187 and 197, with the isotopic distribution around the molecular species (two ions 2 Da apart in a ratio of 3:1, reflecting the relative abundance of $^{35}$Cl and $^{37}$Cl) showing the former to contain chlorine.

Were these to be true unknowns, then product-ion scans could be carried out on these molecular species in an attempt to obtain further structural information.

# Summary

In this chapter, the main aspects of mass spectrometry that are necessary for the application of LC–MS have been described. In particular, the use of selected-ion monitoring (SIM) for the development of sensitive and specific assays, and the use of MS–MS for generating structural information from species generated by soft ionization techniques, have been highlighted. Some important aspects of both qualitative and quantitative data analysis have been described and the power of using mass profiles to enhance selectivity and sensitivity has been demonstrated.

# References

1. (a) National Institute of Standards and Technology, *NIST/EPA/NIH Mass Spectral Database*, Standard Reference Database 1, National Institute of Standards and Technology, Gaithersburg, MD, 1992; (b) National Institute of Standards and Technology, *NIST 98*, Standard Reference Database 1, Standard Reference Data Program, National Institute of Standards and Technology, Gaithersburg, MD, 1998.

2. Chapman, J. R., *Practical Organic Mass Spectrometry*, 2nd Edn, Wiley, Chichester, UK, 1993.
3. De Hoffmann, E., Charette, J. and Stroobant, V., *Mass Spectrometry – Principles and Applications*, Wiley, Chichester, UK, 1996.
4. Barker, J., *Mass Spectrometry*, 2nd Edn, ACOL Series, Wiley, Chichester, UK, 1999.
5. Russell, D. H. (Ed.), *Experimental Mass Spectrometry*, Plenum Press, New York, 1989.
6. Siuzdak, G., *Mass Spectrometry for Biotechnology*, Academic Press, San Diego, CA, 1996.
7. McLafferty, F. W. and Turecek, F., *Interpretation of Mass Spectra*, 4th Edn, University Science Books, Mill Valley, CA, 1993.
8. Ashcroft, A. E., *Ionization Methods in Organic Mass Spectrometry*, RSC Analytical Spectroscopy Monograph, The Royal Society of Chemistry, Cambridge, UK, 1997.
9. McLafferty, F. W. (Ed.), *Tandem Mass Spectrometry*, Wiley, New York, 1983.
10. Busch, K. L., Glish, G. L. and McLuckey, S. A., *Mass Spectrometry/Mass Spectrometry: Techniques and Applications of Tandem Mass Spectrometry*, VCH, New York, 1988.
11. Spengler, B., *J. Mass. Spectrom.*, **32**, 1019–1036 (1997).
12. Trainor, J. and Derrick, P. J., 'Sectors and tandem sectors', in *Mass Spectrometry in the Biological Sciences: A Tutorial*, Gross, M. L. (Ed.), Kluwer Academic Publishers, Dordrecht, The Netherlands, 1990, pp. 3–27.
13. Desaire, H. and Leary, J. A., *J. Am. Soc. Mass Spectrom.*, **11** 1086–1094 (2000).
14. Borchers, C., Parker, C. E., Deterding, L. J. and Tomer, K. B., *J. Chromatogr., A*, **854**, 119–130 (1999).
15. Wehr, T., *LC–GC*, **18**, 406–416 (2000).
16. Steen, R. J. C. A., Bobeldijk, I. and Brinkman, U. A. Th., *J. Chromatogr. A*, **915**, 129–137 (2001).

Chapter 4

# Interface Technology[†]

**Learning Objectives**

- To be aware of the range of interfaces that have been used for LC–MS.
- To understand the principles of operation of each of these interfaces, in particular with regard to the way in which they achieve compatibility between high performance liquid chromatography and mass spectrometry.
- To be aware of the experimental parameters, both for HPLC and MS, that affect the way in which the interface functions and the effect that these have on the analytical information which is generated.
- To be aware of the types of analytical information that can be provided by each of these interfaces.
- To be aware of the strengths and weaknesses of each of these interfaces.

## 4.1 Introduction

The need for a more definitive identification of HPLC eluates than that provided by retention times alone has been discussed previously, as have the incompatibilities between the operating characteristics of liquid chromatography and mass spectrometry. The 'combination' of the two techniques was originally achieved by the physical isolation of fractions as they eluted from an HPLC column, followed by the removal of the mobile phase, usually by evaporation, and transfer of the analyte(s) into the mass spectrometer by using an appropriate probe.

---

[†] A somewhat more descriptive alternative title for this chapter could be 'Getting the Plumbing Right'!

There are a number of potential problems with this approach, as follows:

(i) Any loss of analyte during fraction collection will increase the detection limits. This is of particular importance when there is initially only a small amount of the component(s) present.

(ii) If the chromatographic separation provides well-resolved components, these can usually be isolated separately and spectral data from each component obtained and interpreted. If complete resolution is not achieved, it is not possible to isolate each individual component and the spectra obtained from a single fraction are likely to be from a mixture. These spectra, depending upon the ionization technique used, are difficult to interpret without the use of MS–MS and it must be borne in mind that these techniques were not readily available at the time that fraction collection was the only possible way that HPLC eluates could be studied. Indeed, it may be argued that the advent of soft ionization techniques in combination with HPLC separation was one of the main reasons why MS–MS has become as commonplace as it now is.

(iii) HPLC is particularly applicable to the separation of thermally labile compounds. If the HPLC mobile phase contains a significant proportion of water, i.e. a reversed-phase system is being used, it is likely that heat will be required for its removal and this carries with it the increasing risk that thermal decomposition of the analyte will occur.

(iv) The mass spectrometry ionization technique needs to be chosen carefully. If molecular weight and/or structural information are required, chemical and electron ionization, respectively, are the methods of choice. Since, however, these are both only directly applicable to volatile and thermally stable compounds, i.e. GC- rather than HPLC-compatible analytes, more often than not they are unsuccessful and useful spectra are not obtained. Alternative ionization techniques are often required, e.g. field desorption and fast-atom bombardment [1], and again, at the time fraction collection was carried out routinely, these were not generally available in mass spectrometry laboratories and, in any case, are often limited in the analytical information they can provide.

The direct linking of HPLC to mass spectrometry removes the need for fraction collection and with it the potential problems discussed above. It does, however, introduce a number of other problems which have been mentioned in earlier chapters of this book.

The main one is the incompatibility of HPLC, utilizing flow rates of ml $min^{-1}$ of a liquid, and the mass spectrometer, which operates under conditions of high vacuum. Even if this can be overcome, attention must then be focussed on the ionization of the analyte, bearing in mind the limitations of EI and CI discussed earlier in Chapter 3, and the generation of analytically useful mass spectra.

The history of LC–MS has been recently described [2] and will not be repeated here in detail. It is worthwhile to note, however, the following dates, quoted in this reference, when the interfaces to be described in subsequent parts of this chapter became available commercially:

- Moving-belt interface — 1977
- Direct-liquid-introduction interface — 1980
- Thermospray interface — 1983
- Frit FAB/continuous-flow FAB interface — 1985/1986
- Atmospheric-pressure chemical ionization interface — 1986
- Particle-beam interface — 1988
- Electrospray interface — 1988

In summary, it can be said that prior to the development of the thermospray interface there were an increasing number of reports of the analytical application of LC–MS [3] but in this present author's opinion, based on a number of years of using a moving-belt interface, the technique could not be considered to be 'routine'. The thermospray interface changed this and with the commercial introduction of the combined APCI/electrospray systems in the 1990s the technique, for it now may be considered as a *true hybrid technique*, has reached maturity (although this should not be taken as a suggestion that there will be no further developments).

## 4.2 The Moving-Belt Interface

The first interface to be made available commercially was the moving-belt interface, shown schematically in Figure 4.1.

The operation of the interface may be divided conveniently into four stages, as follows:

(i) application of mobile phase and analyte(s) to a continuously moving belt;

(ii) removal of the mobile phase by passage of the belt under an infrared heater and through a number of differentially pumped regions;

(iii) flash desorption/vaporization of the analyte into the source of the mass spectrometer or, latterly, ionization by a surface technique such as FAB;

(iv) cleaning of the belt with a heater and/or a wash-bath to remove any involatile materials or excess sample prior to the application of further mobile phase and analyte(s) and a repeat of steps (ii)–(iv).

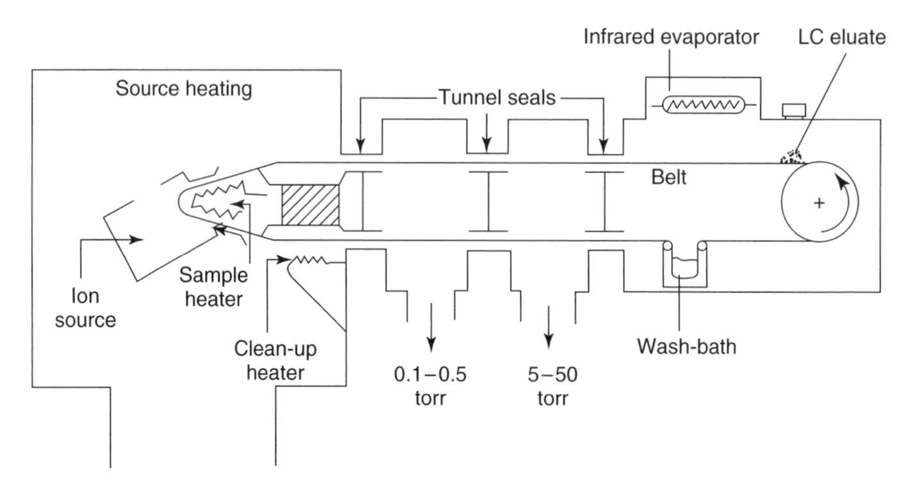


**Figure 4.1** Schematic of a moving-belt LC–MS interface. From applications literature published by Micromass UK Ltd, Manchester, UK, and reproduced with permission.

A uniform deposit of analyte(s) on the belt is required and it is possible to do this with a range of mobile phases and flow rates by a very careful balancing of the rate of solvent deposition, the speed at which the belt moves and the amount of heat supplied by the infrared evaporator.

If the belt moves too quickly, in relation to the rate of deposition, sample will not be deposited on all parts of the belt. This results in the production of an uneven total-ion-current (TIC) trace and a distortion of the mass spectra obtained, with consequent problems in interpretation, particularly if library searching is employed.

An erratic TIC trace is also obtained if the belt is moving too slowly but in these circumstances this is due to the formation of droplets rather than the spreading of mobile phase on the belt. An additional problem encountered when droplets are formed on the belt is that more heat is required to evaporate the solvent and with this comes the increased likelihood of decomposition of any thermally labile compounds that may be present.

When optimum experimental conditions have been obtained, all of the mobile phase is removed before the analyte(s) are introduced into the mass spectrometer for ionization. As a consequence, with certain limitations, it is possible to choose the ionization method to be used to provide the analytical information required. This is in contrast to the other LC–MS interfaces which are confined to particular forms of ionization because of the way in which they work. The moving belt can therefore provide both electron and chemical ionization spectra, yielding both structural and molecular weight information.

Only analytes possessing some degree of volatility, however, can be transferred into the mass spectrometer by direct desorption/vaporization from the belt. While this extends the range of compounds from which EI spectra can be obtained – some compounds which do not provide spectra when a conventional solids probe is used provide spectra when introduced via the moving-belt interface – there is still a significant number of compounds that can be separated by using HPLC but from which spectra cannot be obtained in this way. These less volatile and thermally labile compounds can be studied by using the moving belt with fast-atom bombardment (FAB) ionization which brings about ionization on the surface of the belt.

Only around 10% of the reported uses of the moving-belt interface involved the use of FAB; the vast majority, some 90%, have utilized EI or CI [2].

The moving belt itself is manufactured from either stainless steel or polyimide (a nylon-like polymer), with the latter being more widely used, especially for magnetic-sector instruments operating at high source potentials of around 4 kV. The polyimide belt, initially at least, gives an intense chemical background over a significant part of the $m/z$ range. This necessitates 'conditioning' of the belt, for around 12 h, before use. The changing of a belt therefore results in the interface, and indeed the mass spectrometer to which it is fitted, being put out of use. The equivalent problem in a modern electrospray interface, i.e. a blocked capillary, can be remedied in less than an hour!

The belt suffers from mechanical instability, often causing it to break, usually at the most inconvenient time ('Murphy's Law' – the most important scientific principle in any experimental discipline!). The tunnel seals, used to isolate the differential vacuum regions of the interface, are the most likely places for the belt to snag. Inefficient cleaning of the belt of residual sample and/or inorganic buffers (see below) tends to exacerbate this problem.

A uniform film of analyte, which is required for the production of good quality spectra, can usually be obtained from mobile phases which contain predominantly organic solvents (normal-phase systems). As the percentage of water in the mobile phase increases, however, droplets tend to form on the belt, irrespective of the belt speed. If the belt is not exactly horizontal, and this is often the case, especially after it has been in use for some time, the droplets are likely to roll off the belt and be lost, thus reducing the overall sensitivity of the analysis dramatically.

As mentioned previously, the formation of droplets requires the use of extra heat to effect complete solvent evaporation and with this comes the potential for decomposition of thermally labile materials.

If evaporation is not complete, the majority of the liquid on the belt is not able to pass through the tunnel seals and this leads to a reduction in sensitivity. Any liquid that does pass through is likely to cause significant pressure fluctuations within the mass spectrometer and this will lead to inconsistent performance, particularly in respect of the sensitivity which is obtained. If the heat input can be adjusted to effect complete evaporation of the droplets, an erratic TIC trace is obtained, as discussed previously.

From an analytical perspective, the presence of droplets can also lead to apparent chromatographic peak broadening and a loss of resolution and analytical performance.

A number of solutions to the problem of droplet formation have been proposed and these include both modifications to the interface itself as well as to the chromatographic techniques employed.

The use of the lower flow rates employed with microbore HPLC columns or splitting of the eluate from a conventional column will immediately reduce the volume of liquid being presented to the interface and, while not necessarily totally removing the tendency to form droplets, at least is likely to make the situation more manageable.

The use of smaller volumes of solvent is beneficial, as it is with the direct-liquid introduction (DLI) interface (see Section 4.3 below), in that it reduces the level of mass spectral background arising from the mobile phase and this facilitates the interpretation of the mass spectra obtained. The use of microbore HPLC would be expected to generate sharper chromatographic peaks, thereby increasing the mass flow of analyte to the detector and producing an increase in sensitivity, although the implementation of microbore technology is not trivial. In contrast, the approach of splitting the column flow and allowing only a proportion to the moving belt will reduce the overall sensitivity.

Since droplet formation is a particular problem with aqueous mobile phases, continuous post-column solvent extraction, in which the solutes are extracted into an immiscible organic mobile phase, has been proposed [4]. The mobile phase reaching the belt thus becomes totally organic in nature and is much more easily removed. The major disadvantage of this approach is the possible loss of analyte during the extraction procedure.

The major advance in the way in which column eluate is deposited on the belt was the introduction of spray deposition devices to replace the original method which was simply to drop liquid onto the belt via a capillary tube connected directly to the outlet of the HPLC column. These devices, based on the gas-assisted nebulizer [5], have high deposition efficiencies, transfer of sample can approach 100% with mobile phases containing up to 90% water, and give constant sample deposition with little band broadening.

The use of spray deposition increases the range of solvents which can be used in moving-belt LC–MS and the range of solutes that can be studied by this technique. Since less heat is required to remove the solvent, it is less likely that the solute will be inadvertently removed from the belt or undergo thermal degradation. It is not, however, unknown for particularly volatile and labile analytes to be lost when using spray deposition.

Early versions of this interface did not have any facility to clean the belt after its passage over the sample heater. If the analyte is relatively involatile and/or present in large amounts, a spectrum is sometimes obtained each time the part of the belt on which the sample has been deposited returns to the sample heater. These 'memory effects' are largely removed by incorporating both a second heater and a bath containing an appropriate solvent through which the belt passes.

Reference has been made to the problems associated with the presence of highly involatile analytes. Many buffers used in HPLC are inorganic and thus involatile and these tend to compromise the use of the interface, in particular with respect to snagging of the belt in the tunnel seals. The problem of inorganic buffers is not one confined to the moving-belt interface and, unless post-column extraction is to be used, those developing HPLC methods for use with mass spectrometry are advised to utilize relatively volatile buffers, such as ammonium acetate, if at all possible.

## Summary

The advantages and disadvantages of this type of interface are described below. While the list of disadvantages may appear lengthy, this has to be considered in the light of the alternative interface which was available at the same time, i.e. the DLI interface (see Section 4.3 below), and also compared with what was then possible without the use of an interface.

### *Advantages*

- The interface can be used with a wide range of HPLC conditions, flow rates and mobile phases, both normal and reverse phase, particularly if spray deposition is employed.
- The analyst does have some choice of the ionization method to be used; EI, CI and FAB are available, subject to certain limitations, and thus both molecular weight and structural information may be obtained from the analyte(s) under investigation.

### *Disadvantages*

- An intense chemical background from the material from which the belt is made is often observed in the mass spectra generated by this type of interface unless adequate conditioning is carried out.
- The belt is prone to break during operation.
- Problems may be encountered in the analysis of thermally labile compounds, as heat is required for mobile-phase removal and for the transfer of analyte from the belt into the source of the mass spectrometer, and highly involatile compounds which cannot be desorbed from the belt, unless FAB is used for ionization.

- Mobile phases containing high proportions of water often give droplets on the belt rather than an even film and this may produce an erratic TIC trace and irreproducible mass spectra.
- Memory effects are often experienced, even with the use of a clean-up heater and a wash-bath.
- Surface effects can reduce detection limits.
- The use of involatile buffers must be avoided in order to prevent build-up of material on the belt and its jamming in the tunnel seals.
- This type of interface is not easy to construct and therefore reliance has to be put on commercial manufacturers.

## 4.3 The Direct-Liquid-Introduction Interface

The direct-liquid-introduction (DLI) interface was made available commercially just after the moving-belt interface to which, as no company produced both types, it was an alternative. At this time, therefore, the commercial LC–MS interface used within a laboratory was dictated by the manufacturer of the mass spectrometer already in use unless a new instrument was being purchased solely for LC–MS applications. The development of LC–MS in the early 1980s was such that this was very rare and it was therefore unusual that a scientific evaluation was carried out to assess the ability of a type of interface to solve problems within a particular laboratory.

The direct-liquid-introduction interface is shown schematically in Figure 4.2. This system is effectively a probe, at the end of which is a pinhole of approximately 5 μm diameter, which abuts a desolvation chamber attached to the ion source of the mass spectrometer. The eluate from an HPLC column is circulated

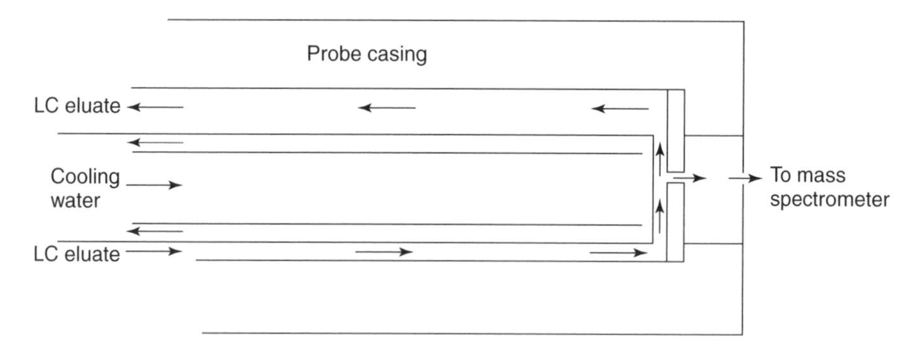


**Figure 4.2** Schematic of a direct-liquid-introduction LC–MS interface. From applications literature published by Agilent Technologies UK Limited, Stockport, UK, and reproduced with permission.

through the probe and as it reaches the pinhole the vacuum in the mass spectrometer draws a proportion of it into the desolvation chamber, and subsequently the source of the mass spectrometer, where analytes are ionized by solvent-mediated chemical ionization processes.

The most important part of this type of interface, from a number of points of view, is the pinhole which, in conjunction with the pumping capacity of the mass spectrometer, controls the flow of eluate into the mass spectrometer. This flow, and therefore the properties of the spray being introduced into the mass spectrometer, is affected by a change in the viscosity of the mobile phase. The use of gradient elution has therefore to be approached with some caution as the sensitivity of the mass spectrometer can change significantly during the course of an analysis.

The maximum flow rate that can be accommodated while still allowing the mass spectrometer to operate is in the range of 10–20 $\mu l\,min^{-1}$. Typical flow rates used in conventional HPLC separations are between 500 and 1000 $\mu l\,min^{-1}$ and therefore only between 1 and 4% of the column eluate, and therefore analyte(s), enter the mass spectrometer source. The sensitivity, or more accurately the *lack* of sensitivity, of the DLI interface is one of its major limitations.

This can potentially be overcome by the use of microbore HPLC columns with flow rates which are directly compatible with mass spectrometer operation, although the necessary decrease in injection volume results in little overall gain in the concentration of sample reaching the mass spectrometer. In addition, at the time that the DLI was available, the use of microbore HPLC, which introduces another set of potential problems related to chromatographic performance, was probably as widespread as the use of LC–MS! It has been assessed [2] that in around 25% of the reported applications of DLI, microbore HPLC has been utilized.

The liquid jet from the DLI probe has to be initiated at atmospheric pressure, i.e. before insertion of the interface into the mass spectrometer, and, for best performance, the spray direction has to be coaxial to the probe. Any deviation from this, however slight, tends to produce changes in the mass spectrum obtained.

One of the major problems with this type of interface, not unsurprisingly, is clogging of the pinhole. For this reason, the HPLC system has to be kept scrupulously clean with solvents being passed through narrow filters to remove any solid particles and in-line filters being incorporated to ensure that column material does not find its way into the probe.

A completely blocked probe is easy to diagnose but of more concern is partial blocking which is not always obvious. Not only will this give reduced sensitivity with less eluate reaching the mass spectrometer, but, if the build-up of material in the pinhole is not even, a non-axial spray is produced which leads to a reduction in sensitivity and changes in the appearance of the mass spectra being generated.

Involatile inorganic buffers, when used as mobile-phase additives, are the prime cause of blocking of the pinhole. The situation can be alleviated either by replacing them by a more volatile alternative, such as ammonium acetate, or by using post-column extraction to separate the analytes from the buffer, with the analytes, dissolved in an appropriate organic solvent, being introduced into the mass spectrometer.

The mobile phase is not totally removed by this interface and consequently ionization by the mass spectrometer is restricted to chemical ionization, which is a relatively high pressure technique. As described earlier in Section 3.2.2, CI produces an adduct of the analyte under investigation with a reactive species generated by the reagent gas which, in the case of the DLI interface, is the vapour from the HPLC mobile phase. The most widely used reagent gases in CI are methane (proton affinity, 129.9 kcal mol$^{-1}$ (543.5 kJ mol$^{-1}$)), isobutane (proton affinity, 162.0 kcal mol$^{-1}$ (677.8 kJ mol$^{-1}$)) and ammonia (proton affinity, 204.0 kcal mol$^{-1}$ (853.6 kJ mol$^{-1}$)) [6]. Methane has a relatively low proton affinity and will consequently protonate the majority of analytes in, what is in CI terms, a relatively energetic process that may well lead to fragmentation of the analyte. Isobutane has a slightly higher proton affinity and will, consequently, protonate fewer compounds. The process, however, is less energetic and, usually, very few fragments ions are formed. Ammonia has a high proton affinity and will only protonate species with an even higher proton affinity but will form adducts, usually with $NH_4^+$, with many compounds. The proton affinities of commonly used reverse-phase solvents are water, 165.2 kcal mol$^{-1}$ (691.0 kJ mol$^{-1}$), methanol, 180.3 kcal mol$^{-1}$ (754.3 kJ mol$^{-1}$) and acetonitrile, 186.2 kcal mol$^{-1}$ (779.2 kJ mol$^{-1}$)) [6], and from these values it can be seen that CI-type processes, known as solvent-mediated CI, are likely to occur in the source of the mass spectrometer, with protonated or adduct species being produced. The formation of adducts is, in itself, not usually a problem and, as we will see later, is a common feature of LC–MS with interfaces, such as thermospray, electrospray and atmospheric-pressure chemical ionization, that operate at relatively high pressures.

There is sometimes a conflict between the need to choose a mobile phase capable of providing adequate separation and one that gives suitable ionization within the mass spectrometer.

It is possible to obtain significant chemical background with the DLI as any impurities present in the mobile phase are being continuously introduced into the mass spectrometer.

From a practical point of view, the DLI, unlike the moving-belt interface, contains no moving parts and is therefore more reliable in operation if adequate precautions are taken to minimize the frequency of the pinhole blocking. In addition, it does not require heat either to remove the mobile phase or to vaporize the analyte into the source of the mass spectrometer. The DLI is, consequently, better for the analysis of thermally labile materials.

### *Summary*

The advantages and disadvantages of this type of interface, particularly in comparison to the moving-belt interface which was available at the same time, are listed below. This was one of the first LC–MS interfaces to be made commercially available and, although used in a number of laboratories, its development was halted prematurely by the introduction of the thermospray interface (as we shall see later).

*Advantages*

- No heat is applied to the interface and it is therefore able to deal with thermally labile materials better than the moving-belt interface.
- The interface contains no moving parts and is cheap and simple to construct and operate and is inherently more reliable than the moving-belt interface.
- Both positive- and negative-ion CI spectra can be generated and the interface provides molecular weight information, plus it can also be used for sensitive quantitative and semi-quantitative procedures.

*Disadvantages*

- Involatile compounds are not usually ionized with good efficiency.
- The pinhole is prone to blockage and therefore the system must be kept completely free of solid materials.
- Only a small proportion of the flow from a conventional HPLC column is able to enter the source of the mass spectrometer and sensitivity is consequently low.
- Ionization is brought about by CI-like processes and structural information is therefore limited unless a mass spectrometer system capable of MS–MS is used.

## 4.4 The Continuous-Flow/Frit (Dynamic) Fast-Atom-Bombardment Interface

The full potential of LC–MS could not be exploited until it was possible to study involatile and thermally labile compounds for which electron and chemical ionization are not appropriate. A relatively small number of reports of the use of the moving-belt interface with fast-atom bombardment ionization for the study of these types of compound have appeared.

The fact that moving-belt interfaces were not offered as accessories by all of the major manufacturers of mass spectrometers, plus the fact that these interfaces could not be easily constructed within the laboratories where the technique was required, necessitated the search for an alternative approach to the study of involatile materials by LC–MS.

In conventional FAB (see Section 3.2.3), the analyte is mixed with an appropriate matrix material and applied to the end of a probe where it is bombarded with a fast-atom, or latterly, a fast-ion, beam. There are two major considerations when linking HPLC to such a system, namely (a) how is the matrix material, which is crucial for effective ionization in conventional FAB, to be incorporated into the system, and (b) how is the flowing HPLC system to be continuously presented to the ionizing beam?

Two forms of interface have been commercially developed [7] which allow analytes in a flowing liquid stream – it has to be pointed out, not necessarily from an HPLC system (see below) – to be ionized by using FAB. These are essentially identical except for that part where the HPLC eluate is bombarded with the heavy-atom/ion beam. Both of these interfaces consist of a probe in the centre of which is a capillary which takes the flowing HPLC eluate. In the continuous-flow FAB interface (Figure 4.3), the column eluate emerges from the end of the capillary and spreads over the probe tip, while in the frit-FAB interface the capillary terminates in a porous frit onto which the atom/ion beam is directed.

Excess mobile phase has to be removed from the probe tip to allow its replenishment and this may be accomplished in a number of ways, for example, with an absorbent pad situated adjacent to the area subjected to atom/ion bombardment.

Typically, flow rates in HPLC are around 1 ml $min^{-1}$, while the vacuum requirements of the mass spectrometer preclude liquid delivery of more than around 15 μl $min^{-1}$ at the probe end. To achieve compatibility therefore requires either the splitting of the flow from a conventional column or the use of some form of HPLC, such as a packed microcolumn, which provides directly compatible flow rates. Whichever of these solutions is employed, the amount of analyte reaching the mass spectrometer, and thus the overall sensitivity of the analysis

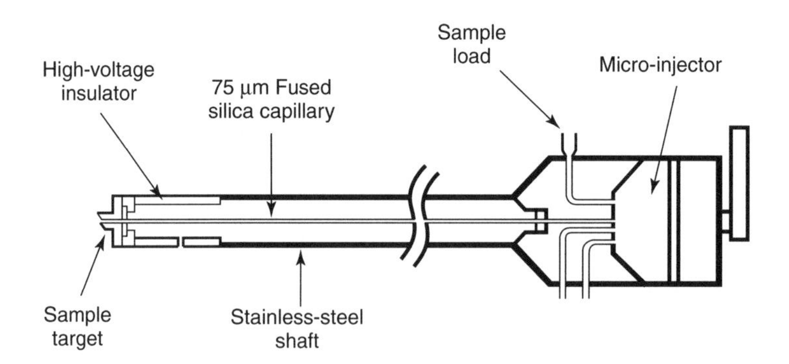


**Figure 4.3** Schematic of a continuous-flow FAB LC–MS interface. From applications literature published by Kratos Analytical Ltd, Manchester, UK, and reproduced by permission of Mass Spectrometry International Ltd.

being carried out, is reduced. The overall sensitivity of dynamic FAB, however, is greater than static FAB and this is attributed to a reduction in the number of analyte–matrix clusters that are formed.

The use of low flow rates introduces two further practical problems. The first is the inability to maintain stable conditions at the end of the probe, hence resulting in fluctuations in ion current, as experienced when droplets are formed on the moving belt. As the liquid emerges onto the probe tip, it experiences the high vacuum and begins to evaporate, with a consequent reduction in the temperature of the probe tip. Sufficient heat must therefore be applied to prevent freezing of the mobile phase and this helps stabilize ion production.

The second is associated with the irreproducibility of the mobile phase gradients that may be formed. The latter is overcome by forming the gradient at a conventional flow rate and then splitting prior to the injector.

The matrix, which in most reported applications appears to be glycerol, may either be incorporated directly into the mobile phase pre-column or added post-column. If added to the mobile phase, its effect on the separation must be considered, while if added post-column, significant peak broadening may be observed.

In static FAB, the analyte is dissolved in the matrix, which fulfils two purposes, namely (i) the transfer of energy from the bombarding ion/atom beam to the analyte, and (ii) the continual replenishing of the bombarded surface with analyte molecules to allow the maximum number to be ionized. In dynamic FAB, the replenishing of the surface is not quite so important, as the liquid stream continually provides a fresh supply of analyte molecules, and the main purpose of the matrix is to provide energy transfer. In dynamic FAB, it is possible to employ smaller amounts of matrix material without reducing ionization efficiency and it is usual to use approximately 1 wt% matrix for frit-FAB and around 5 wt% for continuous-flow-FAB. The presence of this matrix material does lead to a permanent chemical background being observed in the mass spectra obtained although this is only of importance, as with any background, if it occurs at $m/z$ values also generated by the analytes of interest (Murphy's Law often comes into play here!).

Suppression effects are experienced in static FAB, with signals from more hydrophilic materials being reduced compared to those from hydrophobic components. There are fewer suppression effects in dynamic FAB and this is of benefit when it is not possible to achieve complete chromatographic resolution.

It has been previously noted (see Section 4.2 above) that use of the moving-belt interface allows EI spectra to be obtained from compounds that do not yield spectra when analysis is attempted using a conventional EI probe. The same is true when the dynamic-FAB probe is used in that spectra may be obtained from compounds that do not yield spectra when a static-FAB probe is used. This has been attributed to the presence of the mobile phase.

One area in which the performance of dynamic FAB is inferior to that of static FAB is the molecular weight range of analytes that may be studied. Static FAB

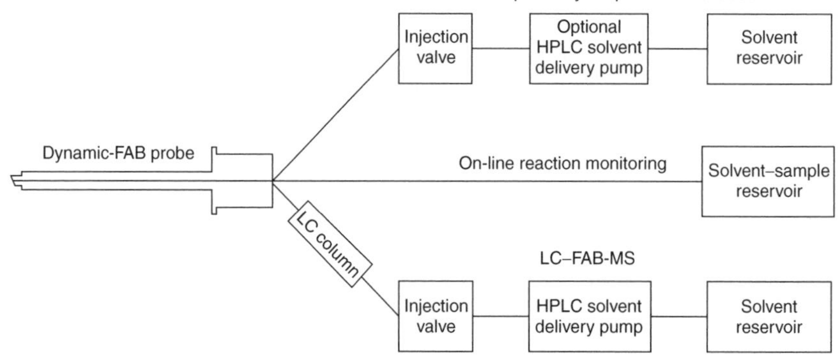


**Figure 4.4** Different configurations of dynamic FAB. From applications literature published by Kratos Analytical Ltd, Manchester, UK, and reproduced by permission of Mass Spectrometry International Ltd.

has a mass range, depending upon the bombarding species and the mass spectrometer being used, in excess of 10 000 Da, while the majority of dynamic FAB applications reported involve analytes with molecular weights below 2000 Da.

While the major interest in dynamic FAB has been when used in conjunction with HPLC, it must be pointed out that the system is capable of being used without a separation column (see Figure 4.4).

The pressure difference between the source of the mass spectrometer and the laboratory environment may be used to draw a solution, containing analyte and matrix material, through the probe via a piece of capillary tubing. When an adequate spectrum of the first analyte has been obtained, the capillary is simply placed in a reservoir containing another analyte (and matrix material) and the process repeated. This may therefore be used as a more convenient alternative to the conventional static FAB probe and this mode of operation may also benefit from the reduction in suppression effects if the analyte is one component of a mixture.

For better precision of quantitative determinations, a loop injector may be incorporated.

This practical set-up may also be used for reaction monitoring by placing the capillary into a reaction mixture and continually acquiring mass spectra, which thus allows the analyst to examine changes in its composition.

## *Summary*

The introduction of the dynamic-FAB LC-MS interface allowed the study of more non-volatile and ionic compounds than was possible with the other interfaces available at that time. It was, however, necessary to attach such an interface to a mass spectrometer with sufficient mass range to allow this potential advantage to be realized. This had significant financial implications on any laboratory contemplating its use and it was therefore usually purchased as an accessory for an existing instrument or as one feature of a multi-purpose instrument.

*Advantages*

- It is possible to study thermally labile materials using this type of interface since the only heat applied to the probe tip is that required to prevent freezing of the mobile phase as it evaporates.
- The interface is directly compatible with the low flow rates used with microbore HPLC and, because of the narrow chromatographic peaks, good sensitivity can be obtained with small amounts of sample.
- Since only a small amount of matrix is required in dynamic FAB, when compared to conventional FAB, less chemical background is observed and, because of the stability of the signal, it is convenient to use computer enhancement to generate good quality mass spectra. For this reason, the dynamic-FAB probe is often used as a convenient alternative to the conventional FAB probe.
- The interface is simple in design and relatively easy to construct.
- Mobile phases with a high percentage of water may be used.

*Disadvantages*

- The presence of the matrix can cause chromatographic problems if added to the mobile phase before the column, especially if this is of small diameter. The low flow rates that are used require an increased concentration of matrix to be present in the mobile phase to ensure an appropriate amount reaches the probe tip.
- If conventional HPLC columns are used, with splitting of the eluate to provide the necessary flow rate, an overall decrease in sensitivity usually results.

## 4.5 The Particle-Beam Interface

Arguably, the ultimate LC–MS interface would be one that provides EI spectra, i.e. a spectrum from which structural information can be extracted by using familiar methodology, and this was one of the great advantages of the moving-belt interface. There is, however, an incompatibility between the types of compound separated by HPLC and the way in which electron ionization is achieved and therefore such an interface has restricted capability, as previously discussed with respect to the moving-belt interface (see Section 4.2 above).

Even so, much effort was put into the development of such a system and this resulted in the introduction of the particle-beam interface, also known as the

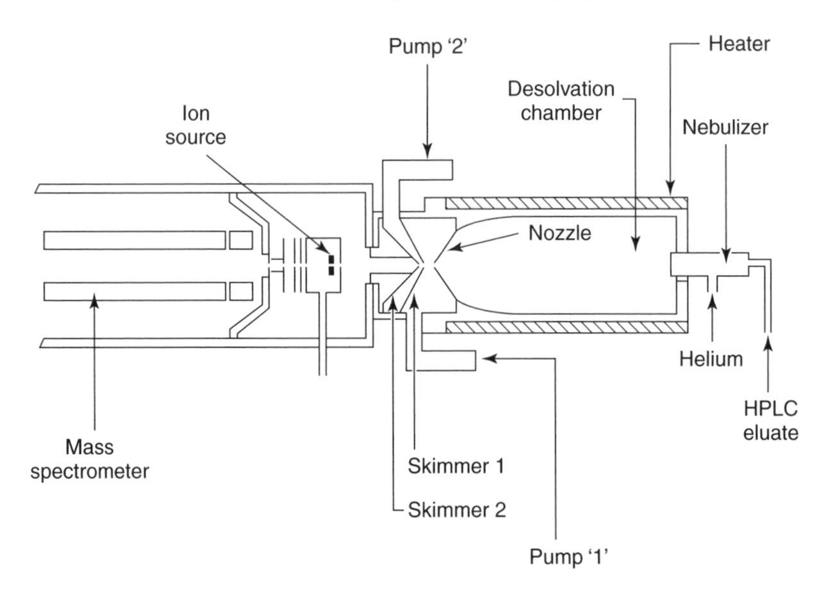


**Figure 4.5** Schematic of a particle-beam LC–MS interface. From applications literature published by Micromass UK Ltd, Manchester, UK, and reproduced with permission.

MAGIC® (**M**onodisperse **A**erosol **G**enerating **I**nterface for **C**hromatography), Thermabeam® and Universal® interfaces.

A general schematic of a particle-beam interface is shown in Figure 4.5. The procedure used with this interface consists of four stages, as follows:

- nebulization of the HPLC eluate;
- desolvation of the droplets so formed;
- removal of the solvent vapour and nebulizing gas;
- ionization of the analyte.

In each of the designs, droplets are formed from the HPLC eluent by passage through a nebulizer; the MAGIC interface employs a pneumatic nebulizer, the Universal interface a thermal nebulizer, and the Thermabeam interface a combined thermal and pneumatic nebulizer. The droplets are then carried, by a high-velocity gas stream, into a desolvation chamber, with the MAGIC and Thermabeam interfaces employing a simple heated metal desolvation chamber, and the Universal interface a desolvation chamber containing a gas diffusion membrane.

The formation of droplets, which range from 50 to 200 nm in diameter, gives a very large surface area from which evaporation may take place rapidly. The desolvation chamber is maintained virtually at ambient temperature by providing sufficient heat to overcome the latent heat of vaporization of the mobile phase. While the volatile components vaporize, the less volatile components, such as the analyte, condense to form sub-micron diameter particles or liquid droplets, with the smaller the particle size, then the lower the temperature required for their subsequent vaporization prior to ionization.

The mixture of sample, vapour and carrier gas finally passes through a two-stage momentum (jet) separator. The latter consists of a number of aligned orifices, the gaps between which are pumped. As the mixture emerges from the first orifice, those components with lower momentum, i.e. the carrier gas and eluent, tend to diverge and are pumped away, while those with higher momentum, i.e. the higher-molecular-weight analytes, pass into the second orifice which forms the inlet to a second similarly configured region where an identical process occurs. Transfer of the higher-momentum particles into the mass spectrometer therefore takes place. The region between the first set of orifices is maintained at between 2 and 10 torr, and that between the second set between 0.1 and 1 torr.

The particles then enter a conventional mass spectrometer source where they are vaporized prior to being ionized using electron impact or chemical ionization. As with other interfaces, this may cause problems during the analysis of thermally labile and highly involatile compounds.

The range of compounds from which electron ionization spectra may be obtained using the particle-beam interface is, like the moving-belt interface, extended when compared to using more conventional methods of introduction, e.g. the solids probe, or via a GC. It is therefore not unusual for spectra obtained using this type of interface not to be found in commercial libraries of mass spectra.

Examples of such compounds include anionic surfactants whose analysis had previously been limited to desorption techniques such as FAB and thermospray but which yielded interpretable EI spectra when using a particle-beam interface [8]. A number of explanations have been put forward to explain this capability [9], including the volatility of the analytes being enhanced by the addition of surface energy during aerosol generation and the size of the droplets being such that evaporation of intact ions directly into the gas phase is favoured.

Thermally labile compounds may also be studied – for example, the EI spectra from the condensation products from the reaction between dimedone and substituted phenylbenzopyrans obtained via a particle-beam interface show less thermal degradation than do the mass spectra obtained when using a direct-insertion probe [10].

The molecular weight limit of the particle-beam interface is around 1000 Da.

The most important feature of any interface which is capable of allowing an EI spectrum to be produced is that the mobile phase is totally removed so that the spectra obtained may be attributed solely to the analyte. Whether or not this is accomplished depends upon the composition of the mobile phase, its flow rate and the conditions employed within the interface, i.e. temperature, nebulizing gas flow, etc.

The particle-beam interface gives optimum performance at flow rates of between 0.1 and 0.5 ml $min^{-1}$. These rates are directly compatible with 2 mm diameter HPLC columns and have been employed in many reported applications. It is possible to employ 4.6 mm columns, either by using a reduced flow rate, although this will have some effect on the chromatographic performance, or by post-column splitting of the flow. The percentage of water in the mobile phase is a dominant factor in determining both the interface operating conditions and the flow that may be accepted. The more water present, then the lower the mobile phase flow that may be accommodated. Since the optimum interface conditions are dependent upon the mobile phase composition, performance is expected to vary during gradient elution.

The effect of the variation in mobile phase composition can be overcome by the use of post-column on-line extraction to remove water from the mobile phase and thus produce a 100% organic mobile phase and this is also likely to bring about an increase in overall sensitivity.

Even though the mobile phase is largely removed, EI spectra generated by a particle-beam interface do tend to show a mobile-phase-associated low-mass background that can mask ions from the analyte. This does not help interpretation and is of particular importance when library searching is being used. The use of RICs, as described above in Section 3.6.2, should be applied to associate low-mass ions with either the analyte or the background. The presence of background may increase the amount of analyte required to give a mass spectrum of adequate quality for interpretation. Normally between 10 and 100 ng of analyte is required to give such a spectrum, although the amount is dependent upon the compound involved and the interface parameters. It has been found that more intense responses are obtained when analytes are present in HPLC eluates containing a low proportion of water.

There are two notable features of the quantitative performance of this type of interface. It has been found that non-linear responses are often obtained at low analyte concentrations. This has been attributed to the formation of smaller particles than at higher concentrations and their more easy removal by the jet separator. Signal enhancement has been observed due to the presence of (a) coeluting compounds (including any isotopically labelled internal standard that may be used), and (b) mobile-phase additives such as ammonium acetate. It has been suggested that ion–molecule aggregates are formed and these cause larger particles to be produced in the desolvation chamber. Such particles are transferred to the mass spectrometer more efficiently. It was found, however, that the particle size distribution after addition of ammonium acetate, when enhancement was observed, was little different to that in the absence of ammonium acetate when no enhancement was observed.

Since the carrier effect is not general for all analytes and all additives, quantitative studies using the particle-beam interface should only be carried out after a very careful choice of experimental conditions and standard(s) to be used, with isotopic-dilution methodology being advocated for the most accurate results.

**DQ 4.1**

The particle-beam interface provides EI spectra from HPLC eluates and this is of great advantage over other interfaces which provide only molecular weight information. Why, then, is it of advantage to be able to generate CI spectra from the particle-beam interface?

*Answer*

*The advantage of being able to generate a CI spectrum by using the particle-beam interface is no different than when other sample introduction systems are used. As discussed earlier in Chapter 3, EI is a relatively high energy process which often leads to the production of unstable molecular ions which fragment before they are detected, i.e. do not appear in the mass spectrum. In these circumstances, the molecular weight of the analyte concerned cannot be determined directly and the generation of a CI spectrum is often the only way in which this important piece of information may be obtained.*

## Summary

The particle-beam interface has been developed primarily to provide EI spectra from HPLC eluates but may be combined with other ionization techniques such as CI. If quantitative studies are being undertaken, a detailed study of experimental conditions should be undertaken. Isotope-dilution methodology is advocated for the most accurate results.

*Advantages*

- A number of thermally labile and relatively involatile compounds which do not yield EI spectra when using more conventional inlet methods do so when introduced via the particle-beam interface.

*Disadvantages*

- The sensitivity of the particle-beam interface is dependent not only on the specific analyte but also on the experimental conditions employed. Detection limits are invariably higher than are desirable.
- Neither extremely volatile or extremely involatile compounds are ideal for investigation using the particle-beam interface.
- The performance of the particle-beam interface deteriorates as the percentage of water in the HPLC mobile phase increases.

# 4.6 The Thermospray Interface

Although each of the previously described interfaces has advantages for particular types of analyte, there are also clear limitations to their overall performance. Their lack of reliability and the absence of a single interface that could be used for the majority of analytes did nothing to advance the acceptance of LC–MS as a routine technique. Their application, even with limitations, did, however, show very clearly the advantages that were to be gained by linking HPLC to MS and the efforts of many to find the 'ideal' LC–MS interface were intensified.

The thermospray interface overcame many of the problems encountered with the moving-belt and direct-liquid-introduction interfaces and with the advent of this, LC–MS became a routine analytical tool in a large number of laboratories. This was reflected in the fact that this was the first type of interface made available commercially by the majority of the manufacturers of mass spectrometers.

It was also the first of a number of interfaces, with the others being electrospray and atmospheric-pressure chemical ionization, in which ionization is effected directly from solution within the interface itself, i.e. the mass spectrometer was not used to produce ions from the analyte simply to separate them according to their $m/z$ ratios.

Ionization involving these interfaces may be considered to comprise the following four stages, with the differences between the techniques being associated with the ways in which these stages are carried out:

- the formation of droplets from the HPLC eluate;
- charging of these droplets;
- desolvation of the droplets;
- the formation of ions from the analyte.

Thermospray may be defined as the controlled, partial or complete, vaporization of a liquid as it flows through a heated capillary tube.

A thermospray system is shown schematically in Figure 4.6. This consists of a heated capillary through which the LC eluate flows, with the temperature of this capillary being carefully controlled to bring about around 95% vaporization of the liquid. The vapour so produced acts as a nebulizing 'gas' and aids the break-up of the liquid stream into droplets.

The droplets so formed undergo desolvation as they traverse a heated region of the interface and ions are formed from analytes contained in the liquid stream by means of ion–molecule reactions, cf. chemical ionization, and/or ion-evaporation processes (see Section 4.7.1 below), depending upon the properties of both the liquid stream and the analyte.

When buffer is present in the mobile phase and the analytes are largely ionic in nature or give preformed ions in solution, the ion-evaporation mechanism is

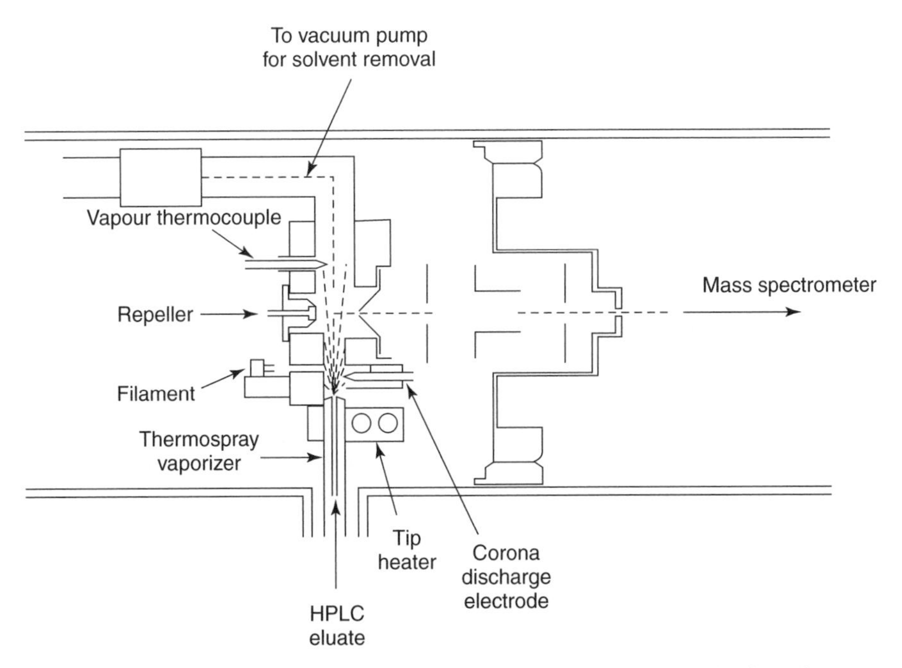


**Figure 4.6** Schematic of a thermospray LC–MS interface. From applications literature published by Vestec (Applied Biosystems), Foster City, CA, and reproduced with permission.

thought to be important. In the majority of applications, however, it is considered that ions are formed in solvent-mediated CI processes, either by gas-phase ion–molecule reactions or by proton transfer at the interface of the liquid droplet and the gas phase. If complete vaporization of the mobile phase stream takes place within the capillary, this latter mechanism is not possible and so provides a further reason why it is essential to control carefully the heat input to the capillary.

Gas-phase ion–molecule reactions involving ionic analytes would lead to their neutralization and, consequently, these have to be minimized. This requires careful control of both the composition and the concentration of any buffer used.

The ions formed are then directed though a sampling cone at 90° to the direction of vapour flow – to minimize the chances of blocking of the entrance to the mass spectrometer – into the source of the mass spectrometer, while the vast majority of the vapour generated by the mobile phase is removed by a pump directly opposite the capillary.

Slightly downstream of the end of the capillary are usually to be found a filament and/or a discharge electrode, which provides secondary methods of ionization, while opposite or slightly downstream of the sampling cone is a repeller or retarding electrode. Their use will be described in more detail later.

Control of the vaporization process, i.e. the temperature of the capillary, is of crucial importance. Optimum performance is obtained with around 95% vaporization of the liquid stream. Too much heat results in vaporization occurring within the capillary with deposition of analyte and, if operation of the interface under these conditions is continued for any length of time, blockage of the capillary. On the other hand, if insufficient heat is applied to the capillary, vaporization does not occur and liquid flows from it and no spray is obtained. The optimum temperature is dependent on a number of parameters, among the most important being the composition of the mobile phase and its flow rate. Good temperature control is therefore required to obtain the best conditions when gradient elution is employed.

In true thermospray, charging of the droplets is due to the presence of a buffer in the mobile phase. Both positively and negatively charged droplets are formed due to the statistical fluctuation in anion and cation density occurring when the liquid stream is disrupted. As with the interfaces previously described, involatile buffers are not recommended as blocking of the capillary is more likely to occur if temperature control is not carefully monitored and for this reason ammonium acetate is often used.

When buffer is not present, i.e. when normal-phase chromatography is being used, thermospray ionization is not possible and a filament or discharge electrode is used to generate a plasma in which CI-type processes can occur. In addition to allowing ionization under these conditions, it is found that the ionization of compounds may be enhanced under conditions in which true thermospray can operate.

The filament operates in the same way as a filament in chemical ionization by generating reactive species from solvent molecules in the high-pressure region of the source. These ionize the analyte by ion–molecule reactions (see Section 3.2.2 above). The discharge electrode, which may also provide more stable conditions when the mobile phase contains a very high proportion of water, provides the electrons required to generate the reactive species by means of a continuous gas discharge.

The CI reagent-gas plasma is either generated by thermospray ionization if buffer is present or, as described above, by the filament or discharge electrode.

In many cases, an ion from the protonated molecule $(M + H)^+$ is the only ion observed in a thermospray spectrum but if ammonium acetate buffer is used, depending upon the relative proton affinities of the species present, an ammonium adduct $(M + NH_4)^+$ may be the predominant ion. In addition, clusters may be formed with components of the mobile phase. Although the thermospray ionization process involves less energy than conventional CI, and very little intense fragmentation is usually observed, the presence of ions due to the elimination of small molecules, e.g. water, methanol and ketene, is not unknown. These latter ions are usually of relatively low intensity when compared to the protonated or ammoniated species but the mass differences involved normally aid the assignment of the molecular weight.

The sampling cone is orthogonal to the direction of vapour and ion flow and ions are therefore extracted into the high-vacuum part of the mass spectrometer. The extraction efficiency, particularly at high mass, can be improved by incorporating an electrode in the region of the sampling cone. If located directly opposite the sampling cone, this is known as a 'repeller', and if slightly downstream of the sampling cone, an 'ion retarder'. The term repeller will be used henceforth for both positions of the electrode. The voltage applied to the repeller is of importance as it has been found that at high (this being a relative term) potentials, fragmentation of ions can occur. Since thermospray ionization inherently produces only molecular species, the ability to bring about fragmentation, and thus derive structural information from an analyte, is of great advantage.

Fragmentation occurs because the repeller voltage increases the kinetic energy of the ions, not only making collision-induced dissociation (CID) more likely but also allowing endothermic ion–molecule and solvent-switching reactions to occur.

Although theoretically an alternative to MS–MS, there is a fundamental difference which may result in a spectrum generated using the repeller being more difficult to interpret. This is because the repeller causes fragmentation of all of the ions present in the source, e.g. $(M + H)^+$, $(M + NH_4)^+$, solvent adducts, etc., and the relationship between the ions formed is not always clear. It must also be remembered that repeller-induced fragmentation, like MS–MS, does not produce an EI spectrum and while interpretation may be attempted by using the same principles, library matching, unless against repeller-fragmented spectra, is not valid.

It should also be noted that the effect of the repeller is both compound-dependent and mode-, i.e. thermospray, filament or discharge, dependent. It is necessary, prior to developing any analysis, therefore, to determine the optimum experimental conditions for the analyte(s) of interest.

## Summary

The introduction of the thermospray interface provided an easy-to-use LC–MS interface and was the first step in the acceptance of LC–MS as a routine analytical technique. It soon became the most widely used LC–MS interface of those available in the mid to late 1980s.

### *Advantages*

- The interface was much easier to use than the (then) available alternatives.
- The interface can operate under a wide range of HPLC conditions, flow rates of 1 $ml\,min^{-1}$ may be used, and mobile phases containing high percentages of water can be accommodated.

- The interface can allow unequivocal determination of molecular weight as thermospray spectra usually contain ions simply from molecular species with little fragmentation being observed.
- The addition of a discharge electrode and a filament to the thermospray source widens the range of compounds that may be studied and HPLC solvents that may be accommodated. Optimum ionization conditions for a particular compound need to be determined empirically and it is essential that switching between the possible ionization modes may be accomplished easily and quickly.
- Since the sample is ionized directly from solution it is protected from heat and many thermally labile analytes may be studied with little or no degradation.
- The sensitivity of the interface is compound-dependent but generally high sensitivity is possible by using one of the ionization methods available (thermospray, filament and discharge).

*Disadvantages*

- If a buffer is present in the HPLC mobile phase, and this is essential for true thermospray ionization, it should ideally be volatile and this may necessitate modifying existing HPLC methodology.
- Decomposition of some thermally labile analytes is observed.
- The interface is not suitable for high-molecular-weight (>1000 Da) analytes.
- The reproducibility of analytical results is affected by a number of experimental parameters and is sometimes difficult to control.
- The formation of adducts may confuse the assignment of molecular weight.
- Usually little structural information is immediately available and repeller-induced fragmentation or MS–MS is required to generate this. Spectra generated by repeller-induced fragmentation, because fragmentation of all ionic species generated in the interface occurs, are often difficult to interpret.
- The interface generates a significant amount of solvent-associated chemical noise at low mass which makes this region unusable for analytical purposes. Since the ions generated from the analyte are molecular species, this is not usually a problem but must be considered when spectra are being interpreted.

## 4.7 The Electrospray Interface

High performance liquid chromatography is an effective technique for the separation of compounds of high molecular weight. There are, however, two major problems with the use of mass spectrometry for the study of this type of molecule and these have severely limited the application of LC–MS. Specifically, these are as follows:

(a) The inability to ionize, in an intact state, many of the labile and/or involatile molecules involved.
(b) Should ionization be possible, the lack of appropriate hardware to allow the mass analysis and efficient detection of the ions of high $m/z$ ratio involved.

The problem of the successful ionization of thermally labile molecules has been addressed by the introduction of 'energy-sudden' techniques, such as fast-atom bombardment (FAB), which rely on the fact that energy may be provided to the molecule so rapidly that desorption takes place before decomposition may occur.

The use of the dynamic-FAB probe (see Section 4.4 above) has allowed the successful coupling of HPLC to this ionization technique but there is an upper limit, of around 5000 Da, to the mass of molecules which may be successfully ionized. Problem solving, therefore, often involves the use of chemical methods, such as enzymatic hydrolysis, to produce molecules of a size more appropriate for ionization, before applying techniques such as peptide mapping (see Section 5.3 below).

The mass range requirement invariably means that FAB is used in conjunction with a magnetic sector instrument. Conventional detectors, such as the electron multiplier, are not efficient for the detection of large ions and the necessary sensitivity is often only obtained when devices such as the post-acceleration detector or array detector are used. Instruments capable of carrying out high-mass investigations on a routine basis are therefore costly and beyond the reach of many laboratories.

Electrospray is an ionization method that overcomes both of the problems previously described.

A liquid, in which the analyte(s) of interest have been dissolved, is passed through a capillary, at atmospheric pressure, maintained at high voltage. The liquid stream breaks up with the formation of highly charged droplets which are desolvated as they pass through the atmospheric-pressure region of the source towards a counter electrode. Desolvation is assisted by a stream of a drying gas, usually nitrogen, being continually passed into the spraying region. Analyte ions are obtained from these droplets (see Section 4.7.1) which then pass through two differentially pumped regions into the source of the mass spectrometer. A schematic of an electrospray system is shown in Figure 4.7.

Since ionization takes place directly from solution, thermally labile molecules may be ionized without degradation.

In contrast to most other ionization methods, the majority of ions produced by electrospray are multiply charged. This is of great significance as the mass spectrometer measures the $m/z$ (mass-to-charge) ratio of an ion and the 'mass' range of an instrument may therefore be effectively extended by a factor equivalent to the number of charges residing on the analyte molecule, i.e. an ion of $m/z$ 1000 with

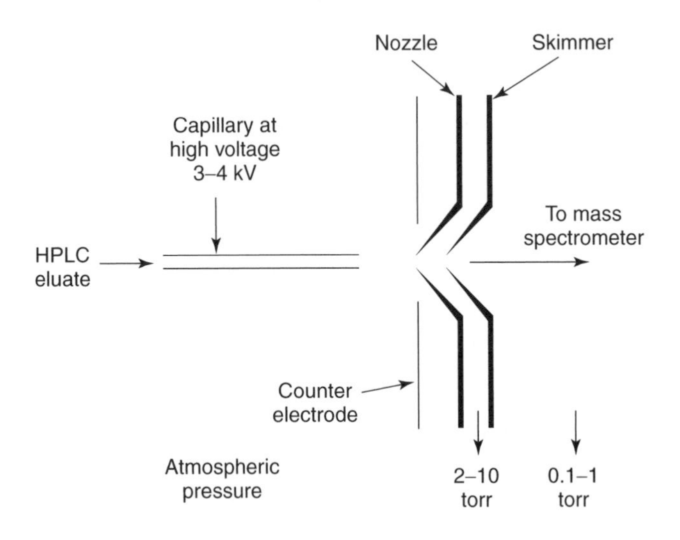


**Figure 4.7** Schematic of an electrospray LC–MS interface. From applications literature published by Micromass UK Ltd, Manchester, UK, and reproduced with permission.

20 charges is derived from a compound with a molecular weight of 20 000 Da. A high molecular weight is not, however, a prerequisite for ionization by electrospray and, as will be described in detail in Chapter 5, the technique is equally applicable to relatively polar, low-molecular-weight (<1000 Da) analytes.

## *4.7.1 The Mechanism of Electrospray Ionization*

Electrospray ionization occurs by the same four steps as listed above for thermospray (see Section 4.6). In contrast to thermospray, and most other ionization methods used in mass spectrometry, it should be noted that electrospray ionization unusually takes place at atmospheric pressure. A similar process carried out under vacuum is known as *electrohydrodynamic ionization* and gives rise to quite different analytical results. This technique has not been developed into a commercial LC–MS interface and will not be considered further.

As described previously, electrospray spectra are produced by passing a liquid stream through a metal capillary maintained at high voltage (typically 3–4 kV for the production of positive ions; slightly less, and of opposite polarity, for the production of negative ions). This high voltage disperses the liquid stream, forming a mist of highly charged droplets that undergo desolvation during their passage across the source of the mass spectrometer. As the size of the droplet reduces, a point is reached (within 100 μs) at which the repulsive forces between charges on the surface of the droplets are sufficient to overcome the cohesive forces of surface tension. A 'Coulombic explosion' then occurs, producing a number of smaller droplets with a radius approximately 10% of that of the parent droplet.

A series of such explosions then takes place until a point is reached at which ions of the appropriate analytes dissolved in these droplets are produced and are transferred through a series of focusing devices (lenses) into the mass spectrometer.

There are alternative explanations for the actual mechanism by which these ions are produced, e.g. the ion-evaporation [11] and charge-residue models [12], and these have been debated for some time.

According to the **ion-evaporation** model, the droplets become smaller until a point is reached at which the surface charge is sufficiently high for direct ion evaporation into the gas phase to occur. In the case of the **charge-residue model**, repeated Coulombic explosions take place until droplets are formed that contain a single ion. Evaporation of the solvent continues until an ion is formed in the vapour phase.

The lack of a definitive explanation, however, does not affect our ability to appreciate the analytical capabilities of the technique, the HPLC characteristics that will affect the production of ions by the electrospray process and the mass spectra that may be obtained.

The selectivity (separation capability) of an HPLC system is dependent upon the combination of mobile and stationary phases. Since ions are being generated directly from the mobile phase by electrospray, its composition, including the identity and concentration of any buffer used, and its flow rate are important considerations.

Desolvation of the droplets formed, and thus ionization of analytes, is favoured by the initial production of small droplets. For this reason, a mobile phase with high surface tension and/or high viscosity should be avoided – pure water is therefore not a desirable mobile phase. The application of higher voltages to the electrospray needle will result in the production of smaller droplets but will ultimately lead to the formation of a high-voltage discharge rather than the formation of droplets. This parameter should be optimized for a particular mobile phase during the instrumental set-up procedure prior to an analysis.

The buffer concentration also directly affects the size of droplets produced – the higher the buffer concentration, then the smaller they are, and this is desirable. The buffer concentration, however, has an effect on the ionization efficiency and at high buffer concentrations ($>10^{-3}$ M) the relationship between detector response and analyte concentration is not linear. As indicated earlier in Figure 2.6, this situation must be avoided for precise quantitative measurements.

The flow rate of liquid in the HPLC–electrospray system is paramount in determining performance both from chromatographic and mass spectrometric perspectives. The flow rate affects both the size and size distribution of the droplets formed during the electrospray process (not all droplets are the same size) and, consequently, the number of charges on each droplet. This, as we will see later, has an effect on the appearance of the mass spectrum which is generated. It should also be noted that the smaller the diameter of the spraying capillary, then the narrower the droplet size distribution and the more efficient the transfer of sample to the mass spectrometer.

True electrospray is most efficient at flow rates of between 5 and 10 μl $min^{-1}$, which are not directly compatible with the majority of HPLC applications. There are two approaches to providing reduced flow rates of an appropriate magnitude.

HPLC columns with reduced diameters (microbore columns) are now available. The flow rate from such columns required to give a desired flow rate at the same linear solvent velocity (and thus retention time) as a 4.6 mm i.d. column operating at 1 ml $min^{-1}$ is given by the following equation:

$$\text{flow rate} = \left(\frac{r}{4.6}\right)^2 \times 1.0 \qquad (4.1)$$

in which $r$ is the HPLC column radius. This may be used to show that a column of 0.65 mm i.d. is required to provide a flow rate of 20 μl $min^{-1}$.

The use of microbore columns is not yet routine as much more rigorous control of parameters, such as flow rate and instrument dead volume, is required to ensure that degradation of chromatographic performance does not occur. It is therefore experimentally more difficult.

The alternative is to split the flow from a conventional column such that an appropriate proportion goes to the electrospray probe, with the remainder going to a second detector and/or for collection or to waste. It might be thought that column splitting would lead to an unacceptable loss in sensitivity, e.g. reduction of the flow rate from 1 ml $min^{-1}$ to 20 μl $min^{-1}$ requires a 50:1 split. At such flow rates, however, electrospray acts as a concentration-sensitive detector, i.e. the signal intensity is proportional to the **concentration** of analyte in the mobile phase rather than the absolute amount of analyte present. Flow rate changes do not affect the intensity of the signal from the detector and splitting of the flow may therefore be carried out without an accompanying reduction in sensitivity.

These solutions are not always practicable and HPLC flow rates of up to 2 ml $min^{-1}$ may be accommodated directly by the use of electrospray in conjunction with pneumatically assisted nebulization (the combination is also known as Ionspray®) and/or a heated source inlet. The former is accomplished experimentally by using a probe that provides a flow of gas concentrically to the mobile phase stream, as shown in Figure 4.8, which aids the formation of droplets from the bulk liquid, and will allow a flow rate of around 200 μl $min^{-1}$ to be used.

Two designs of electrospray system employing a heated source inlet are shown in Figure 4.9. In the first of these (Figure 4.9(a)), a heated capillary is placed directly in line with the electrospray probe so that droplets produced by the electrospray process enter this on their way to the lens system and the mass spectrometer. In the second design (Figure 4.9(b)), a heated stainless-steel block is used in place of the capillary. The block contains four, relatively wide, channels through which the droplets produced by the electrospray process travel. These four channels each contain a 'kink' as shown, (a) to increase the pathlength of

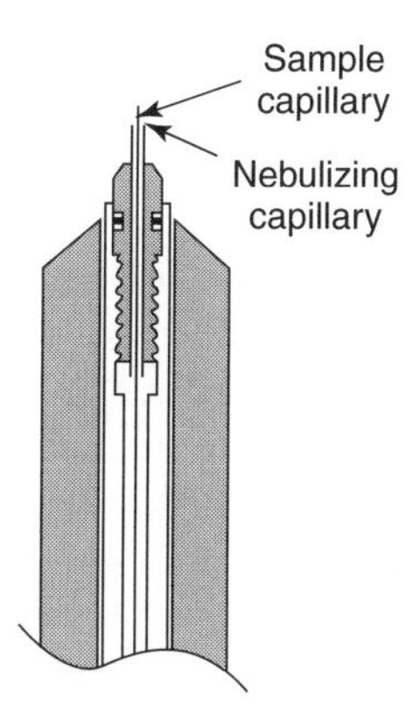


**Figure 4.8** Schematic of an electrospray probe with a concentric flow of nebulizing gas. From applications literature published by Micromass UK Ltd, Manchester, UK, and reproduced with permission.

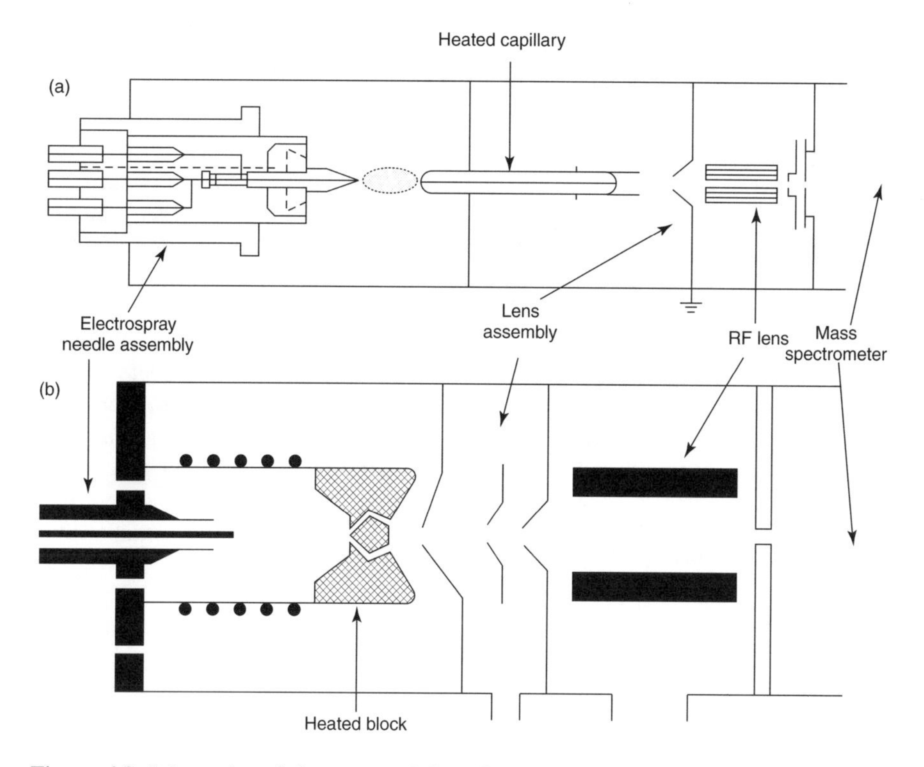


**Figure 4.9** Schematics of electrospray LC–MS interfaces with (a) a heated capillary and (b) a heated block to allow high mobile-phase flow rates. From applications literature published by (a) Thermofinnigan, Hemel Hempstead, UK, and (b) Micromass UK Ltd, Manchester, UK, and reproduced with permission.

the droplets, and thus allow more complete desolvation, and (b) to minimize contamination of the lenses by removing the direct line of sight between them and the droplets as they emerge from the electrospray probe. Systems of this type allow HPLC flow rates of up to 2 ml $min^{-1}$ to be accommodated.

While the major concern has been the compatibility of electrospray with high HPLC flow rates, from a chromatographic standpoint, low-flow regimes are of increasing interest and these require their own special considerations since desolvation of very small droplets may occur too rapidly, hence leading to irregular production of ions. This is as much a problem as the formation of droplets that are too large to undergo sufficient desolvation.

So-called nanoflow systems are now available which operate at flow rates in the nl $min^{-1}$ region. In these, the metal needle of the conventional electrospray system is replaced with a fused silica capillary, often coated with gold to allow the electrospray voltage to be applied, drawn to give a spraying orifice of between 1 and 2 μm in diameter [13]. Approximately 1 μl of solution containing the sample of analyte may be loaded directly into this capillary and, without the application of external pumping, approximately 20 nl $min^{-1}$ is sprayed into the electrospray source. Alternatively, it may be incorporated into an HPLC system as shown in Figure 4.10 [14], where a 75 μm × 15 cm $C_{18}$ column is used at a flow rate of approximately 150 nl $min^{-1}$. In this system, the fused silica capillary had a spray tip of 20 μm i.d. × 90 μm o.d. The use of a low-flow system such as this necessitates a number of practical problems to be addressed. A conventional HPLC pump is used to generate a flow rate of 400 μl $min^{-1}$ which is then split, pre-column, to give the desired flow, a 100 nl loop is employed and, to allow the use of a UV detector in series with the mass spectrometer, a low-flow detector cell is incorporated. In addition to being used with nanoflow HPLC, the interfacing of other separation techniques, such as capillary electrophoresis and capillary electrochromatography, may be accomplished with this type of interface.

At lower flow rates, between 10 and 100 nl $min^{-1}$, the droplets produced have diameters of less than 200 nm and are between 100 and 1000 times smaller than

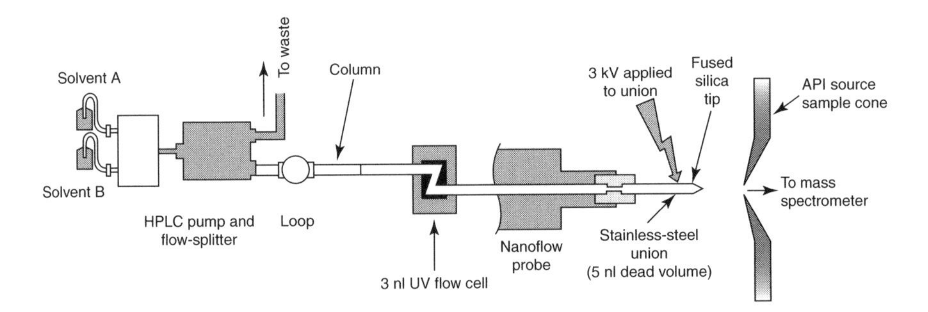


**Figure 4.10** Schematic of a nanoflow LC–MS system. From applications literature published by Micromass UK Ltd, Manchester, UK, and reproduced with permission.

those generated from a conventional electrospray source. The droplets thus have both higher surface-to-volume and charge-to-mass ratios than those produced at higher flow rates and on average are thought to contain a single analyte molecule and that since this molecule is closer to the surface charge it will be easily ionized. Ionization efficiencies approaching 100% can be achieved but under these conditions, however, electrospray behaves as a mass-sensitive detector, i.e. the signal intensity is proportional to the **amount** of analyte reaching the electrospray needle, rather than a concentration-sensitive device [15].

The use of low flow rates allows the production of ions from mobile phases containing high percentages of water which, as noted previously, usually presents a particular problem because of its high surface tension and conductivity. The production of electrospray spectra from aqueous solution is possible at flow rates of less than 50 $\mu l\,min^{-1}$.

### *4.7.2 Sample Types*

The ability to produce ions using electrospray ionization is more reliant on the solution chemistry of the analyte than the other ionization techniques described and this feature may be used by the analyst to advantage. It may also confuse the unwary!

We have previously considered the mechanism of electrospray ionization in terms of the charging of droplets containing analyte and the formation of ions as the charge density on the surface of the droplet increases as desolvation progresses. The electrospray system can also be considered as an electrochemical cell in which, in positive-ion mode, an oxidation reaction occurs at the capillary tip and a reduction reaction at the counter electrode (the opposite occurs during the production of negative ions). This allows us to obtain electrospray spectra from some analytes which are not ionized in solution and would otherwise not be amenable to study. In general terms, the compounds that may be studied are therefore as follows:

(a) ionic compounds that are intrinsically charged in solution;
(b) neutral/polar compounds that may be protonated (for positive-ion mass spectra) or deprotonated (for negative-ion mass spectra) under the solution conditions employed, e.g. appropriate pH;
(c) non-polar compounds that undergo oxidation (positive-ion mass spectra) or reduction (negative-ion mass spectra) at the electrospray capillary tip.

There are a number of properties of the solvent, such as its viscosity, conductivity, surface tension and polarity, that have an effect on the electrospray process.

The pH of the solution is also of critical importance, not only in terms of whether or not ionization occurs, but also, particularly in the case of high-molecular-weight compounds, the appearance of the spectrum (see Section 4.7.3 below). The production of positive ions is favoured at acidic pH but ions have been observed at pHs at which a particular analyte would be expected to be fully deprotonated. The intensity of these ions, it has to be said, is usually significantly less than when produced from solutions of more appropriate pH and it is thought that they are produced in the transition of the analyte from the liquid to the gas phase.

The electrospray process is susceptible to competition/suppression effects. All polar/ionic species in the solution being sprayed, whether derived from the analyte or not, e.g. buffer, additives, etc., are potentially capable of being ionized. The best analytical sensitivity will therefore be obtained from a solution containing a single analyte, when competition is not possible, at the lowest flow rate (see Section 4.7.1 above) and with the narrowest diameter electrospray capillary.

If excess electrolytic materials are present, competition for charge will occur. The efficiency with which the analyte will be ionized depends upon the concentration of each of the species present and also the relative efficiency of the conversion of each to the gas phase.

The ions observed in the mass spectrum may be different to those present in solution owing to processes that occur at the solution/gas phase boundary and ion–molecule reactions that may take place in the gas phase.

### *4.7.3 The Appearance of the Electrospray Spectrum*

Many of the ionization techniques that we have considered thus far are termed 'soft' in that the spectra produced are exceedingly simple, consisting only of a molecular species that enables the molecular weight of the analyte to be determined. Electron ionization, on the other hand, is a more energetic process which often brings about fragmentation of the analyte molecule. An EI mass spectrum therefore usually consists of a significant number of ions of different $m/z$ values that may be related to the structure of the analyte from which it has been generated. More importantly, these spectra are reproducible, and spectra produced in laboratories geographically well separated are likely to be very similar.

An electrospray spectrum is unusual in that while it provides only ions from molecular species, it consists (usually) of ions at a number of $m/z$ values, i.e. it appears more like an EI than a CI spectrum, with the number of these ions increasing with the molecular weight of the analyte. The appearance of the spectrum is much more dependent upon the environment from which it has been produced than those produced using other ionization techniques. The spectra produced from a single analyte under different experimental conditions may therefore vary considerably in appearance.

Electrospray is the softest mass spectrometry ionization technique and electrospray spectra therefore usually consist solely of molecular ions. Electrospray is unique, however, in that if the analyte contains more than one site at which protonation (in the positive-ion mode) or deprotonation (in the negative-ion mode) may occur, a number of molecular ions with a range of charge states is usually observed. For low-molecular-weight materials ($<$1000 Da), the number of sites at which protonation can occur is likely to be small and the electrospray spectrum is likely to be relatively simple and contain only a few ions. As the molecular weight of the analyte increases, particularly if biopolymers such as proteins are involved, the number of sites at which protonation can occur increases and the spectrum becomes more complex.

A typical electrospray spectrum of a high-molecular-weight material, i.e. that of horse heart myoglobin, is shown in Figure 4.11. Each of the ions observed arises from attachment of a different number of protons, and an equivalent number of charges, to the intact molecule.

In contrast to ions generated by other ionization techniques, these ions do not occur at integer $m/z$ values and this 'non-integer' part of the mass is of crucial importance when making a precise determination of the molecular weight of the analyte involved – this being one of the major uses of electrospray ionization.

How then do we determine the molecular weight of an analyte from an electrospray mass spectrum? To do this, we must consider the processes that lead to the formation of the ions observed. Each ion is formed by the addition of a number of species, often protons, to the analyte molecule, with the number of charges residing on that ion being related to the number of species added. The measured $m/z$ of an ion is thus related to the molecular weight of the analyte from which it is generated by the following equation:

$$m/z = (M + nH)/n \tag{4.2}$$

where $M$ is the molecular weight of the analyte, $H$ is the atomic/molecular weight of the adduct species (1 for hydrogen, 18 for the ammonium ion, 23 for

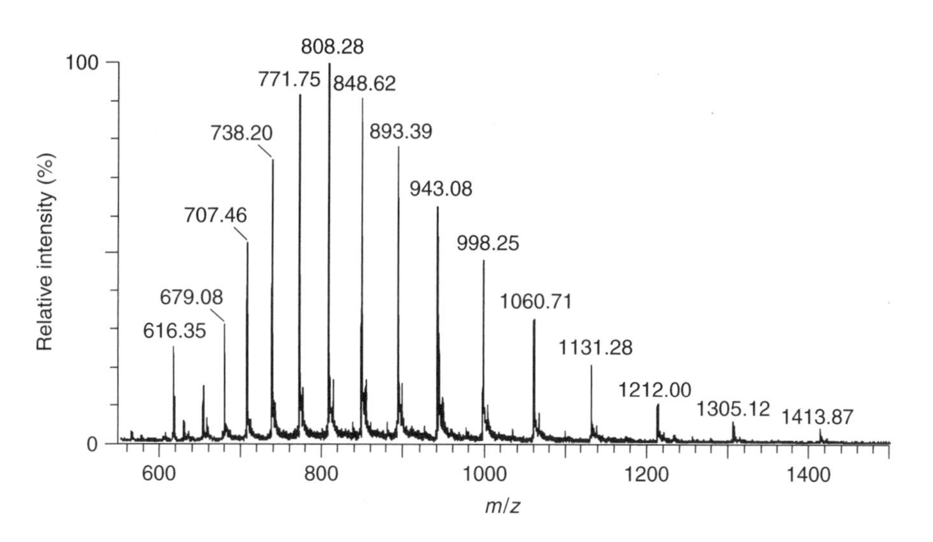


**Figure 4.11** Electrospray spectrum of horse heart myoglobin.

sodium, 39 for potassium, etc,), and $n$ is the number of charges carried by the ion (this is equal to the number of adduct species if these are singly charged).

The only directly measurable variable in this equation is $m/z$ and therefore, in order to determine $M$, we must be able to determine the number of charges on the ion whose $m/z$ value is known. This may be carried out by considering the relationship given in equation (4.2) for two adjacent ions in the spectrum.

For the first ion, whose measured $m/z$ value is $m_1$ we may write the following:

$$m_1 = (M + n_1 H)/n_1$$

which may be rearranged to give:

$$m_1 n_1 = M + n_1 H$$

and finally:

$$M = m_1 n_1 - n_1 H$$

which, if the adduct is a proton, may be simplified to the following:

$$M = m_1 n_1 - n_1 \tag{4.3}$$

For the second ion we may derive a similar expression as follows:

$$M = m_2 n_2 - n_2 \tag{4.4}$$

in which $m_2$ is at a **lower** $m/z$ value than $m_1$.

Equations (4.3) and (4.4) may be combined to give the following:

$$m_1 n_1 - n_1 = m_2 n_2 - n_2$$

Since $m_2 < m_1$ (by definition), $n_2$ must be greater than $n_1$ and it is not unreasonable to assume that $n_2 = n_1 + 1$, i.e. the number of charges on the ions differs by one (this assumption is borne out by the validity of the expression when subsequently applied to all electrospray spectra). We may therefore write the following equation, containing a single unknown $n_1$, which may be solved readily:

$$m_1 n_1 - n_1 = m_2(n_1 + 1) - (n_1 + 1) \tag{4.5}$$

$$m_1 n_1 - n_1 = m_2 n_1 + m_2 - n_1 - 1$$

$$m_1 n_1 - m_2 n_1 = m_2 - 1$$

$$n_1 = (m_2 - 1)/(m_1 - m_2) \tag{4.6}$$

Having calculated the value of $n_1$, the number of charges on one particular ion, the number of charges on each of the ions in the spectrum may be assigned and, by the use of equation (4.3), a number of independent values for the molecular weight of the analyte may be obtained from a single spectrum.

**Table 4.1** Molecular weight of horse heart myoglobin calculated from four adjacent pair of ions observed in its electrospray mass spectrum

| $m_1$ | $m_2$ | $n_1$ (calculated) | $n_1$ | Molecular weight (Da) |
|---|---|---|---|---|
| 893.39 | 848.62 | 18.93 | 19 | 16 955.41 |
| 848.62 | 808.28 | 20.01 | 20 | 16 952.40 |
| 808.28 | 771.75 | 21.10 | 21 | 16 952.88 |
| 771.75 | 738.20 | 21.97 | 22 | 16 956.50 |
| | | *Mean* | | 16 954.30 |
| | | *Standard deviation* | | 1.97 |
| | | *Theoretical molecular weight* | | 16 951.5 |
| | | *Error* | | 0.02% |

If we consider the electrospray spectrum of horse heart myoglobin shown previously (see Figure 4.11), the information given in Table 4.1 may be obtained by application of the above equations. The precision and accuracy of these measurements are also shown in this table.

**SAQ 4.1**

Determine the charge state on the ion of $m/z$ 1060.71 in the mass spectrum shown in Figure 4.11 by using the methodology outlined above. From this, calculate the molecular weight of horse heart myoglobin.

The tedium of carrying out a number of these calculations each time an electrospray spectrum is acquired has been removed by the provision of 'transformation' software with the mass spectrometer data system. This software not only carries out the calculations automatically, but also plots the mass spectrum on a true mass scale.

Figure 4.12 shows the transformed spectrum from the electrospray spectrum shown in Figure 4.11.

Note that for the determination of molecular weight, the charge-state *distribution* is not of great importance as it does not affect the $m/z$ value of the ion involved and thus the calculated molecular weight. If the conformational state of the biopolymer is of interest, however, the distribution of charged states is a fundamental consideration and any parameter likely to change this distribution must be carefully controlled.

The molecular weight of an analyte is therefore arguably the most important single piece of information that may be obtained from an electrospray spectrum and it is therefore important that the term 'molecular weight' is defined unequivocally.

The mass spectrometer allows individual isotopes of an element to be observed. The molecular weight of an analyte must therefore be calculated by using the

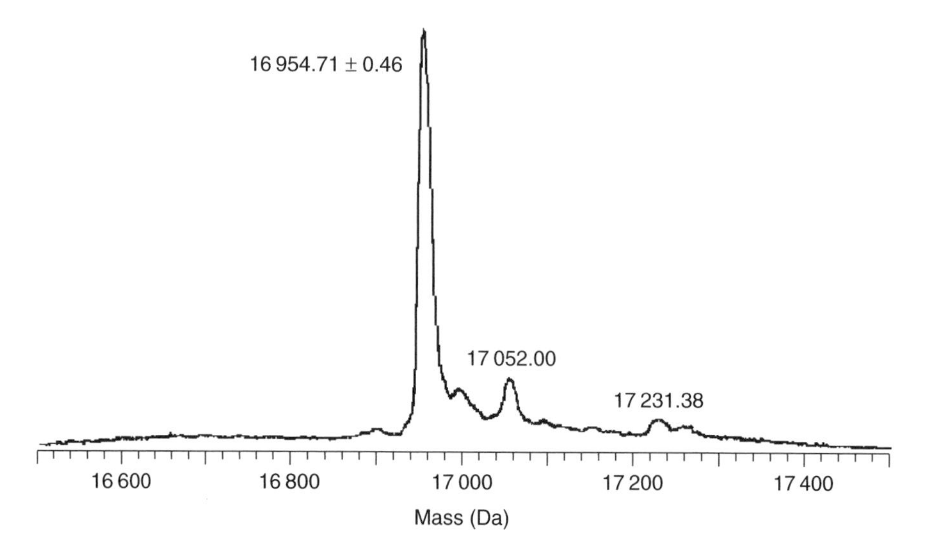


**Figure 4.12** Electrospray spectrum of horse heart myoglobin after transformation.

*individual isotopic masses* rather than the *chemical atomic weight* of the element which is calculated by using the individual masses of the isotopes and their relative abundances. For example, chlorine consists of two isotopes, i.e. $^{35}Cl$, with a mass of 34.9689, and $^{37}Cl$, with a mass of 36.9659, with relative abundances of 75.77 and 24.23%, respectively, which leads to a chemical (average) atomic weight of 35.453. The molecular-ion region of the mass spectrum of an analyte containing a single chlorine atom will not consist of a single ion but of two ions separated by two mass units in the ratio of 74.77:24.23 (3:1) (ignoring the contribution from other species such as $^{13}C$).

From a mass spectrometry perspective, the molecular weight of an analyte is defined as that mass containing the most/more abundant isotope of the elements present, e.g. $C_6H_5Cl = 112$, based on $C = 12$, $H = 1$ and $Cl = 35$.

**Table 4.2** Masses and relative abundances of some isotopes of commonly occurring elements

| Element | Mass (relative abundance) | Mass (relative abundance) | Mass (relative abundance) |
|---|---|---|---|
| Carbon | 12.0000 (98.89%) | 13.0034 (1.11%) | — |
| Hydrogen | 1.0078 (99.985%) | 2.0140 (0.015%) | — |
| Nitrogen | 14.0031 (99.634%) | 15.0001 (0.366%) | — |
| Oxygen | 15.9949 (99.762%) | 17.9992 (0.200%) | — |
| Sulfur | 31.9721 (95.02%) | 32.9715 (0.75%) | 33.9679 (4.21%) |

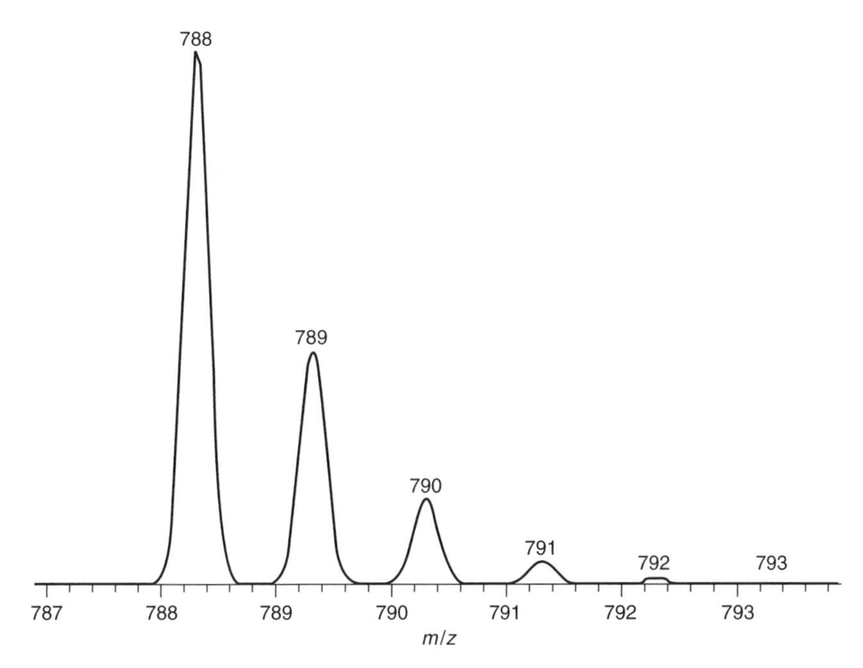


**Figure 4.13** Theoretical molecular-ion region of the mass spectrum of a molecule of composition $C_{35}H_{48}N_8O_{11}S$ at a mass spectrometer resolution of 1500.

The resolution of most mass spectrometers (the ability to separate ions of similar $m/z$ values – see Section 3.3 above) in routine use is sufficient to allow the separation of the ions containing the individual isotopes if low-molecular-weight compounds (<1000 Da) are being studied. This is illustrated in Figure 4.13 which shows the molecular-ion region of a compound having the molecular formula $C_{35}H_{48}N_8O_{11}S$ determined with a mass spectrometer resolution of 1500. The masses of the isotopes present in this molecule are shown in Table 4.2.

**SAQ 4.2**

Using the atomic weights given in Table 4.2, calculate the mass spectral resolution required to separate the molecular ion of atomic composition $C_{35}H_{48}N_8O_{11}S$ from the isotopic peak containing one $^{13}C$ atom and carry out a similar calculation for the ion of composition $C_{284}H_{432}N_{84}O_{79}S_7$ and its single $^{13}C$ satellite.

If, however, we consider a protein of modest size, such as aprotin with a molecular formula of $C_{284}H_{432}N_{84}O_{79}S_7$ at a similar mass spectrometer resolution, the molecular-ion region of its mass spectrum, shown in Figure 4.14, does not show the individual isotopic contributions, a resolution of around 5000 being required for these to be evident (Figure 4.15).

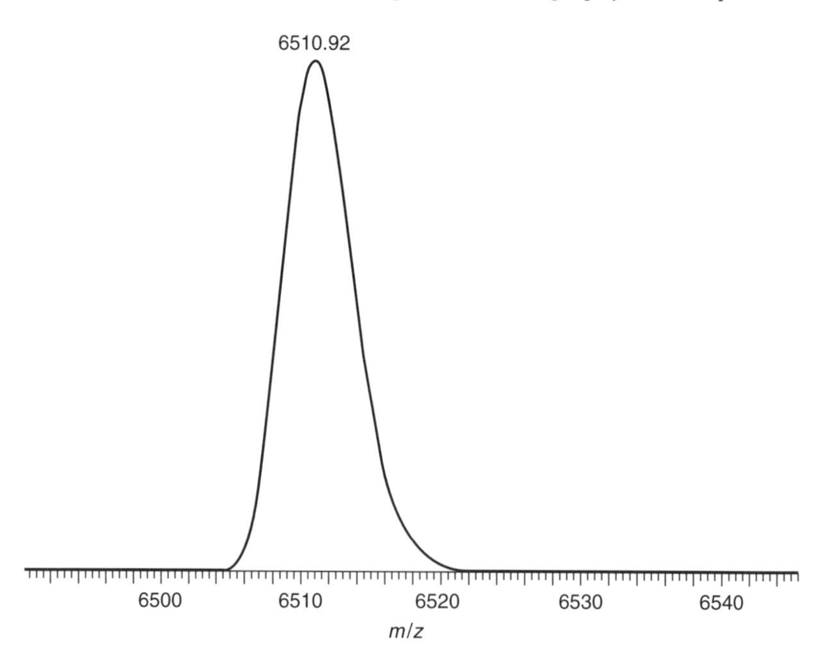


**Figure 4.14** Theoretical molecular-ion region of the mass spectrum of aprotin (molecular weight, 6507) at a mass spectrometer resolution of 1500.

The production of multiply charged ions using electrospray ionization effectively extends the mass range of the mass spectrometer by a factor directly related to the number of charges attached to the analyte molecule. Nature has been kind to us and most high-molecular-weight biopolymers (proteins, oligonucleotides, etc.) give significant ions of $m/z < 2000$ Da, well within the 'mass range' of relatively inexpensive, commercially available, instruments.

What multiple charging does not do, however, is to provide an equivalent increase in the resolution of the mass spectrometer and the resolution required to separate the individual isotopic contributions from a multiply charged species is identical to that required for the corresponding singly charged species. Figures 4.16 and 4.17 show the ions from the 7+ charge state of aprotin at resolutions of 1500 and 5000, respectively.

**SAQ 4.3**

You have calculated the resolution required to separate the molecular ion of atomic composition $C_{284}H_{432}N_{84}O_{79}S_7$ from the isotopic peak containing one $^{13}C$ atom. Carry out a similar exercise to calculate the resolution required to separate the equivalent ions when they each have (a) 5 charges, (b) 7 charges, and (c) 10 charges. Compare the values obtained.

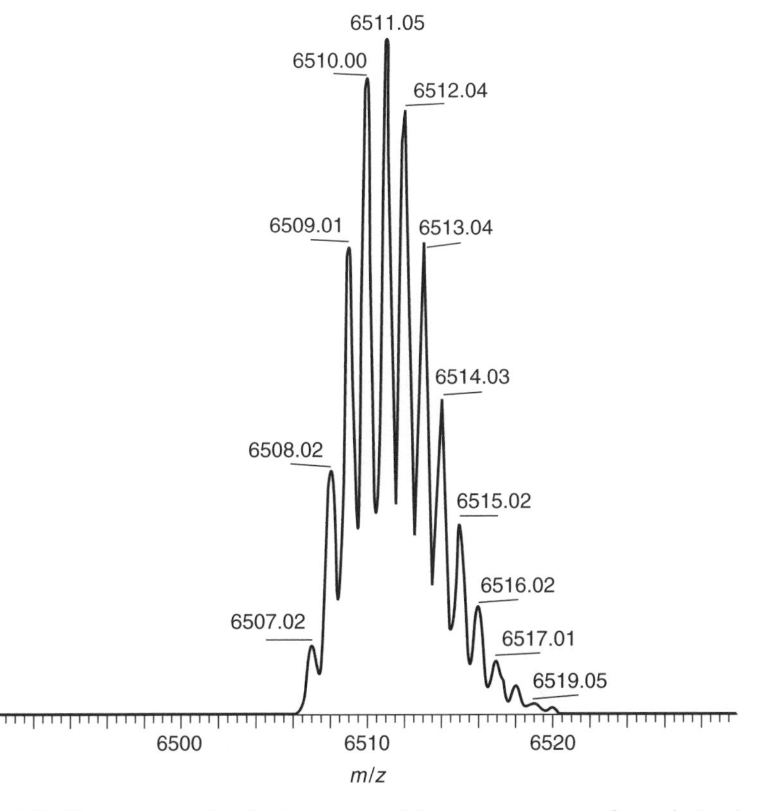


**Figure 4.15** Theoretical molecular-ion region of the mass spectrum of aprotin (molecular weight, 6507) at a mass spectrometer resolution of 5000.

The inability to resolve the individual isotopic contributions has a significant effect on the molecular weight of the analyte that is then calculated. If the measured $m/z$ value shown in Figure 4.16, i.e. 931.14, is used to derive the molecular weight by using equation (4.3), a value of 6510.92 is obtained. This corresponds to the molecular weight calculated by using the average atomic weight of the elements based on the abundance of all of the naturally occurring isotopes, e.g. C is 12.011 Da based on 98.89% $^{12}C$ and 1.1% $^{13}C$. If the monoisotopic ion, namely that corresponding to the presence of the more/most abundant isotope of each element, in Figure 4.17 is used (930.58), a value of 6507.00 is obtained. It is therefore important, when reporting the molecular weight of a compound, to indicate clearly whether this is based on the monoisotopic or average atomic weight of the elements present since these, as demonstrated above, may be significantly different. An apparent variation in the reported molecular weight of a biopolymer determined by using mass spectrometry and more classical methods is often due to this simple ambiguity.

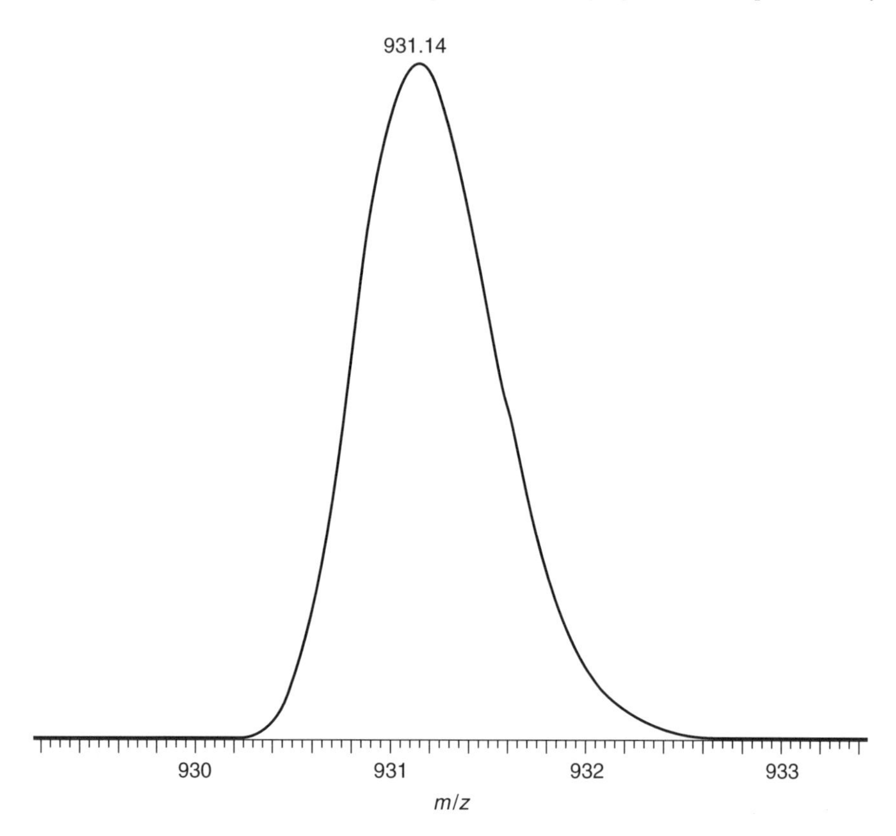


**Figure 4.16** Theoretical mass spectrum of the 7+ charge state of aprotin (molecular weight, 6507) at a mass spectrometer resolution of 1500.

It should also be noted from Figures 4.15 and 4.17 that the most intense ion in the molecular-ion region does not correspond to the species containing only $^{12}C$, i.e. the species used to derive the monoisotopic molecular weight. From simple probability theory, it can be shown that for an ion of $m/z$ $M$ containing $n$ carbon atoms (ignoring any other elements that may be present) that there will be an ion at $m/z$ $(M + 1)$ of intensity, relative to that at $m/z$ $M$, of $n \times 1.1\%$. If the molecule being studied therefore contains more than 91 carbon atoms, and this is not unusual for many biopolymers, the ion at $m/z$ $(M + 1)$ will be of greater intensity than that at $m/z$ $M$. The ion in any mass spectrum used to derive the molecular weight must be chosen correctly to prevent inaccurate results being reported.

For higher-molecular-weight materials, the resolution required to observe the individual isotopic contributions increases proportionally.

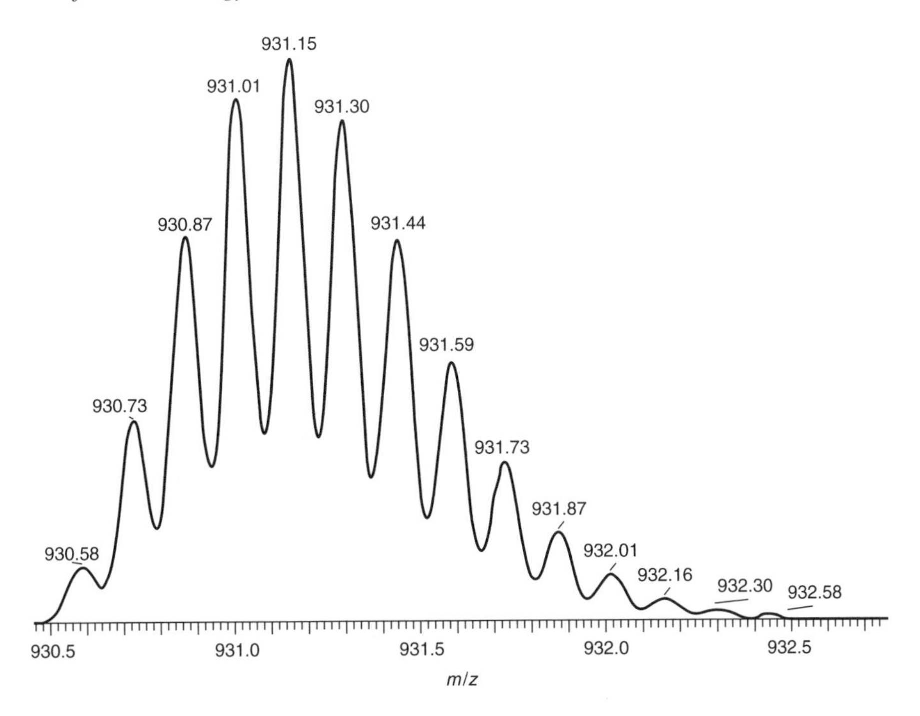


**Figure 4.17** Theoretical mass spectrum of the 7+ charge state of aprotin (molecular weight, 6507) at a mass spectrometer resolution of 5000.

High-resolution mass spectrometers have been used to obtain electrospray spectra and have the added advantage that they allow the direct determination of the charge state of the ions being observed, e.g. if the apparent separation of the $^{12}C$ and $^{13}C$ isotopic contributions is 0.1 Da, the charge state is 10, while if it is 0.05 Da, the charge state is 20, etc.

The suppression effects associated with electrospray ionization have been discussed earlier although if the compounds present are similar in behaviour these may be minimal. The intention, when using chromatography as an introduction device, is to allow individual components to enter the mass spectrometer for analysis. The separation capability of HPLC has been discussed previously and it is not unusual, particularly when complex mixtures are being studied, to encounter electrospray spectra from more than one component.

This sometimes complicates the extraction of molecular weight data as it is not always immediately clear which ions in the spectrum originate from each component. This can be determined by the use of equation (4.6). An example of this is shown in Figure 4.18, which shows the electrospray spectrum from what is apparently a single chromatographic response, while Table 4.3(a) displays the results of applying equation (4.6) to the major ions found in that spectrum. As

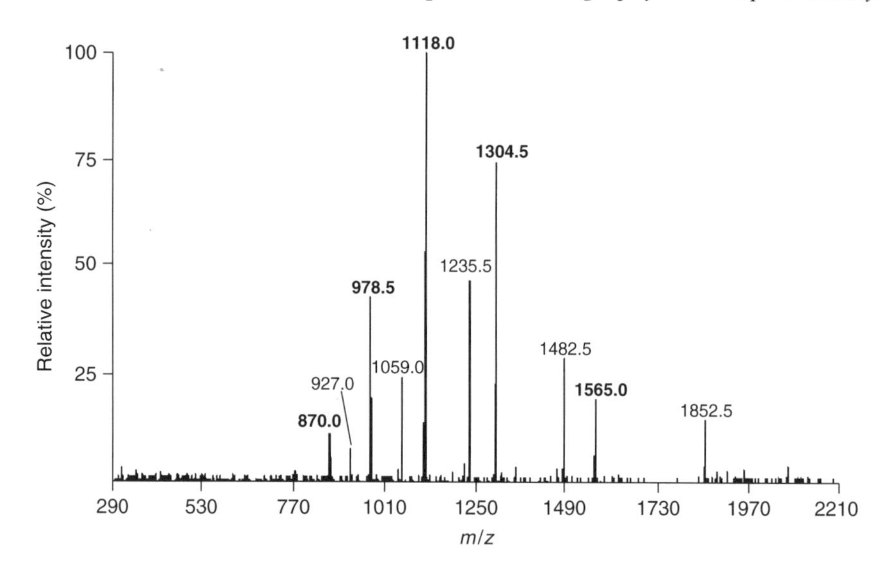


**Figure 4.18** Electrospray spectrum from a single chromatographic response in the LC–MS analysis of a tryptic digest. From applications literature published by SCIEX, Concord, Ontario, Canada, and reproduced by permission of MDS SCIEX, a division of MDS Inc.

**Table 4.3** Charge states calculated from the electrospray spectrum shown in Figure 4.18, assuming (a) all ions form part of the same series, and (b) the presence of two overlapping series containing these ions

| Case | $m_1$ | $m_2$ | $n_1$ (calculated) | Molecular weight (Da) |
|---|---|---|---|---|
| (a) | 927.00 | 870.00 | 15.25 | — |
| | 978.50 | 927.00 | 17.98 | — |
| | 1059.00 | 978.50 | 12.14 | — |
| | 1118.00 | 1059.00 | 17.93 | — |
| | 1235.50 | 1118.00 | 9.51 | — |
| | 1304.50 | 1235.50 | 17.89 | — |
| | 1482.50 | 1304.50 | 7.32 | — |
| | 1565.00 | 1482.50 | 17.96 | — |
| | 1852.50 | 1565.00 | 5.44 | — |
| (b) | 978.50 | 870.00 | 8.01 | 7820.00 |
| | 1118.00 | 978.50 | 7.01 | 7819.00 |
| | 1304.50 | 1118.00 | 5.99 | 7821.00 |
| | 1565.00 | 1304.50 | 5.00 | 7820.00 |
| | 1059.00 | 927.00 | 7.02 | 7406.00 |
| | 1235.50 | 1059.00 | 5.99 | 7407.00 |
| | 1482.50 | 1235.50 | 5.00 | 7407.50 |
| | 1852.50 | 1482.50 | 4.00 | 7406.00 |

can be seen, the values of the charge states calculated do not form a logical series and, in many cases, differ significantly from integer values.

The intensities of the ions observed in an electrospray spectrum usually show an (approximately) Gaussian distribution. The spectrum given in Figure 4.18 does not show this but rather an alternating increase and decrease in intensity and this should immediately alert the analyst to the fact that the spectrum warrants further examination. Table 4.3(b) shows the results obtained when the spectrum is considered to consist of two independent series, with the ions being selected on the basis of their initial increase and subsequent decrease in intensity. As can be seen from this table, the calculated charge states are all very close to an integer value and their values form a logical progression. The molecular weight determination using these values is also shown and provides a consistent value for each series.

The transformation software will carry out a similar analysis and if more than one ion series is present will link the appropriate ions together and produce a 'transformed' spectrum containing the relevant number of molecular species.

An alternative statistical approach to obtain better quality electrospray information is the use of 'maximum entropy' to process the raw data [16]. Maximum entropy differs from other statistical approaches to spectrum deconvolution in that it tries to predict the most likely theoretical signal, or group of signals, that would give rise to the signal observed given certain criteria that cause these signals to differ (to be 'damaged') in appearance from that theoretical signal. In mass spectrometry, the $m/z$ values and the relative intensities of the ions observed at each are considered, with the factors considered to 'damage' the theoretical signal being those that cause an increase in peak width, such as the resolution of the mass spectrometer.

An example of a transformed spectrum and the spectrum resulting from treating the raw data with maximum entropy software is shown in Figure 4.19. As can be seen from this, the spectrum after maximum entropy processing (Figure 4.19(b)) shows the presence of two discrete species that are not obviously visible in the transformed spectrum (Figure 4.19(b)). In this case, the mass difference between the species of masses 15 856.9 and 15 867.0, i.e. 10 Da, is indicative of the difference in their amino acid sequences. There are three pairs of amino acid that differ in molecular weight by 10 Da, namely serine and proline, cysteine and leucine/isoleucine (these have the same molecular weight), and histidine and phenylalanine. The way in which the mass differences of amino acids can be used to derive sequence information will be discussed in greater detail in Section 5.3 below.

### *4.7.4 Structural Information from Electrospray Ionization*

In the previous section, the generation and processing of molecular weight data from electrospray spectra have been described. The other great strength of mass spectrometry is its ability to generate structural information from the analyte

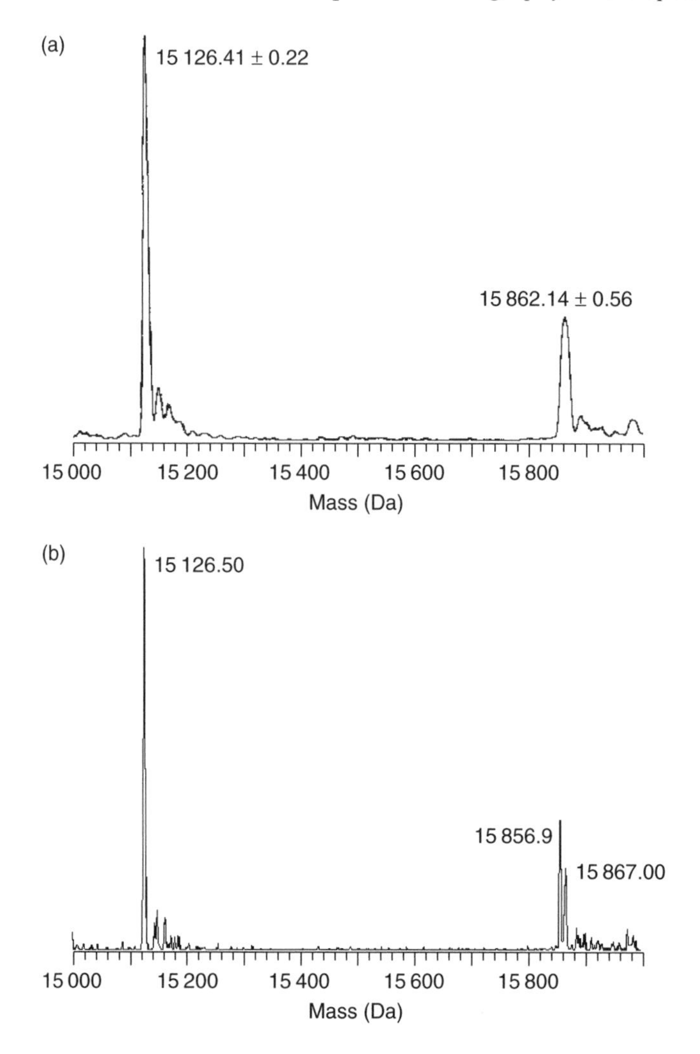


**Figure 4.19** Electrospray spectra of a protein (a) after transformation, and (b) after maximum entropy processing. From applications literature published by Micromass UK Ltd, Manchester, UK, and reproduced with permission.

under investigation. It has already been pointed out that electrospray is the 'softest' ionization process used in mass spectrometry, i.e. the transfer of ions to the gas phase is a low-energy process leading, almost exclusively, to the production of molecular species. There are two ways in which structural information can be generated when electrospray ionization is being used. The first, which may be applied to ions generated by any method of ionization, is tandem mass spectrometry, which was described earlier in Chapter 3.

The second, known variously as 'cone-voltage' or 'in-source' fragmentation, is similar to repeller-induced fragmentation in thermospray (see Section 4.6 above), and is a feature of ionization techniques in which ions are fragmented within the source of the mass spectrometer by the application of a voltage between the 'nozzle' and the 'skimmer' (see Figure 4.7). The application of this voltage increases the velocity of ions exiting the cone into the relatively high pressure region between this and the skimmer (approximately $10^{-2}$ torr; 1 torr = 133.3 Pa). The probability of collision between ions and residual gas molecules is significant and, with the increased energy now possessed by the analyte ions, fragmentation is much more likely.

The amount of energy imparted to the ions by the voltage is proportional to the number of charges on the ion and therefore multiply charged ions often give more fragmentation than their less highly charged counterparts. From an analytical perspective, the fact that ions of different charge states are likely to fragment differently may yield extra, and often, complementary information.

A significant difference between MS–MS and cone-voltage fragmentation is that in the former a specific ion of interest is chosen and its fragmentation studied in isolation. In addition, a number of different MS–MS experiments, yielding different types of information, are available. Cone-voltage fragmentation, on the other hand, is equivalent to one of the MS–MS experiments, i.e. the product-ion scan, and produces fragmentation of all species generated in the source, cf. repeller-induced fragmentation with the thermospray interface. Since a characteristic of electrospray ionization is that it usually produces a number of ions corresponding to different charge states of the analyte molecule, the spectra may be more difficult to interpret. For this reason, MS–MS, if available, is usually the preferred method for obtaining structural information.

The product-ion spectra generated by multiply charged ions differ in one significant respect from those generated by singly charged ions in that the product ions may not be of the same charge state as the precursor ion. The $m/z$ value of the product ion could, therefore, be greater than that of the precursor from which it originated and this must be taken into account when setting up the mass spectrometer to acquire the spectral data. This is also of significance when interpretation of the data is attempted.

Figure 4.20 shows the product-ion spectrum of the doubly charged molecular ion of the peptide glu-fibrinopeptide B at $m/z$ 786 generated by electrospray ionization, displaying a number of ions at $m/z$ ratios above that of the precursor. In this case, because it is known that the precursor is doubly charged, these must be singly charged and the mass differences between them are directly related to the loss of amino acid residues. This spectrum enables the full sequence of the peptide to be confirmed (a discussion of the use of electrospray spectra for deriving sequence information is to be found later in Section 5.3). Were the precursor to have more charges, the charge state of the product ions, and thus the relationship between them, would be less obvious and require a more detailed

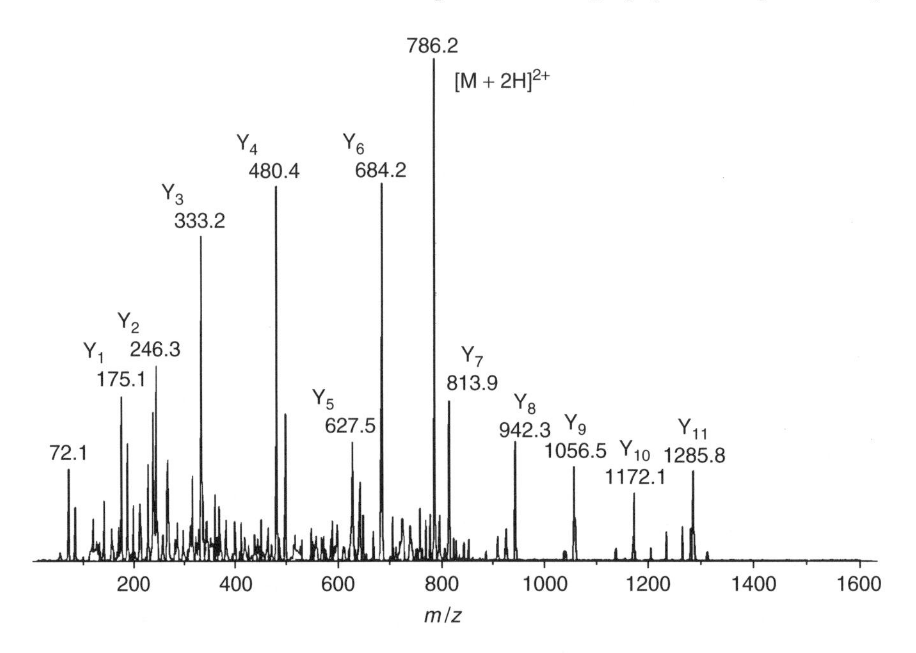


**Figure 4.20** Product-ion MS–MS spectrum from the doubly charged molecular ion of the peptide glu-fibrinogen B generated during electrospray ionization. From applications literature published by Thermofinnigan, Hemel Hempstead, UK, and reproduced with permission.

interpretation. The greater the number of charges on the precursor, then the more difficult is the spectrum likely to be to interpret.

### DQ 4.2

Using Table 5.6 (see p. 156), propose a sequence for glu-fibrinopeptide B based on the product-ion MS–MS spectrum shown in Figure 4.20.

*Answer*

*The ions observed in the spectrum are at* m/z *1285.8, 1172.1, 1056.5, 942.3, 813.9, 684.2, 627.5, 480.4, 333.2, 246.3 and 175.1, thus giving mass difference of 113.7, 115.6, 114.2, 128.4, 129.7, 56.7, 147.1, 147.2, 86.9 and 71.2 Da. By reference to Table 5.6, this is best rationalized in terms of the sequence being Asn–Asp–Asn–Glu/Lys–Glu–Gly–Phe–Phe–Ser–Ala, which is in agreement with the known sequence.*

The implications of charge must also be considered when constant-neutral-loss spectra are obtained because no longer is the loss necessarily of a neutral species, e.g. the loss of neutral CO (mass 28) from a singly charged ion of $m/z$ 250 yields an ion of $m/z$ 222, while from a doubly charged ion of $m/z$ 125 it would yield an ion of $m/z$ 111 (222/2). The loss of $CO^+$ from a doubly charged ion of $m/z$ 125 would yield an ion of $m/z$ 222.

The strength of MS–MS is its ability to select a single ion and study this in isolation and for singly charged ions the resolution of the mass spectrometer is usually sufficient to allow this. For multiply charged ions, as discussed previously, the resolution required to allow the selection of a single ion is often greater than that available from the mass spectrometer being used. What is observed, in these circumstances, is the MS–MS spectrum of the isotopic *cluster* and this needs to be taken into account when interpretation of the resulting spectra is attempted.

Electrospray is an unusual mass spectrometry technique in that it allows the study of the three-dimensional structure of compounds, particularly proteins, in solution as it is believed that this is relatively unchanged when ions are transferred to the vapour phase. This type of application will be discussed in more detail in Chapter 5 but attention is drawn at this point to the previous comments regarding the effect that the HPLC conditions, such as pH, may have on the appearance of an electrospray spectrum and the conformational deductions that may be made from them.

## Summary

The advent of the electrospray interface has allowed the full potential of LC–MS to be achieved. It is now probably the most widely used LC–MS interface as it is applicable to a wide range of polar and thermally labile analytes of both low and high molecular weight and is compatible with a wide range of HPLC conditions.

*Advantages*

- Ionization occurs directly from solution and consequently allows ionic and thermally labile compounds to be studied.
- Mobile phase flow rates from nl min$^{-1}$ to in excess of 1 ml min$^{-1}$ can be used with appropriate hardware, thus allowing conventional and microbore columns to be employed.
- Electrospray ionization, in contrast to the majority of other ionization methods employed in mass spectrometry, produces predominantly multiply charged ions of the intact solute molecule. This effectively extends the mass range of the mass spectrometer and allows the study of molecules with molecular weights well outside its normal range.
- For high-molecular-weight materials, an electrospray spectrum provides a number of independent molecular weight determinations from a single spectrum and thus increased precision.

*Disadvantages*

- Electrospray is not applicable to non-polar or low-polarity compounds.
- The mass spectrum produced from an analyte, in terms of the $m/z$ range of the ions observed and their relative intensities, depends upon a number of factors and spectra obtained using different experimental conditions may therefore differ considerably in appearance. This may or may not have implications on the analytical investigation being undertaken, although the molecular weight information that may be extracted from these spectra, however, is independent of these differences.
- Suppression effects may be observed and the direct analysis of mixtures is not always possible. This has potential implications for co-eluting analytes in LC–MS.
- Electrospray is a soft-ionization method producing intact molecular species and structural information is not usually available. Electrospray sources are capable of producing structural information from cone-voltage fragmentation but these spectra are not always easily interpretable. Experimentally, the best solution is to use a mass spectrometer capable of MS–MS operation but this has not inconsequential financial implications.

## 4.8 The Atmospheric-Pressure Chemical Ionization Interface

The electrospray interface, described in the previous section, enables mass spectra to be obtained from highly polar and ionic compounds.

Atmospheric-pressure chemical ionization (APCI) is another of the techniques in which the stream of liquid emerging from an HPLC column is dispersed into small droplets, in this case by the combination of heat and a nebulizing gas, as shown in Figure 4.21. As such, APCI shares many common features with ESI and thermospray which have been discussed previously. The differences between the techniques are the methods used for droplet generation and the mechanism of subsequent ion formation. These differences affect the analytical capabilities, in particular, the range of polarity of analyte which may be ionized and the liquid flow rates that may be accommodated.

Two APCI systems are shown in Figure 4.22, with the major difference between them being the use of a heated capillary for desolvation and droplet transport in the second of these (Figure 4.22(b)), cf. ESI. Here, the HPLC effluent is passed through a pneumatic nebulizer where the droplets are both generated and desolvated. The spray so formed then passes through a heated region where the vapour is dried. The neutral species thus produced are then passed through a *corona discharge* – the latter occurs when the field at the tip of the electrode is sufficiently high to ionize the gas surrounding it but insufficiently high to cause a

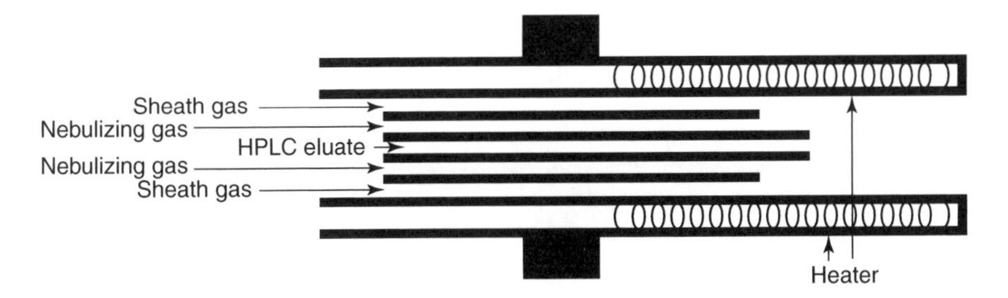


**Figure 4.21** Schematic of an atmospheric-pressure chemical ionization probe. From applications literature published by Micromass UK Ltd, Manchester, UK, and reproduced with permission.

spark – where ionization of the analyte is effected by CI-type processes with the vaporized solvent acting as the reagent gas. The technique is capable of dealing with flow rates between 0.5 and 2 ml min$^{-1}$, so making it directly compatible with 4.6 mm columns, and is much more tolerant to a range of buffers.

The similarity of the hardware required for APCI to that required for electrospray may be seen from comparing Figures 4.9 and 4.22 and makes changing between the techniques convenient from a practical point of view. The complementary nature of the two ionization techniques may, therefore, be readily utilized.

### *4.8.1 The Mechanism of Atmospheric-Pressure Chemical Ionization*

CI is an efficient, and relatively mild, method of ionization which takes place at a relatively high pressure, when compared to other methods of ionization used in mass spectrometry. The kinetics of the ion–molecule reactions involved would suggest that ultimate sensitivity should be obtained when ionization takes place at atmospheric pressure. It is not possible, however, to use the conventional source of electrons, a heated metallic filament, to effect the initial ionization of a reagent gas at such pressures, and an alternative, such as $^{63}Ni$ (a $\beta^-$ emitter) or a corona discharge, must be employed. The corona discharge is used in commercially available APCI systems as it gives greater sensitivity and is less hazardous than the alternative.

The processes leading to the ionization of the analyte at atmospheric pressure are similar to those described previously in relation to CI (see Section 3.2.2 above) in that ions, produced by the interaction of the electrons with the surrounding gas, undergo a number of reactions leading to the generation of reactive ions which interact with the analyte molecules present.

The reagent species in the positive-ion mode may be considered to be protonated solvent ions, and in the negative ion mode $O_2^-$, its hydrates and clusters.

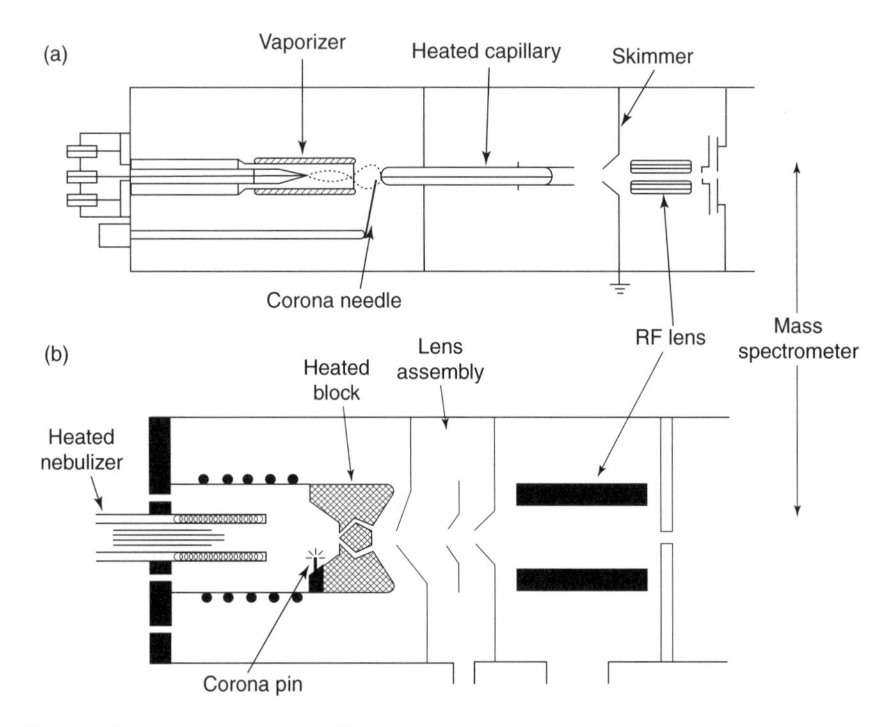


**Figure 4.22** Schematics of two different designs of atmospheric-pressure chemical ionization LC–MS interfaces. From applications literature published by (a) Thermofinnigan, Hemel Hempstead, UK, and (b) Micromass UK Ltd, Manchester, UK, and reproduced with permission.

As with conventional CI, this is a very mild form of ionization leading to molecular species with little or no fragmentation, i.e. $(M + H)^+$ and $(M - H)^-$. This is not, however, always the case. The use of chromatographic modifiers may change the composition of the CI plasma to such a state that, as in CI and thermospray, other ions may be formed, e.g. the presence of ammonium acetate may lead to $(M + NH_4)^+$ and $(M + CH_3COO)^-$ ions in the positive- and negative-ion modes, respectively. The chemistry of the analyte may also have an effect, as has been discussed for ESI, with, for example, the spectra of fullerenes extracted from soot particles yielding an $M^{+\bullet}$ molecular species [17].

In addition to the formation of these ions of direct analytical utility, APCI leads to the formation of ion clusters involving solvent molecules. Since these tend to make interpretation more difficult, they need to be removed and this may be accomplished either by the use of a 'curtain gas' or by cone-voltage fragmentation (see Section 4.7.4 above) which is also applicable to APCI.

From this description, it can be seen that APCI is a gas-phase process and it may seem that this technique is only applicable to the same range of compounds that may be studied by EI and CI, i.e. those that are volatile and thermally stable. APCI is, however, equally applicable to more polar compounds and this has led to its use in combination with HPLC. APCI ionizes most efficiently compounds with low to moderately high polarities and in this respect is complementary to electrospray which gives the best sensitivity for ionic compounds.

The APCI ionization regime is much more harsh that ESI and this precludes its use for the study of large biomolecules, with the mass limit for APCI being generally considered as below 2000 Da. Having said this, as will be shown later, the technique may still be used for the analysis of many thermally labile compounds without their decomposition, and small peptides have been studied.

When ions are present in solution, nebulization produces charged droplets from which ions may be emitted directly into the gas phase by ion evaporation. The formation of ions in this way is by a totally different mechanism to that of APCI and is termed *aerospray* (AS). An LC–MS interface based solely on the aerospray principle alone has not yet been made available commercially. The overall charge on any droplet depends upon the statistical distribution of cations and anions within it – some will have a positive charge and some a negative charge. As with conventional CI, positive-ion formation is favoured by the presence of strong gas-phase acids, and negative-ion formation by the presence of strong gas-phase bases. The production of a distribution of charged droplets applies to thermospray, aerospray and APCI and is one of the major differences between these techniques and electrospray/ionspray.

## Summary

APCI is complementary to electrospray in that it allows spectra to be obtained from non-polar and only slightly polar compounds. There is a great deal of commonality in the hardware used for the production of APCI and electrospray spectra and the combination of the two techniques allows a high proportion of HPLC eluates to be studied.

*Advantages*

- APCI produces ions from solution and while the analyte experiences more heat than with electrospray, compounds with a degree of thermal instability may be studied without their decomposition.
- APCI is best applied to compounds with low to moderately high polarities.
- APCI is a soft ionization technique which usually enables the molecular weight of the analyte under study to be determined.
- APCI is able to deal with flow rates up to 2 ml $min^{-1}$ and is, consequently, directly compatible with 4.6 mm HPLC columns.
- APCI is more tolerant to the presence of buffers in the mobile phase stream than is ESI.

- APCI is more tolerant to changes in experimental conditions than is ESI and a range of mobile phases, including gradient elution, may therefore be accommodated by using a single set of experimental conditions.

*Disadvantages*

- APCI spectra can contain ions from adducts of the analyte with the HPLC mobile phase or organic modifiers, such as ammonium acetate, that may be present. The presence of ions such as $(M + NH_4)^+$ and $(M + CH_3COO)^-$ may hinder interpretation of the spectra obtained.
- Structural information is not usually available unless cone-voltage fragmentation or MS–MS is used.
- APCI is not able to function effectively at very low flow rates.
- APCI is not suitable for analytes that are charged in solution.

# Summary

In this chapter, seven types of LC–MS interfaces have been described and their performance characteristics compared. Any modifications to the HPLC conditions that are required to allow the interface to operate effectively have been highlighted.

Particular emphasis has been placed upon *electrospray* and *atmospheric-pressure chemical ionization* (APCI) which, in addition to being the currently most widely used interfaces, are ionization techniques in their own right.

Electrospray is unusual in that it produces almost exclusively multiply charged ions in a variety of different charge states. The way in which the molecular weight of an analyte may be calculated has been derived. In addition, the appearance of an electrospray spectrum may vary considerably with the conditions in the solution from which it has been generated. For this reason, the mechanisms leading to the production of ions using this technique have been described at some length.

Electrospray and APCI spectra consist predominately of molecular species and the ways in which structural information may be generated from analytes ionized by these techniques have been considered in some detail.

# References

1. Ashcroft, A. E., *Ionization Methods in Organic Mass Spectrometry*, RSC Analytical Spectroscopy Monographs, The Royal Society of Chemistry, Cambridge, UK, 1997.
2. Niessen, W. M. A., *Liquid Chromatography–Mass Spectrometry*, 2nd Edn, Chromatographic Science Series, No. 79, Marcel Dekker, New York, 1999.
3. Yergey, A. L., Edmonds, C. G., Lewis, I. A. S. and Vestal, M. L., *Liquid Chromatography/Mass Spectrometry Techniques and Applications*, Plenum Press, New York, 1990.

4. Karger, B. L., Kirby, D. P., Vouros, P., Foltz, R. L. and Hidy, B., *Anal. Chem.*, **51**, 2324–2328 (1979).
5. Hayes, M. J., Lankmayer, E. P., Vouros, P., Karger, B. L. and McGuire, J. M., *Anal. Chem.*, **55**, 1745–1752 (1983).
6. http://webbook.nist.gov/chemistry/pa-ser.html
7. Caprioli, R. M. (Ed), *Continuous-Flow Fast Atom Bombardment Mass Spectrometry*, Wiley, Chichester, UK, 1990.
8. Solka, B. H., 'Particle beam LC–MS spectra of anionic surfactants', in *Proceedings of the 40th ASMS Conference on Mass Spectrometry and Allied Topics*, Washington, DC, May 31–June 5, 1992, pp. 1464–1465.
9. Sheehan, E. W., Ketkar, S. and Willoughby, R. C., 'Volatility enhancement of nonvolatile solutes by the combination of a heated target and a solvent depleted particle beam', in *Proceedings of the 39th ASMS Conference on Mass Spectrometry and Allied Topics*, Nashville, TN, May 19–24, 1991, pp. 1306–1307.
10. Haddon, W. F. and Harden, L. A., 'Advantages of particle beam sample introduction for analysis of thermally sensitive natural products by mass spectrometry', in *Proceedings of the 39th ASMS Conference on Mass Spectrometry and Allied Topics*, Nashville, TN, May 19–24, 1991, pp. 1316–1317.
11. Iribarne, J. V. and Thompson, B. A., *J. Chem. Phys.*, **64**, 2287–2294 (1976).
12. Schmeizeisen-Redeker, C., Bueftering, L. and Roellgen, F. W., *Int. J. Mass Spectrom. Ion Processes*, **90**, 139–150 (1989).
13. Wilm, M. and Mann, M., *Anal. Chem.*, **68**, 1–8 (1996).
14. Bordoli, R., Hoyes, J., Langridge, J., Chervet, J-P., Vissers, H. and van Veelen, P., 'Nanobore HPLC–MS and HPLC–MS/MS using a nanoflow ESI interface and a Q–Tof hybrid mass spectrometer', Micromass Technical Note 108, Micromass, Manchester, UK, 1998.
15. Abian, J., Oosterkamp, A. J. and Gelpi, E., *J. Mass Spectrom.*, **34**, 244–254 (1999).
16. Buck, B. and Macaulay, V. A. (Eds), *Maximum Entropy in Action*, Oxford University Press, Oxford, UK, 1991.
17. Anacleto, J. F., Quilliam, M. A. and Boyd, R. K., 'Analysis of fullerene soot extracts by liquid chromatography–mass spectrometry using atmospheric pressure ionization', in *Proceedings of the 41st ASMS Conference on Mass Spectrometry and Allied Topics*, San Francisco, CA, May 30–June 4, 1993, p. 1083.

# Chapter 5

# Applications of High Performance Liquid Chromatography–Mass Spectrometry

**Learning Objectives**

- To be aware of the wide range of applications to which LC–MS may be applied and the power of the combined technique.
- To be aware that a number of parameters may affect the data which are obtained and that extensive method development, involving experimental design, may be necessary to obtain optimum performance.
- To understand the circumstances in which particular features of mass spectrometry, such as high-resolution measurements, MS–MS and cone-voltage fragmentation, selected-ion monitoring and selected-decomposition monitoring, may be used to address particular analytical problems.
- **To recognize that the combination of the two techniques is more powerful than either of the techniques in isolation.**

A search of the Science Direct literature database [1] using the simple term 'LC–MS' yielded the number of references shown in Table 5.1.

It is not being suggested that these figures represent the total number of papers published on the subject since the database used does not contain publications from all publishers and it is highly unlikely that all relevant papers will have been indexed using this simple term. It does, however, reflect the significant increase in the number of papers on LC–MS published recently which, in turn, reflects the development of LC–MS into a routine and mature analytical technique.

**Table 5.1** Number of references containing the term 'LC–MS' on the Science Direct database as at 10th December 2002

| Year | Number of references cited |
|---|---|
| 1995 | 116 |
| 1996 | 142 |
| 1997 | 186 |
| 1998 | 269 |
| 1999 | 320 |
| 2000 | 484 |
| 2001 | 507 |
| 2002 (to 10th December) | 1233 |

**DQ 5.1**

What other terms might have been used in a literature search for papers on LC–MS?

*Answer*

*The term 'LC–MS' has been used in this present book to represent the linking of high performance liquid chromatography with mass spectrometry but it is not the only term that is in common usage. Any papers which have been indexed with alternative terms, unless the database search engine has a comprehensive list of synonyms, will not be retrieved. It is therefore worthwhile investigating the efficiency of the system being used by attempting a number of individual searches using alternative terms to determine the level of overlap of the results obtained. In this case, HPLC–MS, LC/MS, liquid chromatography–MS, high performance liquid chromatography–MS, liquid chromatography–mass spectrometry, high performance liquid chromatography–mass spectrometry and even the separate terms HPLC and MS (abbreviated and in full) would be worth trying.*

The application areas for LC–MS, as will be illustrated later, are diverse, encompassing both qualitative and quantitative determinations of both high- and low-molecular-weight materials, including synthetic polymers, biopolymers, environmental pollutants, pharmaceutical compounds (drugs and their metabolites) and natural products. In essence, it is used for any compounds which are found in complex matrices for which HPLC is the separation method of choice and where the mass spectrometer provides the necessary selectivity and sensitivity to provide quantitative information and/or it provides structural information that cannot be obtained by using other detectors.

It is not the intention of this present author to provide, or even attempt to provide, a detailed discussion of all of the applications of LC–MS that have been reported, but rather to discuss a relatively small number of papers which illustrate the way in which LC–MS has been employed to give the required analytical results. In this way, it is hoped that the reader will appreciate the ways in which the technique can be applied and those who are actively involved in the technique may draw parallels that will be of value in considering their own analytical problems.

Equally, it is not intended that every aspect of the papers chosen as examples will be described in detail – the intention is to extract particular aspects of the application of LC–MS from each. Readers requiring further experimental detail and background information should therefore consult the relevant paper and the references therein.

In this book, a number of different LC–MS interfaces have been described, where some of these have been included primarily from an historical standpoint. Currently, the most widely used interfaces are, undoubtedly, the electrospray and APCI interfaces and it is these that will be concentrated upon (a search of the Science Direct database [1] for 2001 using the term 'thermospray', previously the most widely used interface, yielded only one paper).

Many of the recent applications of electrospray ionization, both in isolation and in conjunction with HPLC, have been concerned with the study of biopolymers, such as proteins, glycoproteins and carbohydrates. Readers unfamiliar with biochemistry should consult a general text on the subject for the necessary background information. LC–MS applications may be divided into two categories, i.e. those in which the main task is the identification of analytes and those in which quantitation is required. These may be further subdivided into those involving high-molecular-weight (>2000 Da) compounds such as biopolymers and those involving low-molecular-weight compounds (many drugs and metabolites).

## 5.1 Method Development

In the vast majority of GC–MS applications, the chromatographic conditions employed have little or no effect on the operation of the mass spectrometer. This means that the spectrometer may be 'tuned' for optimum performance and a number of samples containing different analytes can be analysed without operator intervention. This is not the case with LC–MS where the chromatographic conditions will invariably have a significant, compound-dependent, effect on the mass spectrometry conditions required to obtain useful analytical data.

Method development is not always, therefore, a simple task since there are a substantial number of parameters which may influence the final results that are obtained. As a consequence of the number of parameters that may be involved, formal experimental design procedures are increasingly being utilized, indeed are essential, to determine the experimental conditions that give optimum analytical performance.

Experimental design requires the analyst to identify the variables (factors) that are likely to affect the result of the analysis and to carry out experiments which allow those that have an effect on the final outcome to be identified. Having identified the factors of importance, experimental design finally allows the precise experimental conditions that give the 'best' result to be determined. 'Best' is, of course, a subjective term and what is meant by the term will vary from experiment to experiment and must be carefully defined in the context of the determination being carried out. Examples include the conditions required to give complete chromatographic resolution of the analytes present, the minimum analysis time for a complex mixture or the most intense detector signal from some or all of the analytes.

When an analytical method is being developed, the ultimate requirement is to be able to determine the analyte(s) of interest with adequate accuracy and precision at appropriate levels. There are many examples in the literature of methodology that allows this to be achieved being developed without the need to use complex experimental design simply by varying individual factors that are thought to affect the experimental outcome until the 'best' performance has been obtained. This simple approach assumes that the optimum value of any factor remains the same however other factors are varied, i.e. there is no 'interaction' between factors, but the analyst must be aware that this fundamental assumption is not always valid.

As an analytical method becomes more complex, the number of factors is likely to increase and the likelihood is that the simple approach to experimental design described above will not be successful. In particular, the possibility of interaction between factors that will have an effect on the experimental outcome must be considered and factorial design [2] allows such interactions to be probed.

Factors may be classified as 'quantitative' when they take particular values, e.g. concentration or temperature, or 'qualitative' when their presence or absence is of interest. As mentioned previously, for an LC–MS experiment the 'factors' could include the composition of the mobile phase employed, its pH and flow rate [3], the nature and concentration of any mobile-phase additive, e.g. buffer or ion-pair reagent, the make-up of the solution in which the sample is injected [4], the ionization technique, spray voltage for electrospray, nebulizer temperature for APCI, nebulizing gas pressure, mass spectrometer source temperature, cone voltage in the mass spectrometer source, and the nature and pressure of gas in the collision cell if MS–MS is employed. For quantification, the assessment of results is likely to be on the basis of the selectivity and sensitivity of the analysis, i.e. the chromatographic separation and the maximum production of molecular species or product ions if MS–MS is employed.

**SAQ 5.1**

For which types of compound is APCI likely to be the most appropriate ionization technique and for which is electrospray likely to be more effective?

Sets of experimental conditions are then defined that allow the relationship between the factors of interest and the experimental outcome to be probed. The data generated by this exercise are then analysed, usually using appropriate software if the effects of a number of factors are being investigated, to allow the set of experimental conditions giving the 'best' result to be determined.

### 5.1.1 The Use of Experimental Design for Method Development

A method for the development of a generic LC-electrospray–MS method for the analysis of acidic compounds using experimental design has been reported [5]. From an HPLC perspective, this type of analysis often requires the use of an ion-pairing reagent to obtain separation; however, many of these, such as tetraalkylammonium ions, are involatile and have undesirable effects on the performance of the mass spectrometer and more volatile alternatives have to be found – in this case, triethylamine was used.

The factors chosen for study were the concentration of the ion-pairing reagent, the solution pH ('quantitative' factors) and the acid chosen for pH adjustment (formic, acetic, propionic and trifluoroacetic acids) ('qualitative' factor). The effect of these factors was assessed by using responses that evaluated both the HPLC (the number of theoretical plates and the retention time) and MS performance (the total peak area and peak height) for each of the four analytes studied, i.e. 1-naphthyl phosphate (1), 1-naphthalenesulfonic acid (2), 2-naphthalenesulfonic acid (3) and (1-naphthoxy)acetic acid (4).

The method development consisted of three stages. In the first, a number of experiments covering a wide range of ion-pair reagent concentrations and solution pHs were carried out with each of the four acids used for pH adjustment. The cone voltage and electrospray capillary voltage were optimized for each analyte for each mobile phase composition and it was found that the gain in sensitivity obtained in this process was 'minimal', although the effect was not the same for all of the analytes studied. The results from these experiments enabled a number of general observations to be made. For example, the concentration of the ion-pair reagent had a greater effect on the peak height obtained in the mass spectrometer than did the solution pH, while neither the concentration of the ion-pair reagent nor the solution pH had a great effect on the HPLC performance except in the case of one of the analytes. It was also clear that the worst MS response for most of the analytes was obtained when trifluoroacetic acid was used for pH adjustment and so this acid was removed from further consideration.

In the second stage, a number of simple experiments were carried out over a more limited range of the ion-pair reagent concentrations and solution pH values to obtain better definition of the ranges to be investigated in the final optimization procedure. These results highlighted one of the problems encountered when a number of compounds are to be determined in a single analysis in that they do not all behave in the same way. In this case, for compound (1) the number of theoretical plates increased with increasing pH while the retention time decreased – the latter effect being observed with all four compounds. The number of theoretical plates obtained in the analysis of compounds (2) and (3) did not vary with pH when propionic acid was used for pH adjustment but decreased with increasing pH when formic acid was used, and showed a marked reduction when acetic acid was used at pH 5.5. When propionic acid was used for pH adjustment with compound (4), it resulted in a reduction in theoretical plates as the pH increased while with formic acid the opposite behaviour was observed; when acetic acid was used, the number of theoretical plates did not vary greatly. With such complex behaviour, it is not surprising that experimental design is necessary before optimum experimental conditions can be obtained.

For the final optimization, a modified factorial design involving three concentration levels of triethylamine and three pH levels was used. From these results, it was clear that the optimum conditions for the analysis of the carboxylic acid were so different from those required for the other compounds studied that it was not sensible to attempt to analyse all four together and indeed that carboxylic acids were better analysed by using conventional reversed-phase HPLC than by using ion-pairing.

### *5.1.2 The Choice of Electrospray or APCI*

The vast majority of LC–MS analyses currently in use employ either electrospray ionization or APCI. In the previous example, electrospray ionization was employed because of the highly polar nature of the analytes but, as discussed above in Sections 4.7 and 4.8, this and APCI are, to a large extent, complementary, with APCI being used for low- to medium-polarity analytes and electrospray for medium- to high-polarity analytes. There are many compounds, therefore, for which the 'best' ionization technique is not immediately obvious and their relative merits must be investigated.

Picric, picramic and isopicramic acids and dipicrylamine (hexyl) have been studied by LC–MS using electrospray and APCI, employing the volatile ion-pairing reagent tributyl ammonium formate [6]. Qualitatively, the spectra obtained using both techniques gave $[M - H]^-$ species as the most intense ions in their spectra but APCI gave more fragment ions that were useful for identification purposes. Conversely, electrospray was found to be significantly more sensitive and therefore more appropriate for quantitative applications.

**DQ 5.2**

Why was a volatile ion-pairing reagent used in this work?

*Answer*

*An* ***involatile*** *ion-pairing reagent would be deposited in the electrospray interface and lead to a reduction in performance. Some interfaces have been specifically designed to minimize this by removing the line-of-sight between the spray and the entrance to the mass spectrometer, and are thus more tolerant to involatile buffers. The performance of the interface will be improved by the use of* ***volatile*** *alternatives.*

In this study, the effect of mobile-phase flow rate, or more accurately, the rate of flow of liquid into the LC–MS interface, was not considered but as has been pointed out earlier in Sections 4.7 and 4.8, this is of great importance. In particular, it determines whether electrospray ionization functions as a concentration- or mass-flow-sensitive detector and may have a significant effect on the overall sensitivity obtained. Both of these are of great importance when considering the development of a quantitative analytical method.

A detector may be classified as either mass-flow- or concentration-sensitive, depending upon the changes in response observed when a sample is introduced at different flow rates. A mass-flow-sensitive detector will give an increased peak height as the flow rate is increased although, as the peak width will decrease, the area remains constant. As the flow rate increases, a concentration-sensitive device will give constant peak height but, as the peak width will again decrease, a decrease in peak area. The effect on detection limits is therefore not only dependent upon the mode in which the detector is operating but also on whether quantitation is based on peak height or peak area measurements.

The flow dependence requires investigation as part of the method validation because, as discussed in Section 4.7.1, electrospray ionization may be either concentration- or mass-flow-sensitive. APCI is considered to be mass-flow-sensitive. 'Low' and 'high' are both subjective terms and require investigation as part of method validation.

The issue of flow rate is of particular importance when a method is being developed to determine more than one analyte since the dependency of signal intensity on flow rate is likely to be different for each. This is demonstrated in the development of an LC–MS method for the analysis of a number of pesticides [3], the structures of which are shown in Figure 5.1. Initial experiments to determine the MS–MS transitions to monitor, shown in Table 5.2, and the optimum collision cell conditions were carried out by using flow-injection analysis.

Figure 5.2 shows the variation of electrospray signal intensity, based on both height and area, with flow rate for seven of these compounds when using a mobile phase of 97% methanol/3% water containing 1 mM ammonium acetate. Three compounds show constant peak height and decreasing peak area at flow rates up to 600 $\mu l\,min^{-1}$ and thus display concentration-sensitive behaviour. At higher flow rates, they show mass-flow sensitivity, with both peak height and peak area decreasing with increasing flow rate. The remainder show mass-flow sensitivity throughout the flow rate range investigated. At 600 $\mu l\,min^{-1}$, the signal intensity, based on peak area, is between 5 and 40% of that at 200 $\mu l\,min^{-1}$, while based on peak height, with the exception of chlorpyrifos, all compounds show greater than 60% of their signal intensity at 200 $\mu l\,min^{-1}$. When APCI is used for ionization (see Figure 5.3), seven of the compounds show mass-flow-sensitive characteristics over some range of flows with approximately constant peak area and increasing peak heights. Two analytes, diuron and alachlor, show immediate and significant reductions in both peak area and peak height when flows are

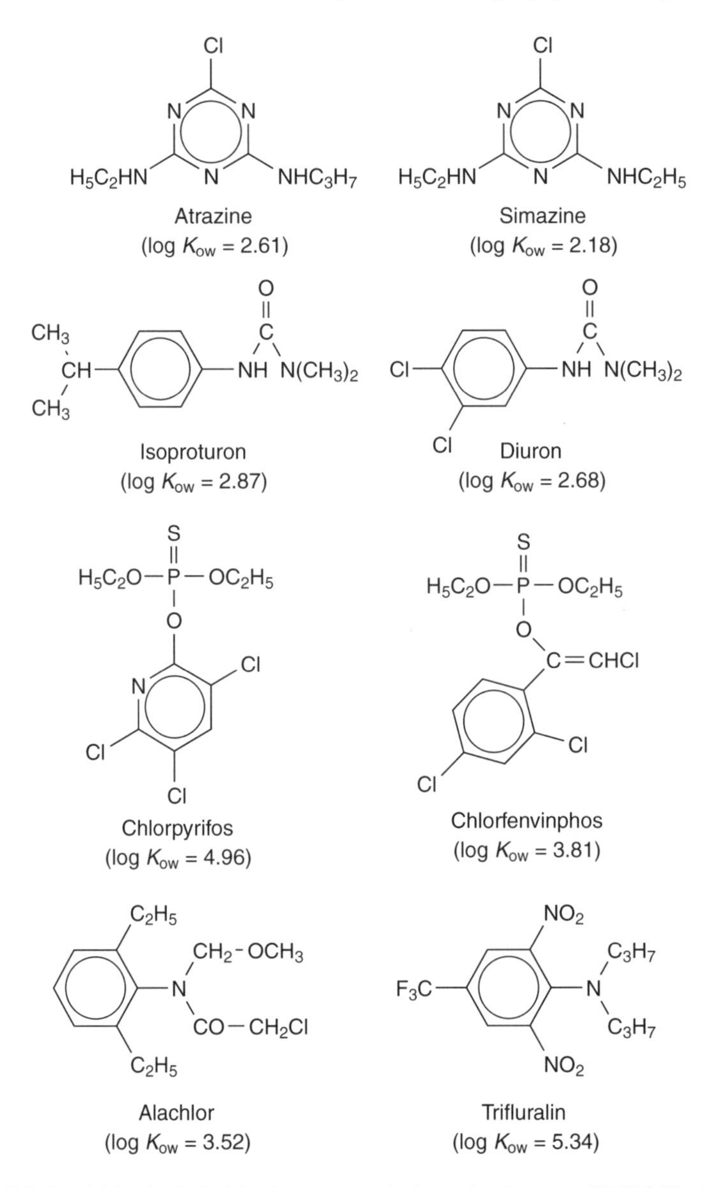


**Figure 5.1** Pesticides included in the systematic investigations on APCI-MS signal response dependence on eluent flow rate; the parameter $K_{ow}$ represents the distribution coefficient of the pesticide between *n*-octanol and water. Reprinted from *J. Chromatogr., A*, **937**, Asperger, A., Efer, J., Koal, T. and Engewald, W., 'On the signal response of various pesticides in electrospray and atmospheric pressure chemical ionization depending on the flow rate of eluent applied in liquid chromatography–mass spectrometry', 65–72, Copyright (2001), with permission from Elsevier Science.

**Table 5.2** Selected-reaction monitoring (SRM) transitions used for MS–MS detection of the pesticides studied in the systematic investigations on APCI–MS signal response dependence on eluent flow rate. Reprinted from *J. Chromatogr., A*, **937**, Asperger, A., Efer, J., Koal, T. and Engewald, W., 'On the signal response of various pesticides in electrospray and atmospheric pressure chemical ionization depending on the flow rate of eluent applied in liquid chromatography–mass spectrometry', 65–72, Copyright (2001), with permission from Elsevier Science

| Compound | SRM (precursor/product ion) |
|---|---|
| Isoproturon | 207/72 |
| Diuron | 233/72 |
| Atrazine | 216/174 |
| Simazine | 202/132 |
| Chlorfenvinphos | 359/99 |
| Chlorpyrifos | 350/198 |
| Alachlor | 270/162[a] |
| | 238/162[b] |
| Trifluralin | 336/202 |

[a] 'Turbo ionspray'.
[b] Using a heated nebulizer.

increased above 200 μl min$^{-1}$. From an analytical perspective, the flow rates which give maximum sensitivity for these two analytes are totally incompatible with the flow rates required for the others. Increasing the aqueous content of the mobile phase brings about a decrease in sensitivity at lower flow rates but the response behaviour is similar to that described above.

**SAQ 5.2**

A solution containing 0.5 mg ml$^{-1}$ of an analyte gives a detector response (based on peak height) of 48 ± 3 arbitrary units when analysed by LC–MS at a flow rate of 0.75 ml min$^{-1}$. At a flow rate of 1.00 ml min$^{-1}$, the detector response was 49 ± 3 arbitrary units. Is the mass spectrometer behaving as a concentration- or mass-flow-sensitive detector?

The variation in signal intensity can be reduced [3] by the use of flow programming. At a constant flow rate of 200 μl min$^{-1}$ (Figure 5.4(a)), signals from chlorpyrifos and trifluralin, using APCI, are observed although the analysis time is approaching 20 min and the chromatographic peak shape is not ideal. At a

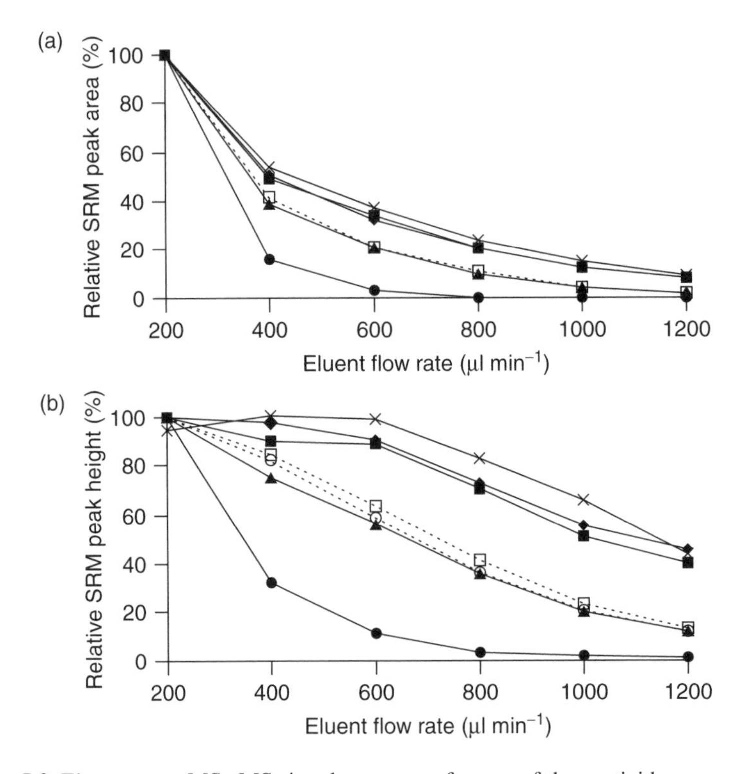


**Figure 5.2** Electrospray-MS–MS signal response of seven of the pesticides versus eluent flow rate, based on (a) peak area, and (b) peak height: ♦, atrazine; ■, simazine; ▲, diuron; ×, isoproturon; □, chlorfenvinphos; ●, chlorpyrifos; ○, alachlor. Reprinted from *J. Chromatogr., A*, **937**, Asperger, A., Efer, J., Koal, T. and Engewald, W., 'On the signal response of various pesticides in electrospray and atmospheric pressure chemical ionization depending on the flow rate of eluent applied in liquid chromatography–mass spectrometry', 65–72, Copyright (2001), with permission from Elsevier Science.

constant flow rate of 1200 $\mu l\,min^{-1}$ (Figure 5.4(b)), the analysis time is much reduced and the chromatographic peak shape is improved. As expected from the flow rate variation discussed above, however, the signals from chlorpyrifos and trifluralin are much reduced. Figure 5.4(c) shows the TIC trace obtained when a flow rate of 1200 $\mu l\,min^{-1}$ was used for the first five minutes of the LC–MS analysis and 200 $\mu l\,min^{-1}$ from that point onwards, with this providing a reduced analysis time, good chromatographic peak shape and optimum intensity for all of the components of the mixture.

The authors of this paper indicated that the parameters which had an effect on signal intensity included the specific design and manufacturer of the LC–MS

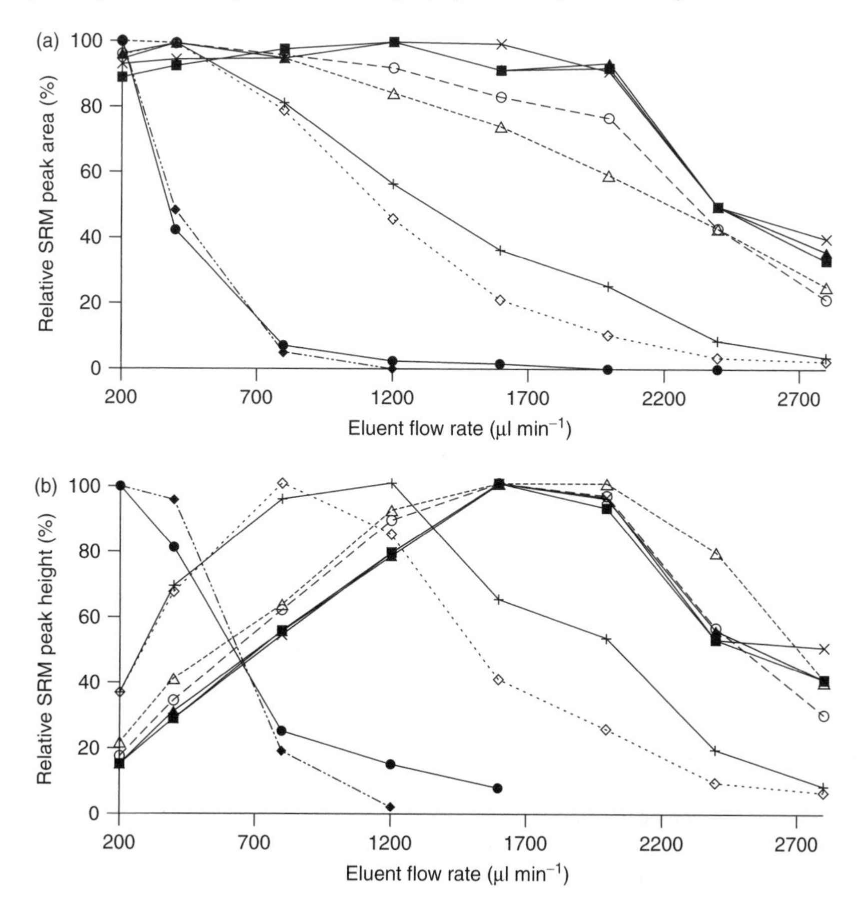


**Figure 5.3** APCI-MS–MS signal response of the eight pesticides versus eluent flow rate, based on (a) peak area, and (b) peak height: ■, atrazine; ▲, simazine; ×, isoproturon; ◇, diuron; ○, chlorfenvinphos; ●, chlorpyrifos; +, alachlor; △, diuron (−); ♦, trifluralin. Reprinted from *J. Chromatogr., A*, **937**, Asperger, A., Efer, J., Koal, T. and Engewald, W., 'On the signal response of various pesticides in electrospray and atmospheric pressure chemical ionization depending on the flow rate of eluent applied in liquid chromatography–mass spectrometry', 65–72, Copyright (2001), with permission from Elsevier Science.

interface, the pressures of the nebulizing and drying gases used in that interface, the mass spectrometer source temperature and the eluent composition. They do also comment, not unexpectedly perhaps, that the precise behaviour observed 'may be specific to their system' but the possibility of flow-rate dependence of analyte signal should be considered during method development.

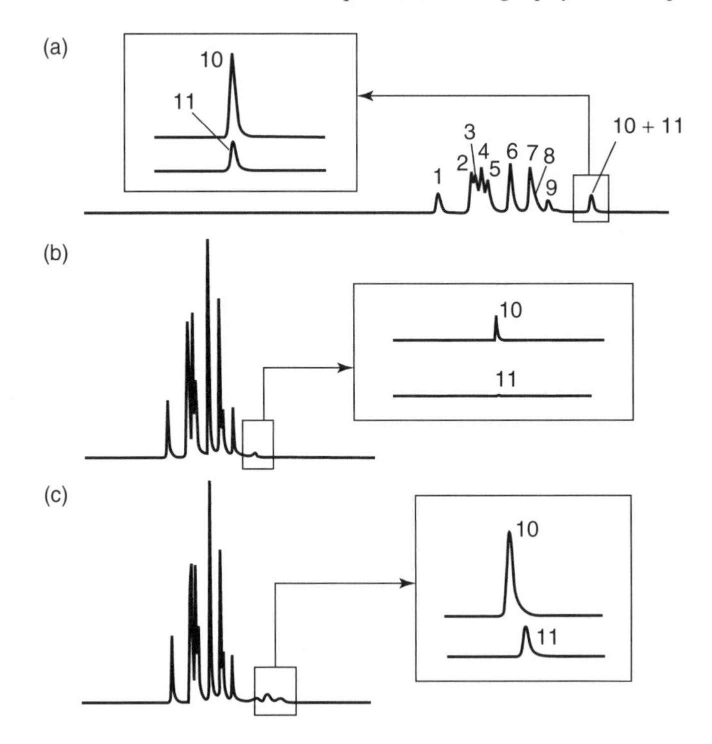


**Figure 5.4** Summed multiple-reaction monitoring traces and extracted selected-reaction monitoring traces for chlorpyrifos (10) and trifluralin (11) from the LC-APCI–MS–MS analysis of a standard pesticide mixture obtained (a) at a constant mobile-phase flow rate of 200 $\mu l\,min^{-1}$, (b) at a constant mobile-phase flow rate of 1200 $\mu l\,min^{-1}$, and (c) by using a programmed mobile-phase flow rate of 1200 $\mu l\,min^{-1}$ from 0–5.1 min and 200 $\mu l\,min^{-1}$ from 5.1–10 min. Reprinted from *J. Chromatogr., A*, **937**, Asperger, A., Efer, J., Koal, T. and Engewald, W., 'On the signal response of various pesticides in electrospray and atmospheric pressure chemical ionization depending on the flow rate of eluent applied in liquid chromatography–mass spectrometry', 65–72, Copyright (2001), with permission from Elsevier Science.

In the previous example, the behaviour of an analyte was investigated by using both electrospray and APCI.

From the foregoing discussion it may seem that a complex experimental design must be carried out before any analysis is attempted. While it is certainly a necessity/advantage that the analyst has some understanding of the effect that a parameter is likely to have on the experimental outcome, many analyses, particularly those involving mixtures containing relatively few components at relatively high concentrations, will be accomplished successfully on the basis of a simple study of selected experimental variables.

# 5.2 The Molecular Weight Determination of Biopolymers

The polarity and thermal instability of biopolymers, together with the almost exclusive formation of singly charged ions renders APCI an inappropriate ionization technique for their study. Much of the early work involving electrospray ionization, on the other hand, was connected with the analysis of this type of molecule, in particular, determining the molecular weight of proteins for which it is particularly effective.

**SAQ 5.3**

At what $m/z$ value would the singly charged molecular ion of a compound with molecular formula $C_{284}H_{432}N_{84}O_{79}S_7$ be observed?

## *5.2.1 Electrospray Spectra of Co-Eluting Components*

A potential problem encountered in these determinations is the ion suppression encountered in the presence of polar/ionic interfering materials which compete with the analyte(s) for ionization (see Section 4.7.2 earlier). Many of these analyses therefore involve some degree of off-line purification and/or an appropriate form of chromatography. Since it is not unusual to encounter closely related compounds that are not easily separated, it is also not unusual to employ both of these approaches, as in the following example. This illustrates the use of HPLC as a method of purification and demonstrates that highly efficient separations are not always required for valuable analytical information to be obtained.

Part of the characterization of a protein encountered in the attempted development of a vaccine to the hepatitis E virus involved the determination of its molecular weight [7].

The protein was purified by a dialysis procedure, denatured and analysed by sodium dodecyl sulfate–polyacrylamide gel electrophoresis (SDS–PAGE). Western blotting[†] indicated that the protein of interest consisted of two components, one of which increased in concentration as the purification proceeded. The authors initially suggested that this could be due to the presence of a number of species produced by modification of the amino acid side-chains, for example, by glycosylation, or by modification of the *C*- or *N*- terminus.

The 'purified' protein was subjected to reversed-phase HPLC analysis by using a $150 \times 1$ mm $C_{18}$ column with gradient elution from 0.1% aqueous trifluoroacetic acid (TFA) to 0.1% TFA in acetonitrile, over a period of 55 min, at a flow

[†] This is a means of transferring protein bands from an electrophoresis gel onto a fixing medium for further analysis.

rate of 50 μl min$^{-1}$. The column eluate was combined with methoxyethanol/iso-propanol (1:1 vol/vol), also at 50 μl min$^{-1}$, in a post-column mixing chamber, and 10 μl min$^{-1}$ of this was taken to an electrospray system – 100 μl min$^{-1}$ being too great a rate to use directly. The mass spectrometer, scanning from $m/z$ 500 to $m/z$ 2000 at 1.5 s per spectrum, was operating in the positive-ion mode.

The function of the post-column solvent addition was not discussed but is presumably to overcome, in the early stages of the gradient elution, the need to spray a largely aqueous mobile phase.

The TIC trace from this analysis, shown in Figure 5.5, exhibits a maximum at ca. 19 min, and a representative electrospray spectrum is illustrated in Figure 5.6. Transformation of the latter produces the spectrum presented in Figure 5.7 which indicates the presence of two species with relative molecular masses (RMMs) of 56 548.5 and 58 161.4 Da. These masses are lower than the value of 59.1 kDa calculated from previously obtained sequence information.

Experimentation showed that the protein was not glycosylated and that the sequence at the $N$-amino acid terminus corresponded to that expected. The $C$-terminus sequence, however, did not correspond to that predicted and these data were interpreted in terms of the presence of a heterogeneous, truncated, protein. A study of the tryptic digest fragments from this protein with matrix-assisted laser desorption ionization (MALDI) with post-source decay enabled the authors to suggest the positions at which the parent protein had been truncated.

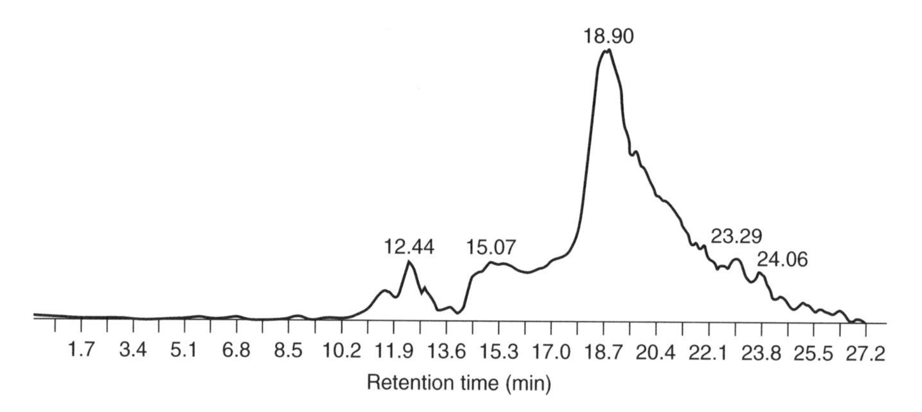


**Figure 5.5** TIC trace from the LC–MS analysis of a purified recombinant 62 kDa protein using a $C_{18}$ microbore 50 × 1 mm column and a flow rate of 50 μl min$^{-1}$. The starting buffer (buffer 'A') was 0.1% TFA in water, while the gradient buffer (buffer 'B') consisted of 0.1% TFA in acetonitrile–water (9:1 vol/vol). The running conditions consisted of 0% 'B' for 5 min, followed by a linear gradient of 100% 'B' for 55 min. Reprinted from *J. Chromatogr., B*, **685**, McAtee, C. P., Zhang, Y., Yarbough, P. O., Fuerst, T. R., Stone, K. L., Samander, S. and Williams, K. R., 'Purification and characterization of a recombinant hepatitis E protein vaccine candidate by liquid chromatography–mass spectrometry', 91–104, Copyright (1996), with permission from Elsevier Science.

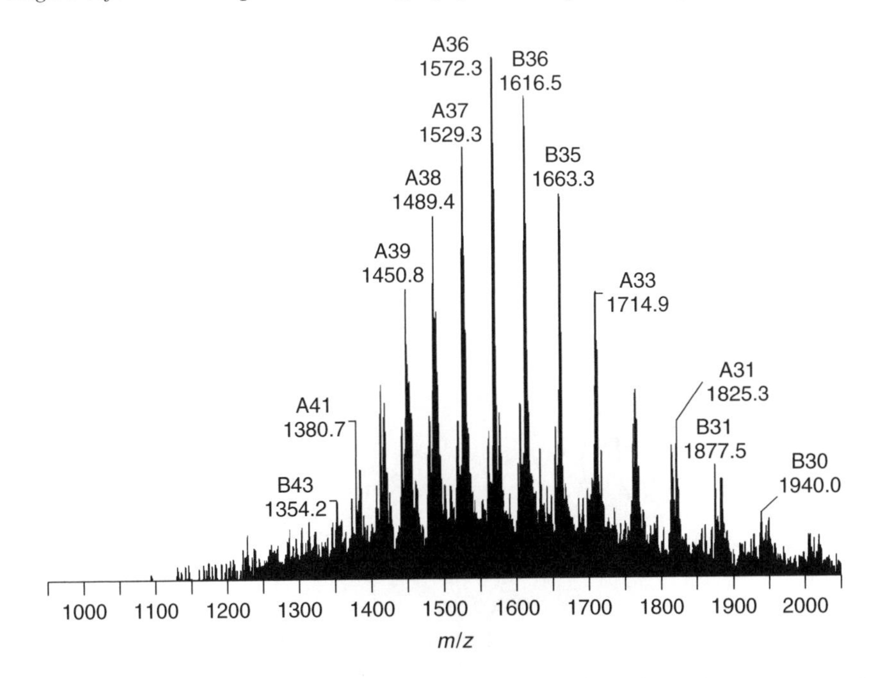


**Figure 5.6** Positive-ion electrospray spectrum obtained from the major component in the LC–MS analysis of a purified recombinant 62 kDa protein using a $C_{18}$ microbore 50 × 1 mm column and a flow rate of 50 $\mu l\,min^{-1}$. The starting buffer (buffer 'A') was 0.1% TFA in water, while the gradient buffer (buffer 'B') consisted of 0.1% TFA in acetonitrile–water (9:1 vol/vol). The running conditions consisted of 0% 'B' for 5 min, followed by a linear gradient of 100% 'B' for 55 min. Reprinted from *J. Chromatogr., B*, **685**, McAtee, C. P., Zhang, Y., Yarbough, P. O., Fuerst, T. R., Stone, K. L., Samander, S. and Williams, K. R., 'Purification and characterization of a recombinant hepatitis E protein vaccine candidate by liquid chromatography–mass spectrometry', 91–104, Copyright (1996), with permission from Elsevier Science.

### *5.2.2 The Use of Selected-Ion Monitoring to Examine the Number of Terminal Galactose Moieties on a Glycoprotein*

**DQ 5.3**

What is selected-ion monitoring and what are the advantages of using this technique?

*Answer*

*Selected ion monitoring is a technique in which only the ions at a limited number of* m/z *ratios chosen to be characteristic of the analyte(s)*

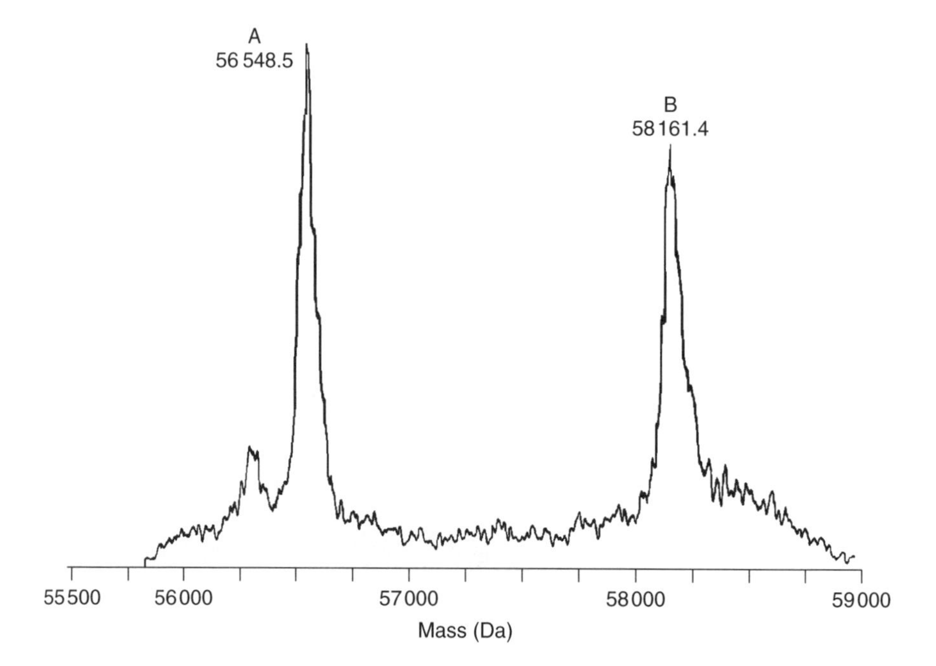


**Figure 5.7** Transformed electrospray spectrum of that shown in Figure 5.6. Reprinted from *J. Chromatogr., B*, **685**, McAtee, C. P., Zhang, Y., Yarbough, P. O., Fuerst, T. R., Stone, K. L., Samander, S. and Williams, K. R., 'Purification and characterization of a recombinant hepatitis E protein vaccine candidate by liquid chromatography–mass spectrometry', 91–104, Copyright (1996), with permission from Elsevier Science.

> *of interest are monitored, rather than ions at all* m/z *ratios. In this way, both the selectivity – analytes which do not produce ions at these* m/z *ratios will not be detected – and sensitivity – more time is spent monitoring each ion and therefore more ions of these* m/z *ratios reach the detector – of an analysis are enhanced.*

Once the molecular weight of a particular species has been determined and its significance ascribed, it is not always necessary to continue to acquire mass spectra over the full mass range to provide the required analytical information. Indeed, as discussed earlier in Section 3.5.2.1, there are often significant benefits to be gained by only monitoring a relatively small number of ions generated by an analyte.

Rituximab is a recombinant mouse/human chimeric monoclonal antibody whose *in vitro* activity varies with the number of terminal galactose moieties glycosylated to the peptide backbone at residue asparagine 301 [8]. The ability to monitor the levels of each discrete species present would allow the manufacturing process to be controlled to provide maximum yield of the desired product and a product of consistent composition.

To allow all culture production to be controlled, a method for rapid analysis is required. Prior to development of an LC–MS method, the analysis was both complex and time-consuming, involving the purification of a relatively large amount of the antibody using affinity chromatography, enzymatic release, and subsequent derivatization of the oligosaccharides and their analysis by using capillary electrophoresis.

Chemical reduction of the antibody results in the production of both 'light' and 'heavy' chains, with the heavy chains showing the different levels of glycosylation that are of interest. The HPLC system used to separate the light and heavy chains consisted of a Poros R1/H 100 × 2.1 mm column maintained at 60°C. Gradient elution was used from 90% of a 2% acetic acid solution (solvent A):10% acetonitrile/2-propanol (70:30 vol/vol) (solvent B) to 25% solvent A:75% solvent B over 30 min at a flow rate of 0.5 ml $min^{-1}$.

The HPLC separation of the light and heavy chains, together with their electrospray spectra, are shown in Figure 5.8.

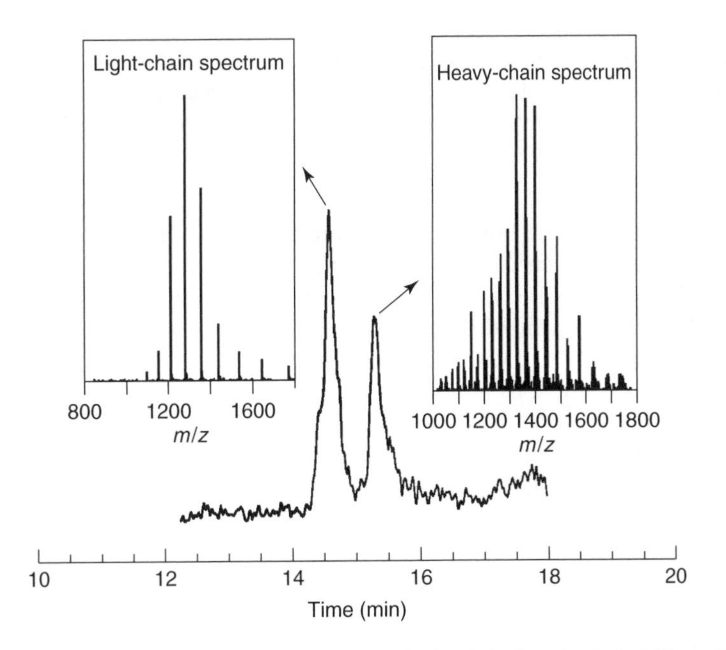


**Figure 5.8** TIC trace and electrospray spectra obtained during the LC–MS analysis of the light- and heavy-chain antibody fragments of reocombinant rituximab. Reprinted from *J. Chromatogr., A*, **913**, Wan, H. Z., Kaneshiro, S., Frenz, J. and Cacia, J., 'Rapid method for monitoring galactosylation levels during recombinant antibody production by electrospray mass spectrometry with selective-ion monitoring', 437–446, Copyright (2001), with permission from Elsevier Science.

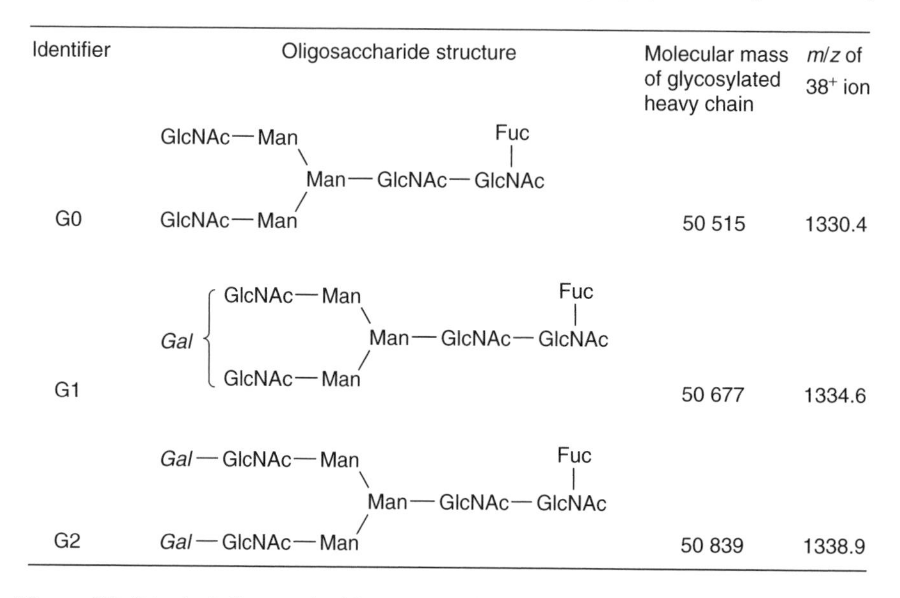


**Figure 5.9** Principal oligosaccharide structures found on recombinant rituximab. Reprinted from *J. Chromatogr., A*, **913**, Wan, H. Z., Kaneshiro, S., Frenz, J. and Cacia, J., 'Rapid method for monitoring galactosylation levels during recombinant antibody production by electrospray mass spectrometry with selective-ion monitoring', 437–446, Copyright (2001), with permission from Elsevier Science.

Transformation of the raw spectrum from the heavy-chain component shows the presence of three species with molecular weights of 50 517.2, 50 676.8 and 50 840.2 Da, which were rationalized in terms of the oligosaccharide structures shown in Figure 5.9 – the differences of 162 Da corresponding to a dehydrated galactose residue ($C_6H_{12}O_6$ – $H_2O$).

Although the HPLC separation of these three species has not been achieved, the ions generated in their electrospray spectra, i.e. those with $38^+$ charges, occurring at $m/z$ 1330.4, 1334.6 and 1338.9, respectively, may be separated by the mass spectrometer and this allows the independent but simultaneous quantitation of all three species, as shown in Figure 5.10. The successful application of this methodology relies upon the fact that the ions to be monitored are sufficiently well resolved and that there is no interference from other ions present. While the mass spectrometer resolution will clearly affect this capability, it should be remembered that as the number of charges on an ion increases then the separation of the isotopic contributions, on the $m/z$ scale, decreases. Better 'resolution' may therefore be obtained by choosing to monitor a lower charge state of the species of interest although this will have implications for the intensity of the signal to be monitored. Changes in the cone-voltage, which change the relative proportions of the different charge states, may be utilized to optimize performance.

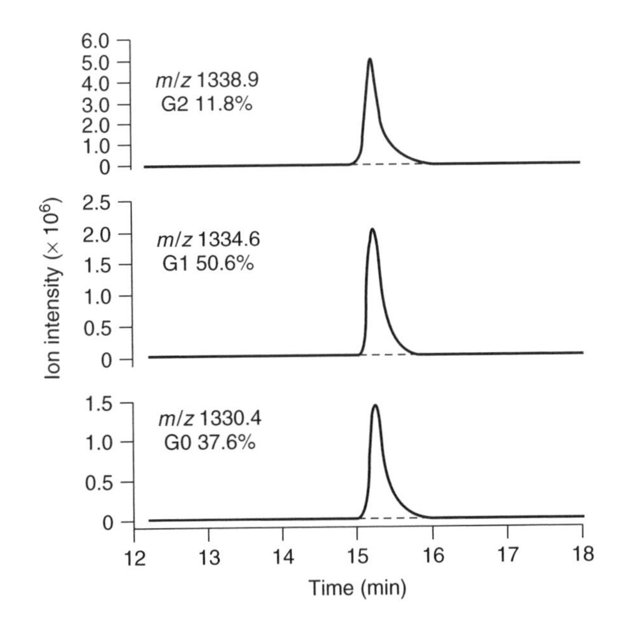


**Figure 5.10** Reconstructed ion chromatograms used to quantify the levels of the principal oligosaccharides found in rituximab process samples. Reprinted from *J. Chromatogr., A*, **913**, Wan, H. Z., Kaneshiro, S., Frenz, J. and Cacia, J., 'Rapid method for monitoring galactosylation levels during recombinant antibody production by electrospray mass spectrometry with selective-ion monitoring', 437–446, Copyright (2001), with permission from Elsevier Science.

## *5.2.3 The Effect of Mobile-Phase Additives and Cone-Voltage*

As part of their method development, these same authors studied the effect of a number of experimental variables on the HPLC separation and the mass spectral quality and it is worthwhile considering their results, reproduced in Table 5.3, in some detail.

It is not unexpected that the choice of the type of HPLC column, the temperature at which it is maintained and the choice of buffer added to the mobile phase are likely to have an effect on the separation obtained, even for a simple mixture as in this case.

The effect of the buffer on the efficiency of electrospray ionization was mentioned earlier in Section 4.7.1. This is a good example of the dramatic effect that this may have – good chromatographic separation and ionization efficiency with formic, acetic and propionic acids, and good separation, although with complete suppression of ionization, with trifluoroacetic acid (TFA), the additive used for the protein application described previously. Post-column addition of propionic acid to the mobile phase containing TFA has been shown to reduce, or even

**Table 5.3** Effect of experimental parameters on the LC–MS analysis of recombinant antibodies. Reprinted from *J. Chromatogr., A*, **913**, Wan, H. Z., Kaneshiro, S., Frenz, J. and Cacia, J., 'Rapid method for monitoring galactosylation levels during recombinant antibody production by electrospray mass spectrometry with selective-ion monitoring', 437–466, Copyright (2001), with permission from Elsevier Science

| Method variable | Test conditions | Results |
|---|---|---|
| Acidic modifier | Trifluoroacetic acid | Ionization completely suppressed |
| | Formic acid | Good separation/ionization |
| | Propionic acid | Good separation/ionization |
| | Acetic acid | Best separation/ionization |
| Column chemistry | Silica | Poor separation and recovery |
| | Polymeric | Good separation and recovery |
| Column temperature | Ambient | Poor separation and peak shape |
| | 60°C | Good separation and peak shape |
| Fragmentor voltage | Over 220 V | Collision-induced dissociation effect |
| | Under 90 V | Poor ionization efficiency |
| | 110–150 V | Good ionization efficiency |
| Data acquisition | Scan mode | Erratic component quantitation |
| | Selected-ion-monitoring mode | Reproducible component quantitation |

remove, the suppression effect of the latter but may be a less convenient solution than using an alternative additive if adequate ionization and peak shape can be obtained.

The results obtained with different fragmentor voltages, referred to in this present text as cone-voltages (see Section 4.7.4 above), clearly show the importance of optimizing this parameter. The voltage used will depend upon the analytical information required. A sufficiently high voltage is required to provide adequate ionization of the analyte, and thereafter qualitative applications require adequate fragmentation to be observed, while quantitative applications benefit from fragmentation being minimized.

**DQ 5.4**

What is meant by the term 'cone-voltage fragmentation' and what information may it provide?

*Answer*

*Atmospheric-pressure chemical ionization (APCI) and electrospray ionization are both* ***soft*** *ionization techniques which give rise, almost exclusively, to the production of molecular species. Structural information,* *which is of value for many applications, is therefore rarely obtained. Cone-voltage fragmentation is a technique in which a voltage is applied to the electrospray or APCI source to induce fragmentation of the ions present and thus provide structural information.*

It must be reiterated that the conditions which provide the 'best' analysis will vary from application to application and the type of study carried out by these authors should be undertaken to determine optimum experimental conditions (see Section 5.1 above). Selected-ion monitoring was used for the quantitation to provide enhanced sensitivity, as outlined previously in Section 3.5.2.1.

The accuracy and precision of the determinations were investigated. Recovery was found to be 101 $\pm$ 2.0% for a range of volumetrically mixed samples and the relative standard deviation (RSD), for a standard injected 23 times over a period of 4.5 months, was found to be 1.1%. It should be noted that the performance of a method for samples based on standard materials may not be attainable when 'real' samples are being determined and further method development may be required.

# 5.3 Structure Determination of Biopolymers

There are four levels of structure observed in proteins [9] and it is important that once the molecular weight of a protein has been determined that these are also elucidated in order that the way in which the protein functions may be understood.

These four levels are as follows:

(1) The *primary* structure – the sequence of peptide-bonded amino acids in the protein chain and the location of any disulfide bridges.

(2) The *secondary* structure – the regular, recurring, arrangements in space of adjacent amino acid residues in the protein chain, e.g. $\alpha$-helix, $\beta$-sheet, etc.

(3) The *tertiary* structure – the spatial arrangement of all of the amino acids in a single protein chain, i.e. the three-dimensional structure of the protein.

(4) The *quaternary* structure – the spatial relationship between the protein chains in a multimeric protein.

These are shown pictorially in Figure 5.11.

MS may be used to investigate all of these but by far the most common application of LC–MS is concerned with the determination of amino acid sequences.

## *5.3.1 Amino Acid Sequencing of Proteins*

Amino acid sequencing may be carried out in a number of ways. The most widely used is the *Edman degradation procedure* in which phenylisothiocyanate is used to react with the amino acid residue at the amine end of the protein chain. This derivatized residue is removed from the remainder of the protein and converted to a phenylhydantoin derivative which is identified by using, for example, HPLC.

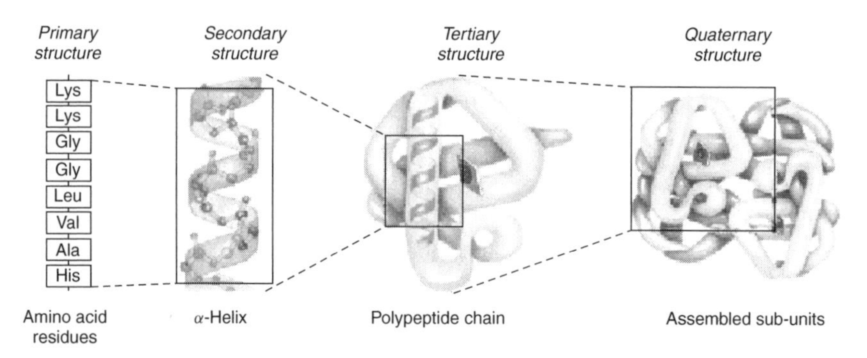


**Figure 5.11** The four levels of protein structure. From Lehninger, A.L., Nelson, D. and Cox, M.M. *Principles of Biochemistry*, 3rd Edn, Worth Publishers, NJ, 2000, and reproduced with permission.

A series of these procedures are then carried out on the remaining protein, thus enabling each 'new' terminal amino acid residue to be identified.

It is not practical to sequence more than about 50 amino acid residues on a single protein in this way and larger proteins need to be broken down into polypeptides with more appropriate lengths to allow complete sequencing to be carried out. This 'shortening' of the polypeptide chain may be carried out using chemical or enzymatic methods, cf. hydrolysis.

## *5.3.2 The Use of Enzymes for Amino Acid Sequencing*

Enzymes are now being used in conjunction with LC–MS to provide sequence information as an alternative to the Edman degradation procedure.

This is a task for which electrospray ionization is well suited although, as discussed earlier in Section 4.7, this is a soft ionization technique that yields almost exclusively molecular ions with little fragmentation and consequently, in the case of biopolymers, little sequence information directly.

As discussed above in Section 4.7.4, there are two ways in which ions generated by electrospray ionization may be caused to fragment, i.e. (a) cone-voltage/in-source fragmentation (CVF), and (b) MS–MS. Problems occur, however, when these methodologies are applied to ions generated by high-molecular-weight compounds – CVF brings about the fragmentation of all species within the source of the mass spectrometer, while MS–MS spectra of highly charged species are often complex. It is not usually possible, therefore, to obtain useful sequence information by the analysis of mass spectra generated directly from the parent biopolymer.

**DQ 5.5**

What is meant by the term 'MS–MS' and what information can it provide? What are the advantages of MS–MS over cone-voltage fragmentation (CVF)?

*Answer*

*MS–MS is a term that covers a number of techniques in which two stages of mass spectrometry are used to investigate the relationship between ions found in a mass spectrum. In particular, the product-ion scan is used to derive structural information from a molecular ion generated by a soft ionization technique such as electrospray and, as such, is an alternative to CVF. The advantage of the product-ion scan over CVF is that it allows a specific ion to be selected and its fragmentation to be studied in isolation, while CVF bring about the fragmentation of all species in the ion source and this may hinder interpretation of the data obtained.*

As with the Edman degradation approach, the parent biopolymer is cleaved into more appropriately sized molecules by using an appropriate enzyme – the term 'enzyme digestion' is often used – or by using chemical hydrolysis. This results in the formation of a number of smaller polypeptides – the larger the original protein, then the more of these there are likely to be – and these are separated, usually by HPLC, and molecular weight and sequence information generated from each of them. A careful analysis of the sequence information from each of these polypeptides, in particular, looking for sequences that partially overlap, allows sequence information from the parent protein to be derived. This process, known as 'peptide mapping', requires the use of a computer and appropriate software has been developed for this purpose.

A number of enzymes are in common use and each of these cleaves the polypeptide backbone adjacent to a particular amino acid residue. The one used for a particular investigation is therefore chosen for the specificity with which it will cleave the polypeptide backbone of the protein being studied. A number of the enzymes used for this purpose are shown in Table 5.4.

In order to obtain complete sequence information for a particular protein, it may be necessary to carry out digestion with more than one enzyme.

**Table 5.4** Enzymes used for the cleavage of proteins for sequence investigations and the corresponding amino acid residues at which they break the peptide backbone

| Enzyme | Cleavage site |
|---|---|
| Trypsin | The carboxy terminus sides of lysine (*except* when proline is attached) and arginine |
| Chymotrypsin | The carboxy terminus sides of phenylalanine, tryptophan and tyramine |
| Asp-*N*-protease | The amino terminus sides of aspartic acid and glutamic acid |
| Pepsin | The amino terminus sides of phenylalanine, tryptophan and tyramine |
| Cyanogen bromide | The carboxy terminus side of methionine |
| Endoproteinase Lys-C | The carboxy terminus side of lysine, *except* when proline is attached to lysine |

The specificity of the enzyme is important since it determines both the number and the size of the polypeptides that are produced. If it is too specific, a small number of (possibly) high-molecular-weight polypeptides are likely to be obtained and the original problem of being unable to generate interpretable sequence information from relatively high-molecular-weight compounds remains. If the enzyme is insufficiently specific, a large number of low-molecular-weight fragments may be generated and, while these are likely to be relatively easy to sequence, the relationship of these sequences to that of the parent protein is not easy to derive. Access to the amino acid composition of the parent protein, i.e. the number of each particular amino acid residue present, may give a useful indication of the number of polypeptides that are likely to be generated by a particular enzyme while a knowledge of the amino terminus residue is also likely to be of advantage.

### 5.3.3 The Mass Spectral Fragmentation of Peptides

Before discussing the application of LC–MS to the determination of sequence information, it is necessary to consider the ways in which polypeptides fragment in the mass spectrometer [10]. These may be considered in terms of (a) breaking of the polypeptide backbone at various locations, with charge retention either on that part of the molecule containing the intact carboxy or the amino function, and (b) fragmentation of the substituent groupings attached to the polypeptide backbone. Fragmentations are described by using a notation consisting of a letter and a number, where the letter indicates the particular bond that has been cleaved, and the number indicates the number of the amino acid residue in the polypeptide chain that has been cleaved. These fragmentations, and the corresponding nomenclature, are shown in Table 5.5. In fact, formation of the fragment ions in some cases involves proton-transfer reactions as well as bond scission and the structures and the nomenclature of the ions formed in this way are shown in Figure 5.12. In addition to these, immonium and acyl ions, as shown in Figure 5.13, may also be produced.

When extracting sequence information from mass spectra, not only is the $m/z$ value at which the ions occur of importance since these provide an indication of the amino acid composition of the peptide giving rise to the ion, but so is the mass difference between adjacent ions. This indicates the particular amino acid residue that has been lost and thus provides the sequence information required. The mass differences arising from each of the amino acids are shown in Table 5.6.

### 5.3.4 Confirmation of Amino Acid Sequence Using the Analysis of LC–MS Data from an Enzyme Digest of a Protein

The sequence of a globular protein was confirmed by a combination of enzymatic digestion and HPLC with both Fourier-transform infrared spectroscopy (LC–FTIR spectroscopy) and mass spectrometry [11].

**Table 5.5** Nomenclature of the ions formed in the mass spectral fragmentation of polypeptides. From Chapman, J. R. (Ed.), *Protein and Peptide Analysis by Mass Spectrometry*, Methods in Molecular Biology, Vol. 61, 1996. Reproduced by permission of Humana Press, Inc.

| Structure | Ions | Fragmentation pattern |
|---|---|---|
| $[H_2N-CH(R1)-CO-NH-CH(R2)-CO-[NH-CH(R)-CO]_{n-3}-NH-CH(Rn)-C(O)-OH]H^+$ (cleavage sites: $a_1$, $b_1$, $c_1$; $x_{(n-1)}$, $y_{(n-1)}$, $z_{(n-1)}$) | a, b and c | Polypeptide chain fragmentation with charge retention on the *N*-terminus; the series is numbered from the amino terminus |
| | x, y and z | Polypeptide chain fragmentation with charge retention on the *C*-terminus; the series is numbered from the carboxy terminus |
| $H_2N-CH(R1)-CO-NH-CH(R2)-CO-NH-CH(CHR'R'')$ (cleavage sites: $a_x$, $d_x$; H transfer) | d | Side-chain fragmentation with charge retention on the *N*-terminus |

*(continued overleaf)*

**Table 5.5** (*continued*)

| Structure | Ions | Fragmentation pattern |
|---|---|---|
| 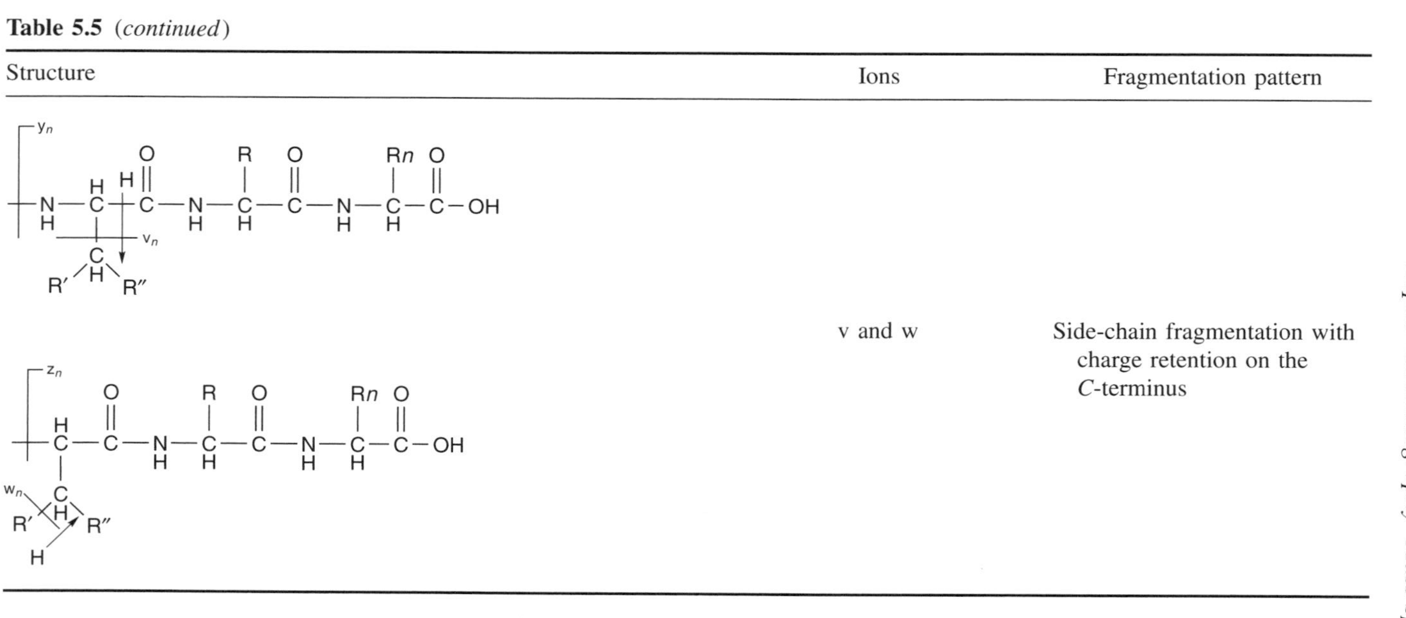  | v and w | Side-chain fragmentation with charge retention on the *C*-terminus |

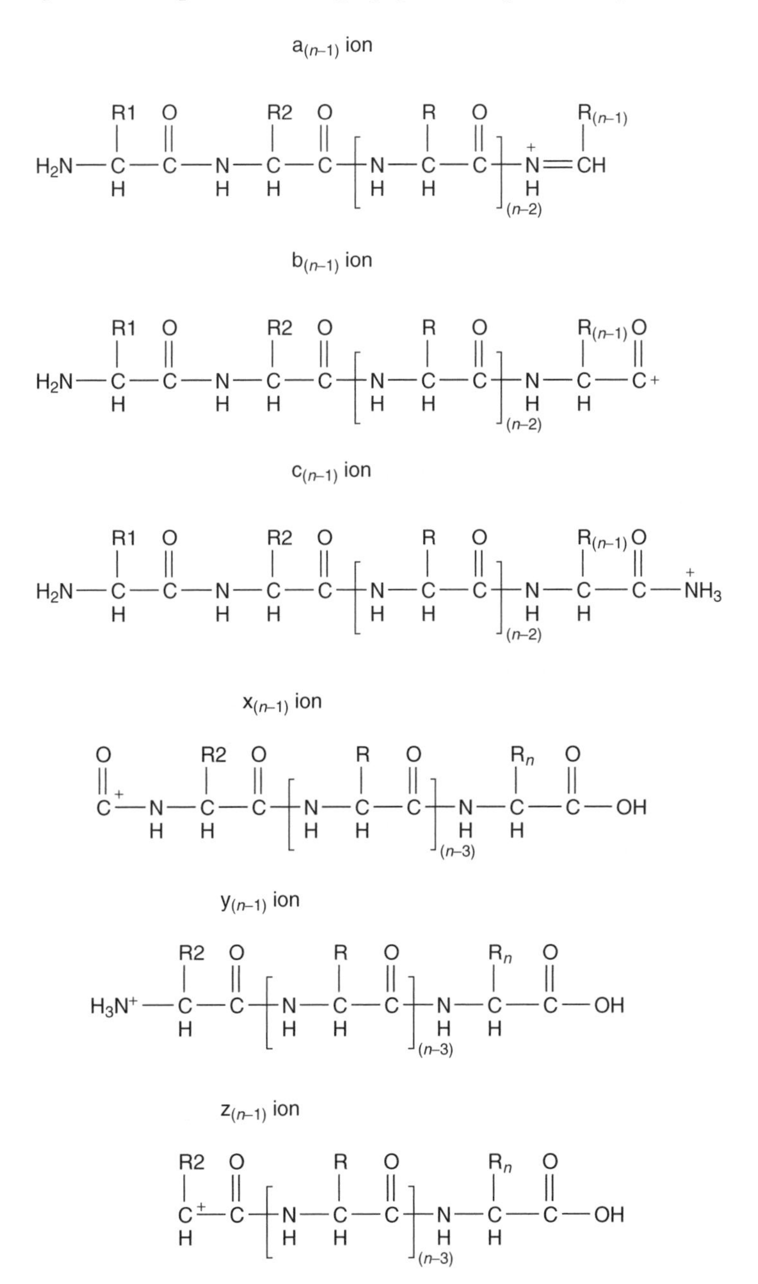


**Figure 5.12** Structures and nomenclature of the ions formed in the mass spectral fragmentation of peptides which involve scission of the polypeptide backbone. From Chapman, J. R. (Ed.), *Protein and Peptide Analysis by Mass Spectrometry*, Methods in Molecular Biology, Vol. 61, 1996. Reproduced by permission of Humana Press, Inc.

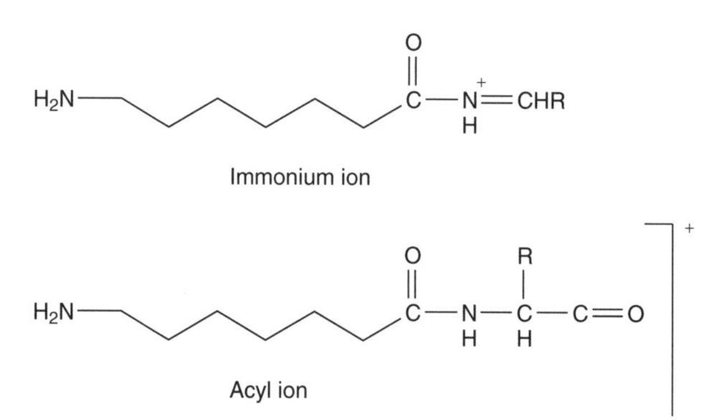


**Figure 5.13** Structures of the immonium and acyl ions formed in the mass spectral fragmentation of peptides. From Chapman, J. R. (Ed.), *Protein and Peptide Analysis by Mass Spectrometry*, Methods in Molecular Biology, Vol. 61, 1996. Reproduced by permission of Humana Press, Inc.

**Table 5.6** Codes used for amino acid residues and the corresponding mass losses observed from each in mass spectra used for sequence determination. From Chapman, J. R. (Ed.), *Protein and Peptide Analysis by Mass Spectrometry*, Methods in Molecular Biology, Vol. 61, 1996. Reproduced by permission of Humana Press, Inc.

| Amino acid | Code | Monoisotopic mass (Da) |
|---|---|---|
| Glycine | Gly (G) | 57.021 |
| Alanine | Ala (A) | 71.037 |
| Serine | Ser (S) | 87.032 |
| Proline | Pro (P) | 97.053 |
| Valine | Val (V) | 99.068 |
| Threonine | Thr (T) | 101.048 |
| Cysteine | Cys (C) | 103.009 |
| Isoleucine | Ile (I) | 113.084 |
| Leucine | Leu (L) | 113.084 |
| Asparagine | Asn (N) | 114.043 |
| Aspartic Acid | Asp (D) | 115.027 |
| Glutamine | Gln (Q) | 128.059 |
| Lysine | Lys (K) | 128.095 |
| Glutamic Acid | Glu (E) | 129.043 |
| Methionine | Met (M) | 131.040 |
| Histidine | His (H) | 137.059 |
| Phenylalanine | Phe (F) | 147.068 |
| Arginine | Arg (R) | 156.101 |
| Tyrosine | Tyr (Y) | 163.063 |
| Tryptophan | Trp (W) | 186.079 |

The LC–FTIR part of the study was concerned with the verification of the presence of particular functional groups within each of the tryptic fragments and will not be considered any further here.

The proteolytic digestion of $\beta$-lactoglobulin was carried out with trypsin which, as indicated in Table 5.4 above, is expected to cleave the polypeptide backbone at the carboxy-terminus side of lysine (K) and arginine (R). On this basis, and from the known sequence of the protein, nineteen peptide fragments would be expected, as shown in Table 5.7. Only 13 components were detected after HPLC separation and, of these, ten were chosen for further study, as shown in Table 5.8.

**Table 5.7** Theoretically predicted polypeptides from the trypsin digestion of $\beta$-lactoglobulin ($\beta$LG)[a]. Reprinted from *J. Chromatogr., A*, **763**, Turula, V. E., Bishop, R. T., Ricker, R. D. and de Haseth, J. A., 'Complete structure elucidation of a globular protein by particle beam liquid chromatography–Fourier transform infrared spectrometry and electrospray liquid chromatography–mass spectrometry – Sequence and conformation of $\beta$-lactoglobulin', 91–103, Copyright (1997), with permission from Elsevier Science

| Theoretical fragment number | Amino acid residues | Sequence | Calculated molar mass[a] (Da) |
|---|---|---|---|
| T1 | 1–8 | LIVTQTMK | 993.3 |
| T2 | 9–14 | GLDIQK | 673.4 |
| T3 | 15–40 | VAGT*W**Y*SLAMAASDISLLDAQSAPLR | 2707.4 |
| T4 | 41–47 | V*Y*VEELK | 986.1 |
| T5 | 48–60 | PTPEGDLEILLQK | 1452.7 |
| T6[c] | 61–69 | WENDE**C**AQK | 1122.5 |
| | 61–70 | WENDE**C**AQKK | 1250.6 |
| T7 | 70 | K | 147.1 |
| T8 | 71–75 | IIAEK | 573.4 |
| T9 | 76–77 | TK | 248.2 |
| T10 | 78–83 | IPAV*F*K | 674.4 |
| T11 | 84–91 | LDAINENK | 916.5 |
| T12 | 92–100 | VLVLDTD*Y*K | 1065.6 |
| | 92–101 | VLVLDTD*Y*KK | 1192.7 |
| T13 | 101 | K | 147.1 |
| T14[c] | 102–124 | *Y*LLF**C**MENSAEPEQSLV**C**Q**C**LVR | 2675.2 |
| T15 | 125–135 | TPEVDDEALEK | 1245.6 |
| T16 | 136–138 | *F*DK | 409.2 |
| T17 | 139–141 | ALK | 331.2 |
| T18 | 142–148 | ALPMHIR | 837.5 |
| T19[c] | 149–162 | LSFNPTLQEEQ**C**HI | 1658.8 |

[a]Showing the monoisotopic masses of fragments $MH^+$ of $\beta$-lactoglobulin A: cysteine residues are shown in bold script, i.e. cysteine-**C**: aromatic residues are underlined, i.e. phenylalanine-*F*, tyrosine-*Y*, and tryptophan-*W*.

[b] $m/z$ expected for singly charged species.

[c]*S*-Carboxymethylated cysteine residue corresponds to a mass increase of 43 Da. Fragment 6 – one *S*-carboxymethylated cysteine residue $\beta$LG A: 61–69, 1165.6 Da; 61–70, 1293.6 Da. Fragment 14 – three *S*-carboxymethylated cysteines $\beta$LG A: 102–104, 2804.2 Da. Fragment 19 – one *S*-carboxymethylated cysteine residue: 149–162, 1701.8 Da.

**Table 5.8** Polypeptides detected during the LC–electrospray-MS analysis of the tryptic digest from $\beta$-lactoglobulin ($\beta$LG). Reprinted from *J. Chromatogr., A*, **763**, Turula, V. E., Bishop, R. T., Ricker, R. D. and de Haseth, J. A., 'Complete structure elucidation of a globular protein by particle beam liquid chromatography–Fourier transform infrared spectrometry and electrospray liquid chromatography–mass spectrometry – Sequence and conformation of $\beta$-lactoglobulin', 91–103, Copyright (1997), with permission from Elsevier Science

| Chromatographic peak | Retention time (min) | *m/z* | Ion assigned[a] | Fragment and residue number[b] | Sequence |
|---|---|---|---|---|---|
| 1 | 22.5 | 933.3 | $MH^+$ | [T1] 1–8 | LIVTQTMK |
| 2 | 23.2 | 818.4 | Measured $MH_2{}^{+2}$ | [T15 + T16] 125–138 | TPEVDDEALEKFDK |
| | | 1635.8 | Actual | | |
| 3 | 24.0 | 696.0 | Measured $Na^+$ | [T10] 78–83 | IPAVFK |
| | | 674.4 | Actual | | |
| 4 | 25.1 | 1065.3 | $MH^+$ | [T12] 92–100 | VLVLDTDYK |
| 5 | 25.8 | 690.0 | $MH^+$ | 15–20 | VAGTWY |
| 7 | 27.8 | 836.5 | Measured $MH^+$ | [T18] 142–148 | ALPMHIR |
| | | 837.0 | Actual | | |
| 8 | 30.1 | 1015.5 | Measured $MH_2{}^{+2}$ | 21–40 | SLAMAASDISLLDAQSAPLR |
| | | 2030.6 | Actual | | |
| 9 | 30.9 | 771.9 | Measured $MH_3{}^{+3}$ | [T4 + T5] 41–60 | VYVEELKPTPEGDLEILLQK |
| | | 2313.3 | Actual | | |
| 10 | 31.5 | 762.0 | Measured $MH^+$ | 27–33 | SDISLLD |
| | | 762.4 | Actual | | |

[a] Actual – indicates the calculated $m/z$ value for the ion of the singly charged species ($MH^+$).
[b] Fragments with no number are unexpected from the digestion with trypsin.

It is important to consider why the number of peptide fragments observed may differ from that predicted theoretically.

It could be that not all of the expected fragments were produced during the digestion process or that certain of them were produced in insufficient quantity to be detected. Although the digestion process involves the reduction of disulfide bridges and an unfolding of the protein in a process intended to make all of the appropriate amino acid residues accessible for reaction, this is not always achieved. Examination of Tables 5.7 and 5.8 shows that T15 and T16 are not observed but the combined fragment T15 + T16 is, and similarly T4 and T5 are not observed, although T4 + T5 is.

The amino acid sequence may be such that enzymatic cleavage at different residues may give rise to identical fragments, thus reducing the number of discrete polypeptides expected. Examination of Table 5.7 shows that, in this case, fragments T7 and T13 are identical.

It could also be the case that the HPLC conditions employed did not give complete resolution of all of the tryptic fragments or indeed did not separate some of the fragments produced. One of the reasons given in the cited reference [11] for choosing not to study some of the fragments was that they 'nearly co-elute with the proteolytic agent' and therefore presumably did not give mass spectra of adequate quality. If resolution is incomplete, the possibility of suppression of ionization during the electrospray process must also be considered.

The way in which mass spectral data are acquired may preclude detection of certain polypeptides. In this case, the $m/z$ range monitored was from 600 to 1200 and examination of Table 5.7 shows that six of the expected fragments have molecular weights below the lower limit for acquisition. In addition, it must be remembered that electrospray ionization gives rise to multiply charged ions and the singly charged species is not always observed. A number of further peptides with molecular weights in excess of 600 Da may therefore not give rise to intense ions in the $m/z$ range being monitored.

The unseparated digest mixture was studied directly by mass spectrometry using matrix-assisted laser desorption ionization (MALDI) and this showed six of the polypeptides detected by LC–MS and three of the expected polypeptides that had not been detected by LC–MS. In contrast, MALDI did not show three polypeptides observed by LC–MS.

The presence of three polypeptides in Table 5.8 that were not predicted from the relationship between the amino acid sequence and the enzyme used for digestion is worthy of note when interpretation of data of this sort is undertaken. The MALDI data showed six further 'unexpected' polypeptides, none of which were detected in the LC–MS data!

Although the use of MALDI *per se* is not covered in this present text, the data cited here clearly show that it is complementary to LC–MS employing electrospray ionization.

### 5.3.5 Determination of the Amino Acid Sequence of a Novel Protein Using LC–MS Data from an Enzyme Digest

In the example discussed above, the analysts were dealing with a known sequence and were able to identify the 'unexpected' results easily. If a novel protein was under investigation, these anomalies would not have been obvious and interpretation would undoubtedly have been hindered. A slightly different methodology may therefore be needed, as illustrated in the following example.

A protein, designated cyctochrome c″, isolated from the methylotrophic bacterium *Methylophilus methylotrophus*, has been studied extensively because of its unusual properties and was found to have an average molecular mass of 14 293.0 Da and to contain 124 amino acid residues. The *N*-terminal sequence to residue 62 had been determined and the heme binding site had been located at Cys-49 and Cys-52 [12]. Further studies were concerned with determining the remainder of the sequence.

The molecular mass of the protein was redetermined by infusing a 5–10 pmol $\mu l^{-1}$ solution of the protein in 50% aqueous acetonitrile containing 0.2% formic acid at a flow rate of 6 $\mu l\,min^{-1}$ into an electrospray source. The scan rate employed on the mass spectrometer was from $m/z$ 60 to $m/z$ 1800 in 12 s. This is a relatively slow scan speed which will lead to a more precise molecular weight determination. Scan speeds of this order may be, and indeed should be, utilized for infusion experiments if sufficient sample is available but it is unlikely to be feasible when chromatographic separations, particularly those involving capillary columns, are employed because of the restriction imposed by the chromatographic peak width (see Section 3.5.2.1 above).

The raw electrospray spectrum obtained is shown in Figure 5.14. Maximum entropy processing of these data yielded the spectrum given in Figure 5.15, which shows the presence of two species with molecular masses of 14 293.6 and 14 309.6 Da, with the latter being attributed to partial oxidation of the parent protein.

The methodology employed for deriving the sequence of the parent protein may be summarized as follows:

(a) Determination of the amino acid content – in this case, that the protein was found to contain 16 lysine residues, hence indicating that Lys-C endoproteinase is an appropriate enzyme for digestion. The latter cleaves on the carboxyl terminus of a lysine residue except where proline is the adjacent residue (see Table 5.4 above).

(b) Determination of the number of heme residues – in this case, one. Heme is the name given to the iron-containing non-amino acid part of the molecule. In c-type cytochromes, these are usually covalently attached to the polypeptide chain through cysteine residues.

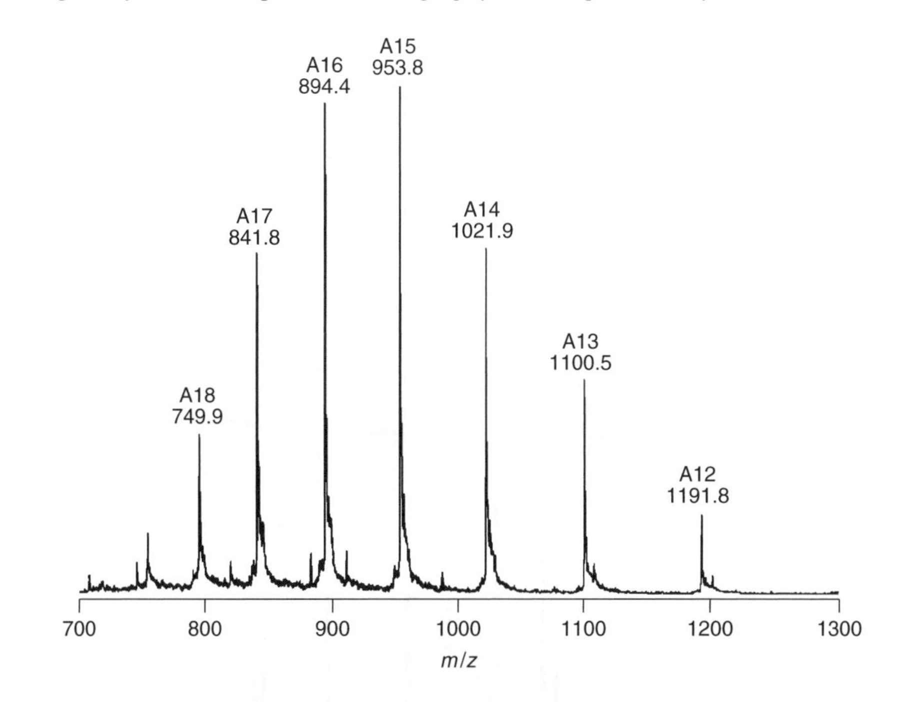


**Figure 5.14** Electrospray spectrum of holocytochrome c″. Reprinted from *Biochim. Biophys. Acta*, **1412**, Klarskov, K., Leys, D., Backers, K., Costa, H. S., Santos, H., Guisez, Y. and Van Beeumen, J. J., 'Cytochrome c″ from the obligate methylotroph *Methylophilus methylotrophus*, an unexpected homolog of sphaeroides heme protein from the phototroph *Rhodobacter sphaeroides*', 47–55, Copyright (1999), with permission from Elsevier Science.

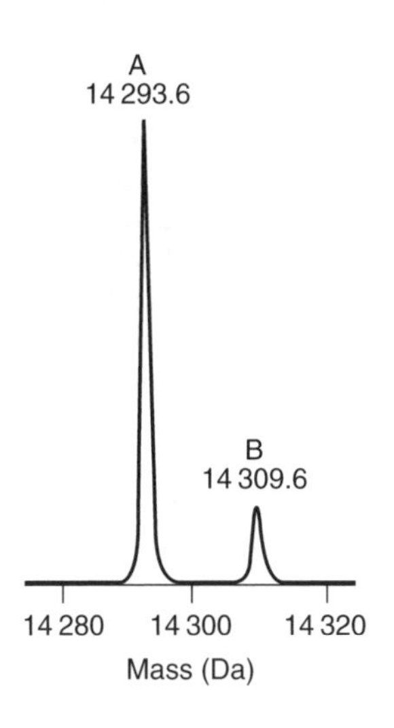


**Figure 5.15** Electrospray spectrum of holocytochrome c″ after application of the maximum entropy algorithm to the spectrum shown in Figure 5.14. Reprinted from *Biochim. Biophys. Acta*, **1412**, Klarskov, K., Leys, D., Backers, K., Costa, H. S., Santos, H., Guisez, Y. and Van Beeumen, J. J., 'Cytochrome c″ from the obligate methylotroph *Methylophilus methylotrophus*, an unexpected homolog of sphaeroides heme protein from the phototroph *Rhodobacter sphaeroides*', 47–55, Copyright (1999), with permission from Elsevier Science.

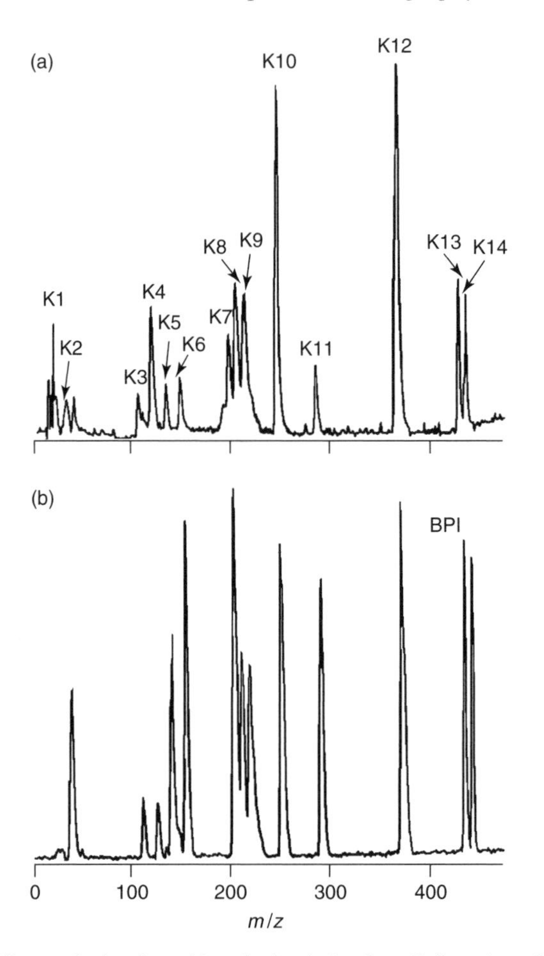


**Figure 5.16** On-line analysis of peptides obtained after Lys-C digestion of holocytochrome c″ showing (a) the UV trace obtained at 220 nm, and (b) the corresponding mass spectral trace from electrospray ionization of the LC eluate. The latter is represented by a modification of an RIC known as the base-peak trace; BPI, base-peak intensity (see text for further details). Reprinted from *Biochim. Biophys. Acta*, **1412**, Klarskov, K., Leys, D., Backers, K., Costa, H. S., Santos, H., Guisez, Y. and Van Beeumen, J. J., 'Cytochrome c″ from the obligate methylotroph *Methylophilus methylotrophus*, an unexpected homolog of sphaeroides heme protein from the phototroph *Rhodobacter sphaeroides*', 47–55, Copyright (1999), with permission from Elsevier Science.

(c) Digestion with Lys-C endoproteinase – this was monitored by the analysis of the reaction mixture, using LC–MS, to determine the optimum time for digestion to generate the required fragments.
(d) The LC–MS analysis of the reaction mixture after complete digestion of the protein.

HPLC separation was carried out by using a 2.1 × 100 mm capillary column with a stationary phase of ODS-AQ $C_{18}$. A flow rate of 200 $\mu l\,min^{-1}$ was employed with a linear gradient from 0 to 35% solvent B over 40 min and then to 100% B over 10 min. (Solvent A was 5% aqueous acetonitrile containing 0.05% trifluoroacetic acid (TFA), while solvent B was 80% aqueous acetonitrile containing 0.05% TFA.)

The results of this analysis are shown in Figure 5.16, which contains the UV trace, at 220 nm, and the equivalent mass spectral trace. In this case, the mass spectral data are represented by a modification of a reconstructed ion chromatogram (RIC) known as the base-peak trace, i.e. a plot of the intensity of the base peak in each scan (the $m/z$ value of this may well vary from scan to scan) as a function of time.

The molecular weights of the fourteen peptides indicated are summarized in Table 5.9. A number of these, i.e. K1, K2, K4, K7 and K10, were rationalized in terms of the expected cleavages within the already known sequence of residues 1 to 62. It is interesting to note that no fragment containing residues 39 and 40 was detected.

**Table 5.9** Peptides detected during the LC–electrospray-MS (LC–ESMS) analysis of the endoproteinase Lys-C digest from native cytochrome c″. Reprinted from *Biochim. Biophys. Acta*, **1412**, Klarskov, K., Leys, D., Backers, K., Costa, H. S., Santos, H., Guisez, Y. and Van Beeumen, J. J., 'Cytochrome c″ from the obligate methylotroph *Methylophilus methylotrophus*, an unexpected homolog of sphaeroides heme protein from the phototroph *Rhodobacter sphaeroides*', 47–55, Copyright (1999), with permission from Elsevier Science

| HPLC fraction | Measured[a] $M_r$ (Da) | Calculated[b] $M_r$ (Da) | Sequence position |
|---|---|---|---|
| K1 | 488.5 | 488.5 | 41–45 |
| K2 | 775.9 | 775.8 | 1–7 |
| K3[c] | 630.6 | 630.7 | 63–68 |
| K4 | 521.6 | 521.7 | 8–11 |
| K5[c] | 867.0 | 866.9 | 88–94 |
| K6 | 894.1 | 894.0 | 81–87 |
| K7 | 824.2 | 824.0 | 33–38 |
| K8[c,d] | 1587.0 | 1586.7 | 95–109 |
| K9[c] | 1335.1 | 1334.6 | 69–80 |
| K10 | 2280.6 | 2280.5 | 12–32 |
| K11[c] | 1743.3 | 1742.9 | 81–94 |
| K12[c,e] | 2302.8 | 2303.3 | 46–62 |
| K13[c] | 1710.4 | 1710.0 | 110–124 |
| K14 | 1383.7 | 1383.6 | 110–121 |

[a]The measured masses are obtained from the LC–ESMS analysis.
[b]The average $M_r$ weights were calculated using 'Biolynx' (Micromass).
[c]Peptides submitted to Edman degradation.
[d]Peptide containing intra-sulfide bridge at positions 96 and 104.
[e]Heme-containing peptide.

Sequence information for the remaining fragments was obtained by Edman degradation (see Section 5.3.1 above) after isolation of the individual peptides using preparative HPLC – the chromatographic resolution being sufficient to allow this, and thus enabled the complete sequence to be determined.

The digestion of the protein, after heme removal, using Glu-C endoproteinase was also carried out. This enzyme cleaves the polypeptide backbone on the carboxyl terminus of a glutamic acid residue and in this case yielded twelve chromatographic responses. Despite two of these arising from unresolved components, molecular weight information was obtained from 15 polypeptides, one of which was the intact protein, covering the complete sequence, as shown in Table 5.10.

Such a process led to the full assignment of the sequence of the cytochrome c″ protein shown in Figure 5.17, in which the polypeptide fragments generated by enzymatic digestion with endoproteinase Lys-C and Glu-C are also indicated.

**Table 5.10** Peptides detected during the LC–electrospray-MS (LC–ESMS) analysis of the Glu-C endoproteinase digest of $HgCl_2$-treated apocytochrome c″. Reprinted from *Biochim. Biophys. Acta*, **1412**, Klarskov, K., Leys, D., Backers, K., Costa, H. S., Santos, H., Guisez, Y. and Van Beeumen, J. J., 'Cytochrome c″ from the obligate methylotroph *Methylophilus methylotrophus*, an unexpected homolog of sphaeroides heme protein from the phototroph *Rhodobacter sphaeroides*', 47–55, Copyright (1999), with permission from Elsevier Science

| HPLC fraction | Measured[a] $M_r$ (Da) | Calculated $M_r$ (Da) | Sequence position |
|---|---|---|---|
| E1 | 647.5 | 647.6 | 1–6 |
| E2 | 602.4 | 602.3[b] | 25–30 |
| E3 | 2498.0 | 2498.2[c] | 47–69 |
| E5 | 1884.6 | 1884.2 | 31–46 |
| E6 | | | |
| A | 602.4 | 602.7 | 85–89 |
| B | 1834.9 | 1835.0[d] | 92–108 |
| C | 1837.6 | 1838.2 | 109–124 |
| E7 | 2142.8 | 2145.5 | 7–24 |
| E8 | | | |
| A | 2469.4 | 2468.8 | 25–46 |
| B | 2554.2 | 2553.9 | 70–91 |
| E9 | 5034.6 | 5034.1[c] | 47–91 |
| E10 | 13 877.0 | 13 876.1[c,d] | 1–124 |
| E11 | 1282.5 | 1282.5 | 109–119 |
| E12 | 3099.6 | 3099.5[d] | 92–119 |

[a]The molecular masses are derived from LC–ESMS.
[b]The sequence was confirmed from a separate LC–ESMS/MS analysis.
[c]One Hg atom ($M_r$ 200.6) is included in the calculated molecular weight.
[d]Peptide containing intra-sulfide bridge at positions 96 and 104.

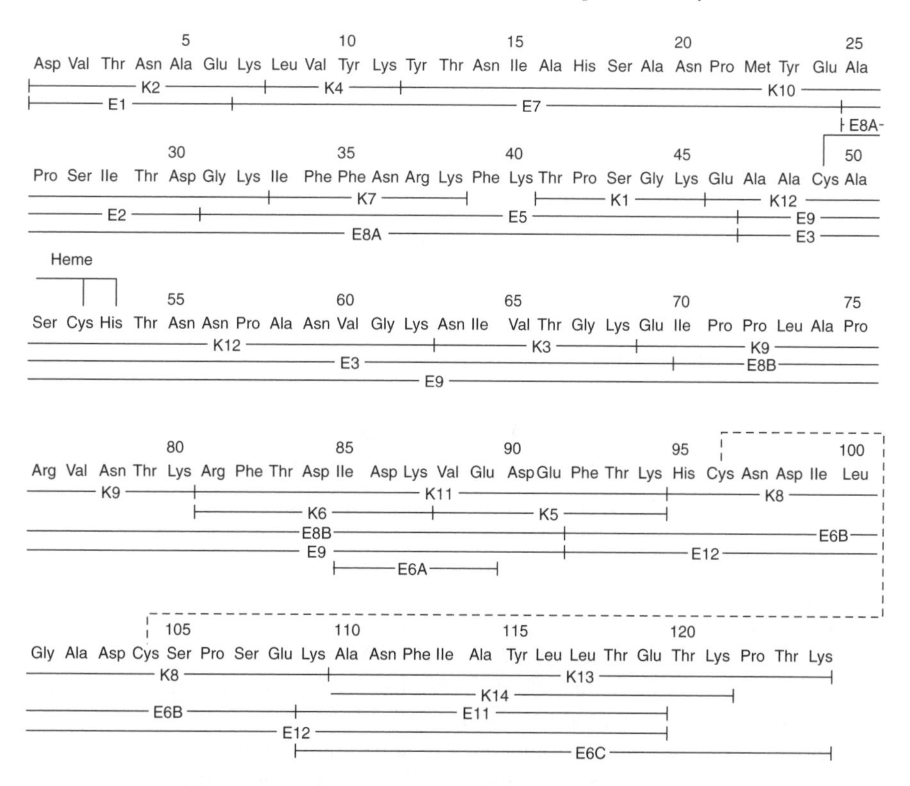


**Figure 5.17** Amino acid sequence of cytochrome c″ from *M. methylotrophus*. The peptides obtained after digestion of the protein with Lys-C endoproteinase are indicated by the letter K, and those after digestion with Glu-C by the letter E. Reprinted from *Biochim. Biophys. Acta*, **1412**, Klarskov, K., Leys, D., Backers, K., Costa, H. S., Santos, H., Guisez, Y. and Van Beeumen, J. J., 'Cytochrome c″ from the obligate methylotroph *Methylophilus methylotrophus*, an unexpected homolog of sphaeroides heme protein from the phototroph *Rhodobacter sphaeroides*', 47–55, Copyright (1999), with permission from Elsevier Science.

In this case, LC–MS was used to provide molecular weight information relating to the peptides generated by enzymatic digestion, while sequence information was from Edman degradation. This approach is often time-consuming and is only possible when the peptides can be isolated in sufficient purity. A different approach to obtaining sequence information is to study the polypeptides using MS–MS and while this was carried out by the authors in the previous example, their data were not reproduced in the paper.

This process, sometimes known as *sequence tagging*, may be a more convenient method of obtaining sequence information but it has to be said that it is unlikely to allow as many residues to be determined as when using the classical Edman approach.

### 5.3.6 Amino Acid Sequencing of Polypeptides Generated by Enzyme Digestion Using MS–MS

The use of MS–MS to provide sequence information has been described [13] for the study of proteins extracted from yeast (*Saccharomyces cerevisiae*). The procedure was somewhat complex and consisted of the following steps:

(a) the extraction of the proteins from a bulk yeast sample;
(b) the separation of the proteins using two-dimensional polyacrylamide gel electrophoresis (2D-PAGE);
(c) excision of the proteins of interest from the gel;
(d) in-gel digestion of the proteins with trypsin;
(e) excision of the resulting peptides from the gel;
(f) separation of the peptides by reversed-phase HPLC and subsequent analysis of the mass spectra from these by using a quadrupole time-of-flight mass analyser in combination with high performance liquid chromatography and tandem mass spectrometry (Q–ToF–LC–MS–MS).

The two-dimensional gel of the yeast extract is shown in Figure 5.18 and clearly illustrates the complexity of the analytical problem.

Matrix-associated laser desorption ionization with a time-of-flight mass analyser (MALDI–ToF) was used to examine the crude tryptic peptide mixture from a number of the proteins, without HPLC separation, to provide a *mass map*, i.e. a survey of the molecular weights of the peptides generated by the digestion process.

**DQ 5.6**

What are the advantages of using MALDI–ToF to examine the crude tryptic peptide mixture?

*Answer*

*MALDI–ToF is a technique that allows the molecular weights of proteins and peptides to be determined. It is less susceptible to suppression effects than electrospray ionization and thus is able to be used for the direct analysis of mixtures. In the case of a crude tryptic digest, MALDI–ToF will provide a molecular weight profile of the polypeptides present without the analysis time being extended by the need to use some form of chromatographic separation.*

Attempts were then made, using these data, to identify the proteins by searching against known peptide databases, such as *ProFound* [14], *PepSea* [15] and *MSFit* [16] that contain the molecular weights of theoretically expected polypeptides obtained from a known protein using a specific enzyme. The molecular weights

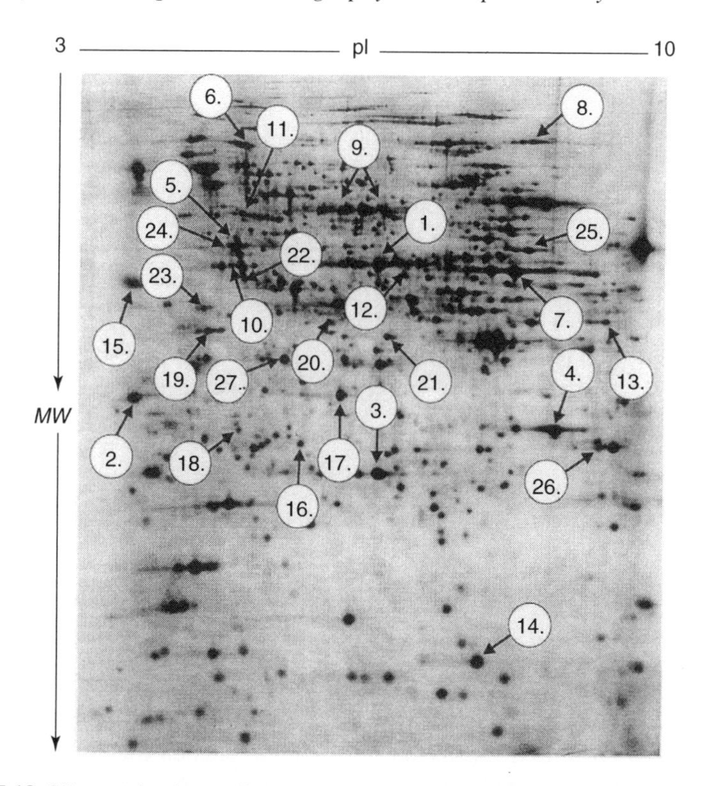


**Figure 5.18** Silver-stained two-dimensional gel of the proteins extracted from the yeast *S. cerevisiae*. From Poutanen, M., Salusjarvi, L., Ruohonen, L., Penttila, M. and Kalkkinen, N., *Rapid Commun. Mass Spectrom.*, **15**, 1685–1692, Copyright 2001. © John Wiley & Sons Limited. Reproduced with permission.

found experimentally are compared with entries in the database and those proteins with the most number of matching fragments form the computer output.

Trypsin digests of 17 proteins were carried out and the peptides produced then studied by using LC–MS–MS. In summary, these produced between one and six peptides which gave some sequence information by MS–MS. Of these peptides, the number of amino acid residues sequenced ranged from three to thirteen.

One protein was identified as a 70 kDa heat-shock protein. It would, theoretically, be expected to give 17 peptides with molecular weights in the range from 900 to 2500 Da – the $m/z$ range scanned using the mass spectrometer. Twelve of these were observed in the MALDI–ToF mass map shown in Figure 5.19, which also shows three peptides produced by incomplete digestion. Five of the theoretically predicted peptides could not be observed, although one of them was detected by using LC–MS–MS and in fact gave an MS–MS spectrum from which a sequence of nine amino acids could be determined. The complementary

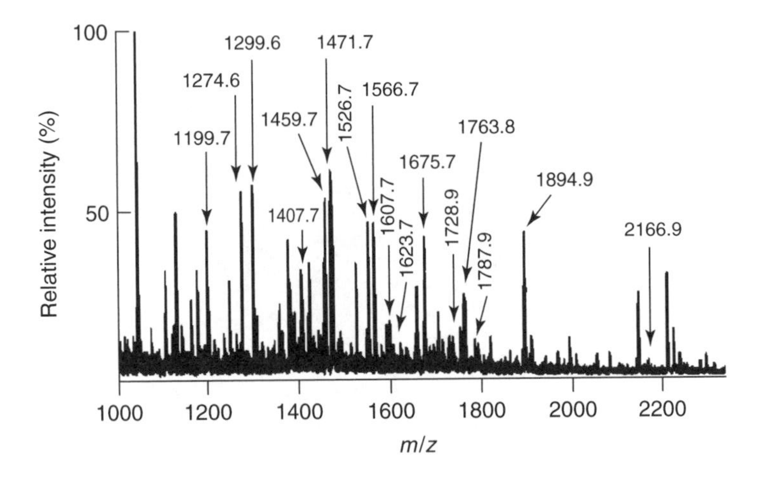


**Figure 5.19** MALDI–ToF mass spectrum, providing a molecular-weight profile of the tryptic peptides derived from spot '22' (see Figure 5.18) of the silver-stained two-dimensional gel of the proteins extracted from the yeast *S. cerevisiae*. From Poutanen, M., Salusjarvi, L., Ruohonen, L., Penttila, M. and Kalkkinen, N., *Rapid Commun. Mass Spectrom.*, **15**, 1685–1692, copyright 2001. © John Wiley & Sons Limited. Reproduced with permission.

nature of MALDI to LC–MS has again been ably demonstrated and shows that thought must be given by the analyst to the technique to be used for a particular investigation.

The HPLC system comprised a 75 μm × 15 cm 'PepMap' column with a linear gradient of acetonitrile/0.1% aqueous formic acid (5 to 50% acetonitrile over 45 min) at a flow rate of 250 nl $min^{-1}$. Positive-ion electrospray ionization was employed using a nanospray interface. MS–MS spectra were acquired over the range $m/z$ 40 to 2000 at a rate of 1 s per scan.

Figure 5.20 shows an MS–MS spectrum produced from a peptide of molecular weight 1782.96 Da. The mass losses observed and corresponding amino acid assignments are shown in Table 5.11.

It may be thought that a small part of the sequence of one peptide is insufficient information on which to base the characterization of a protein but the following is reproduced verbatim from the authors' conclusions: 'The discriminating power of sequence tags for searching in databases is usually high enough to ensure that the partial sequence in an MS–MS spectrum of a single peptide is adequate for protein identification with confidence'.

These authors also mention some shortcomings that should be borne in mind, in particular, that some peptides observed were from the autolysis of trypsin, the digestion agent, and from contaminants such as human keratin, while some peptide ions did not produce interpretable MS–MS spectra.

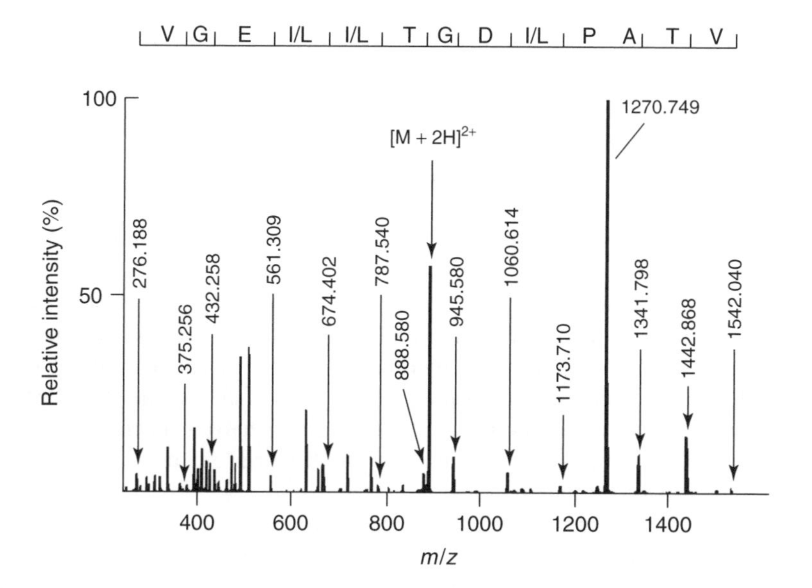


**Figure 5.20** LC–MS–MS spectrum of a tryptic peptide derived from spot '13' (see Figure 5.18) of the silver-stained two-dimensional gel of the proteins extracted from the yeast *S. cerevisiae*, showing the sequence information that may be extracted. From Poutanen, M., Salusjarui, L., Ruohonen, L., Penttila, M. and Kalkkinen, N., *Rapid Commun. Mass Spectrom.*, **15**, 1678–1692, Copyright 2001. © John Wiley & Sons Limited. Reproduced with permission.

**Table 5.11** Sequence information extracted from the LC–MS–MS spectrum shown in Figure 5.20

| *m/z* Observed in mass spectrum | Mass loss (Da) | Amino acid assignment |
|---|---|---|
| 1542.040 | — | — |
| 1442.868 | 99.172 | V |
| 1341.798 | 101.069 | T |
| 1270.749 | 71.050 | A |
| 1173.710 | 97.039 | P |
| 1060.614 | 113.096 | I/L |
| 945.580 | 115.034 | D |
| 888.580 | 57.000 | G |
| 787.540 | 101.040 | T |
| 674.402 | 113.138 | I/L |
| 561.309 | 113.093 | I/L |
| 432.258 | 129.051 | E |
| 375.256 | 57.002 | G |
| 276.188 | 99.068 | V |

### *5.3.7 The Location of Post-Translational Modifications Using LC–MS Data from an Enzyme Digest*

Post-translational modification may affect the biological activity of a protein and the location of such modifications is an extension of sequencing.

From a mass spectrometry perspective, these modifications, such as phosphorylation or glycosylation, manifest themselves as an increase in the molecular weight of both the parent protein and also of the polypeptides (produced by enzymatic digestion) containing the modification.

Assuming the sequence of the parent protein is known, it is not necessary to redetermine the whole sequence merely to locate, and sequence, that/those polypeptide(s) that have undergone modification. This can be done by examination of the total-ion-current (TIC) trace before and after protein hydrolysis for the appearance of new polypeptides or to use mass spectrometry methodology to locate those polypeptides that contain certain structural features. Examples are provided here of both methodologies.

Over the past few years there have been an increasing number of reports of diseases that are becoming resistant to previously effective drug treatments. This resistance is often due to the presence of enzymes that bring about chemical modification of the drug to an inactive form, e.g. $\beta$-lactamase enzymes deactivate $\beta$-lactam antibiotics by their conversion to penicillanic acid.

Inactivation of a $\beta$-lactamase by tazobactam (Figure 5.21) has been investigated [17]. The molecular weight of the $\beta$-lactamase was determined, using electrospray ionization with maximum entropy processing of the raw data (see Section 4.7.3 above), to be 39 851.5 Da and that of the inhibited enzyme to be 39 931.5 Da, an increase of 80 Da. Since tazobactam has a molecular weight of 300 Da, this increase in molecular weight did not allow the nature of the enzyme modification or its location to be determined. Enzymatic digestion of the native protein and the product after inactivation with tazobactam was carried out and the two TIC traces after HPLC separation of the peptide digest are shown in Figure 5.22.

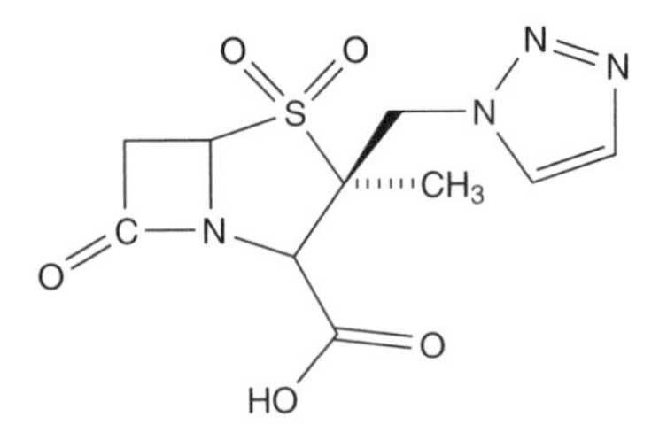


**Figure 5.21** Structure of tazobactam. Reprinted from *Biochim. Biophys. Acta*, **1547**, Bonomo, R. A., Liu, J., Chen, Y., Ng, L., Hujer, A. M. and Anderson, V. E., 'Inactivation of CMY-2 $\beta$-lactamase by tazobactam: initial mass spectroscopic characterization', 196–205, Copyright (2001), with permission from Elsevier Science.

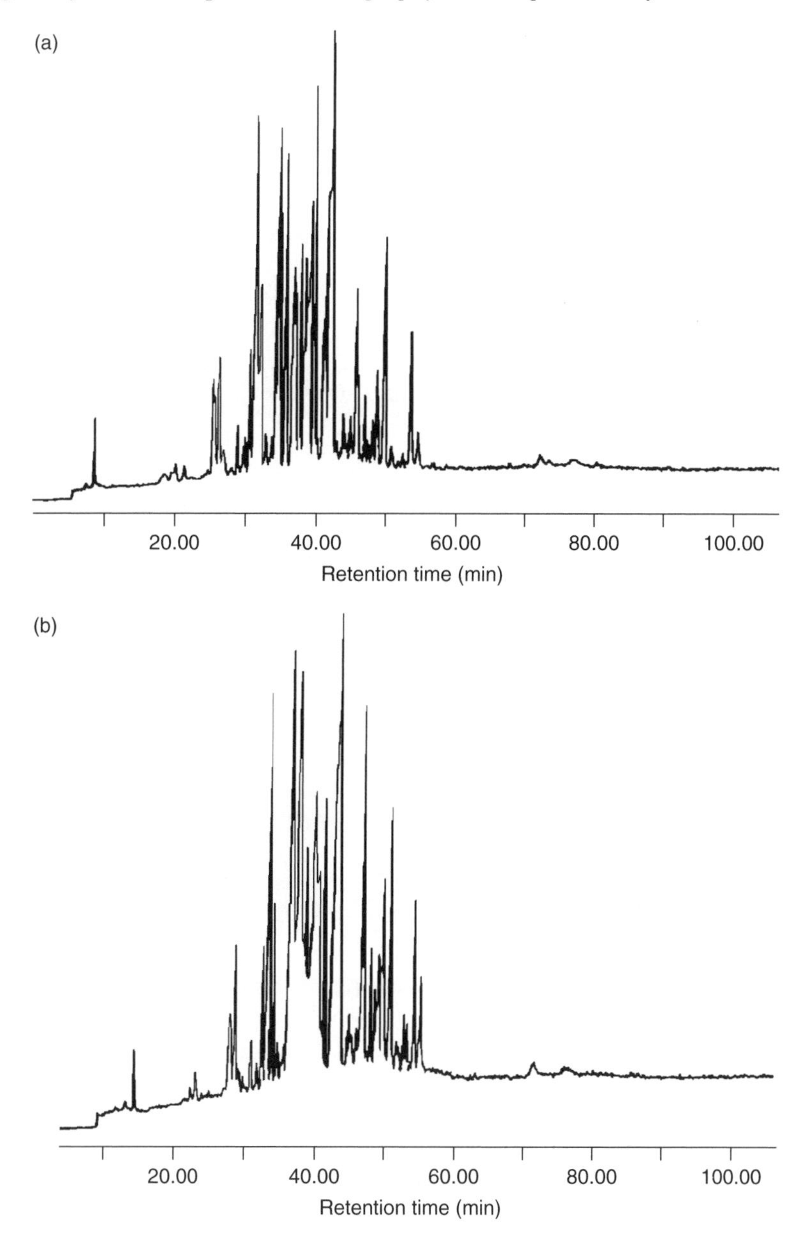


**Figure 5.22** TIC traces obtained from LC–MS analyses of trypsin digests of (a) CMY-2 $\beta$-lactamase, and (b) tazobactam/CMY-2 $\beta$-lactamase. Reprinted from *Biochim. Biophys. Acta*, **1547**, Bonomo, R. A., Liu, J., Chen, Y., Ng, L., Hujer, A. M. and Anderson, V. E., 'Inactivation of CMY-2 $\beta$-lactamase by tazobactam: initial mass spectroscopic characterization', 196–205, Copyright (2001), with permission from Elsevier Science.

The complexity of the traces is such that the differences between them are not immediately obvious and this is not an unusual situation, particularly as the molecular weight of the parent protein increases. A detailed examination of the spectra associated with each of the chromatographic responses may therefore be necessary before the information required by the analyst is obtained.

In this case, one particular chromatographic response, i.e. that at a retention time of 42.12 min in Figure 5.22(b), was found to show significant changes. In the digest of the parent protein, this chromatographic response, the electrospray spectrum of which is shown in Figure 5.23, was found to consist of three peptides with molecular weights of 1825.45 ± 2.66, 2370.51 ± 2.79 and 2578 ± 3.34 Da, the first two of which could be rationalized in terms of amino acid residues 291–309 and 46–67, respectively. The third component, with a molecular weight of 2578 ± 3.34 Da, was not assigned.

The electrospray spectrum from the corresponding chromatographic response in the LC–MS analysis of the tryptic digest of the protein after reaction with the inhibitor is shown in Figure 5.24. In addition to the three species found in the digest of the parent protein, two additional polypeptides, with molecular weights of 2439.36 ± 0.07 and 2457.43 ± 0.02 Da, i.e. 70 and 88 Da above

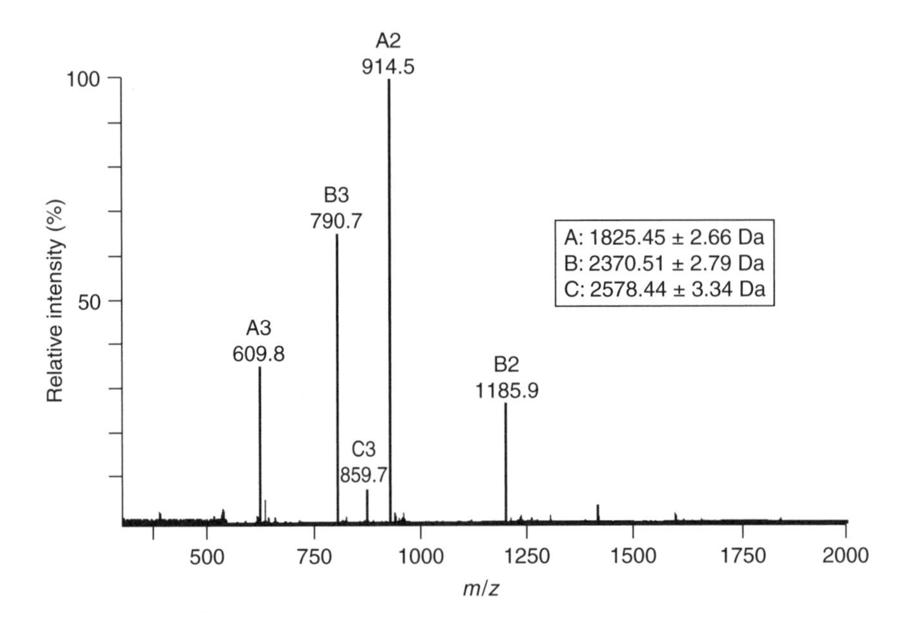


**Figure 5.23** Electrospray mass spectrum of the tryptic peptide with a retention time of 41.81 min from intact CMY-2 $\beta$-lactamase. Reprinted from *Biochim. Biophys. Acta*, **1547**, Bonomo, R. A., Liu, J., Chen, Y., Ng, L., Hujer, A. M. and Anderson, V. E., 'Inactivation of CMY-2 $\beta$-lactamase by tazobactam: initial mass spectroscopic characterization', 196–205, Copyright (2001), with permission from Elsevier Science.

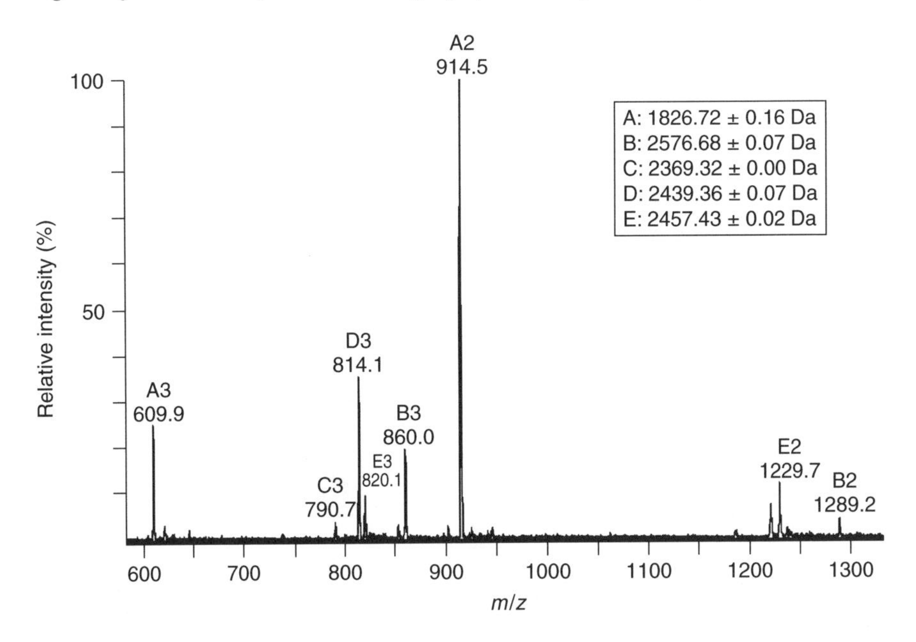


**Figure 5.24** Electrospray mass spectrum of the tryptic peptide with a retention time of 42.12 min from CMY-2 $\beta$-lactamase inhibited by tazobactam. Reprinted from *Biochim. Biophys. Acta*, **1547**, Bonomo, R. A., Liu, J., Chen, Y., Ng, L., Hujer, A. M. and Anderson, V. E., 'Inactivation of CMY-2 $\beta$-lactamase by tazobactam: initial mass spectroscopic characterization', 196–205, Copyright (2001), with permission from Elsevier Science.

that in the parent protein, were found and these were attributed to derivatization with tazobactam specifically at the serine residue at position 64 in the sequence. These were rationalized in terms of hydrolysis of tazobactam during the inhibition process by the mechanism shown in Figure 5.25. The molecular weight of the inhibited enzyme had been determined, after maximum entropy processing of the electrospray spectrum, to have increased by 80 Da, which is not directly in accord with the results of the digestion experiment. This was rationalized in terms of the measurements on the intact protein yielding an unresolved doublet at approximately 39 921.5 and 39 939.5 Da and demonstrates that computer manipulation of raw data, even using sophisticated algorithms, does not always provide unambiguous information.

## *5.3.8 The Location of Post-Translational Modifications Using MS–MS*

The detailed comparison of a complex chromatogram containing a large number of peaks, particularly when some of these peaks contain a number of components,

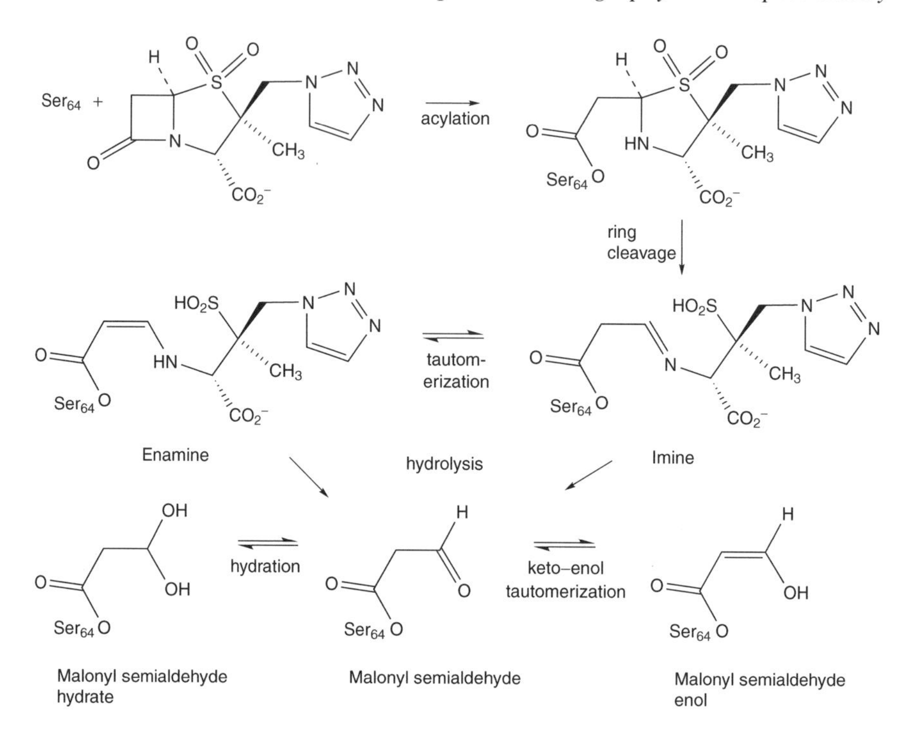


**Figure 5.25** Proposed mechanism for the inactivation of CMY-2 $\beta$-lactamase with tazobactam. Reprinted from *Biochim. Biophys. Acta*, **1547**, Bonomo, R. A., Liu, J., Chen, Y., Ng, L., Hujer, A. M. and Anderson, V. E., 'Inactivation of CMY-2 $\beta$-lactamase by tazobactam: initial mass spectroscopic characterization', 196–205, Copyright (2001), with permission from Elsevier Science.

is likely to be a time-consuming process. The specificity of the mass spectrometer may be used to simplify the analysis by allowing components containing the required structural features to be located.

**SAQ 5.4**

What is a reconstructed ion chromatogram (RIC) and how does it allow a specific compound to be located in the total-ion current (TIC) trace from an LC–MS analysis?

It has been shown [18] that when high cone-voltages are used in conjunction with negative-ion electrospray, phosphopeptides produce diagnostic ions at $m/z$ 63 ($PO_2^-$) and $m/z$ 79 ($PO_3^-$). LC–MS analysis of a trypsin digest of bovine

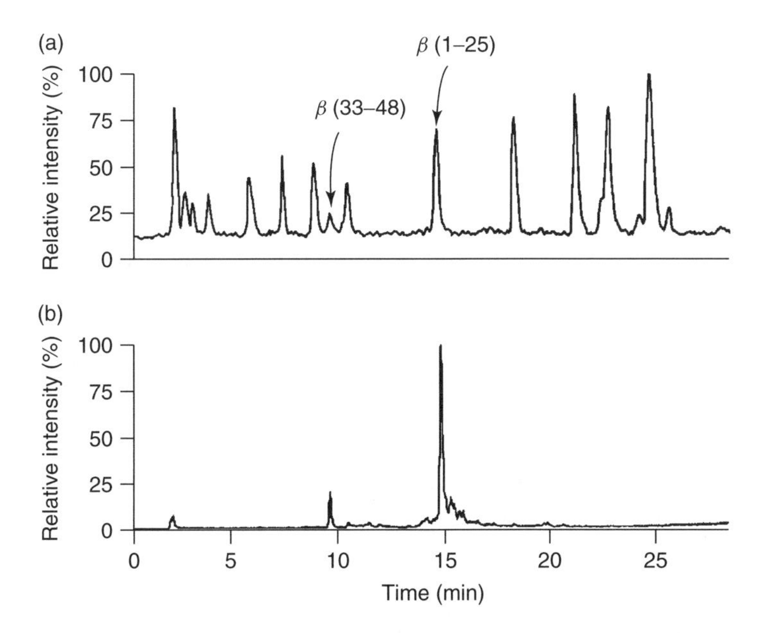


**Figure 5.26** Selective detection of phosphorylated peptides in a tryptic digest: (a) LC–electrospray-MS total-ion-current trace in full-scan positive-ion mode; (b) reconstructed ion chromatogram of $m/z$ 79 ($PO_3^-$) in negative-ion mode with high repeller voltage to induce fragmentation. Reprinted from *J. Chromatogr., A*, **881**, Leonil, J., Gagnaire, V., Molle, D., Pezennec, S. and Bouhallab, S. 'Application of chromatography and mass spectrometry to the characterization of food proteins and derived peptides', 1–21, Copyright (2000), with permission from Elsevier Science.

$\beta$-casein [19] produces a number of peptides, as shown in Figure 5.26(a). A reconstructed ion chromatogram of $m/z$ 79, shown in Figure 5.26(b), allows immediate location of the phosphopeptides present, spectra from which can be studied further in an attempt to generate sequence information and the position of phosphorylation on the parent protein. This simplification has been effected by using conventional scanning of the mass spectrometer which has the added advantage that the data may be reprocessed in other ways should further investigations need to be carried out. A similar procedure to locate glycopeptides using ions of $m/z$ 204 has also been reported [20].

More specific data may be obtained by using MS–MS, where, for example, collision-induced dissociation of molecular ions from electrospray ionization of glycosylated peptides leads to a loss of a sugar residue followed by dehydration – 216 $m/z$ units if the peptide is singly charged, and 108 $m/z$ units if doubly charged [21]. Constant-neutral-loss scanning of the mass spectrometer therefore allows all components of the complex protein digest which fragment in this way to be located. Since this procedure requires the precursor ion(s) to be defined, a full-scan experiment to identify these must also be carried out.

It was found that the experimental conditions for the optimum production of the product ions were not identical for all peptides, in particular, these were influenced by whether the precursor ion was singly or doubly charged. The software with most instruments allows the acquisition time, in this case the time for HPLC separation of the peptides, to be divided into different periods during which the mass spectrometer may be programmed to operate under different sets of conditions, e.g. different mass ranges, cone-voltages, precursor-ion masses, etc. In this case, it was found that three sets of conditions allowed useful data to be obtained, as indicated in Figure 5.27, which shows the full-scan TIC and the TIC trace obtained for constant-neutral losses of 216 $m/z$ units (P1) or 108 $m/z$ units (P2 and P3 – these employed different cone-voltages and mass ranges for acquisition in order to obtain maximum sensitivity). This shows that the digest contains nine glycosylated peptides, the spectra of which can be extracted from the full-scan data and analysed.

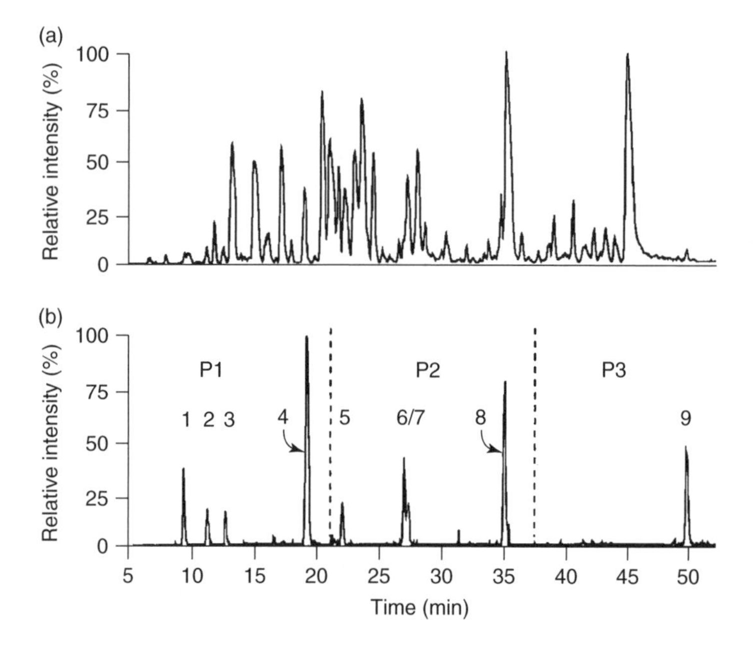


**Figure 5.27** Selective detection of lactolated peptides from a tryptic digest of $\beta$-lactoglobulins by LC–electrospray-MS–MS, showing (a) the total-ion-current trace in full-scan mode, and (b) the total-ion-current trace in neutral-loss-scanning mode. Figure from 'Selective detection of lactolated peptides in hydrolysates by liquid chromatography/electrospray tandem mass spectrometry', by Molle, D., Morgan, F., Bouhallab, S. and Leonil, J., in *Analytical Biochemistry*, Volume 259, 152–161, Copyright 1998, Elsevier Science (USA), reproduced with permission from the publisher.

### *5.3.9 The Analysis of Polysaccharides Present in Glycosylated Proteins*

While the majority of reported work concerning the sequencing of biomolecules by LC–MS has involved proteins, important information may also be obtained from oligosaccharides by employing a similar methodology to that described previously.

Mucins are highly glycosylated proteins, changes in the saccharide parts of which are associated with diseases such as cystic fibrosis and cancers. Oligosaccharides constitute between 50 and 80% of the glycoprotein and a large number of different oligosaccharides, including sulfated species, are present on a single mucin. A number of different analytical approaches have been taken to the study of the glycans (polysaccharides) present on this type of molecule. One approach has involved the separation of glycans chemically released from a mucin as alditols into neutral, sialylated and sulfated species [22]. A study of the sulfated glycans using LC–MS–MS has been reported [23].

The study of sulfated standards enabled certain structural features to be associated with particular ions that appeared in their negative-ion electrospray MS–MS spectra. The following is reproduced from this paper [23]:

> 'The presence of sulfate was shown by the fragment ions at $m/z$ 97 ($HSO_4^-$) and $m/z$ 139 in all spectra. The fragment ion at $m/z$ 199 was found to be useful as a diagnostic ion for locating the sulfate group, indicating sulfate linked to C-4 or C-6 of GalNAc, Gal or GlcNAc but not when attached to the C-3 of Gal or GlcNAc.' 'The location of the sulfate group to a Hex or HexNAc was distinguished by the presence of either $m/z$ 241 [Hex $SO_3]^-$ or $m/z$ 282 [HexNAc–$SO_3]^-$.'

It was also noted that ions at $m/z$ 180 and $m/z$ 513, associated with the attachment of sulfate to C-3 of GlcNAc and C-4 of GalNAc, found in the FAB MS–MS spectra, were also observed in electrospray MS–MS spectra but were of low abundance.

GalNAc, Gal, GlcNAc, Hex and HexNAc are the abbreviations used when discussing sequences of monosaccharides and represent *N*-acetylgalactosamine, galactose, *N*-acetylglucosamine, hexose, i.e. a monosaccharide with six carbon atoms, and *N*-acetylhexosamine, respectively.

The TIC trace from the LC–MS analysis of the sulfated mucin oligosaccharides from the porcine large intestine is reproduced in Figure 5.28 and shows the presence of at least 40 compounds, with the molecular weights of 28 of these which were characterized also being shown. Characterization was not facilitated by the presence of a number of oligosaccharides with the same molecular weight, e.g. four of molecular weight 870 Da, four of molecular weight 1470 Da and five of molecular weight 1016 Da, which differ solely in the sequence of sugars within the saccharide moieties.

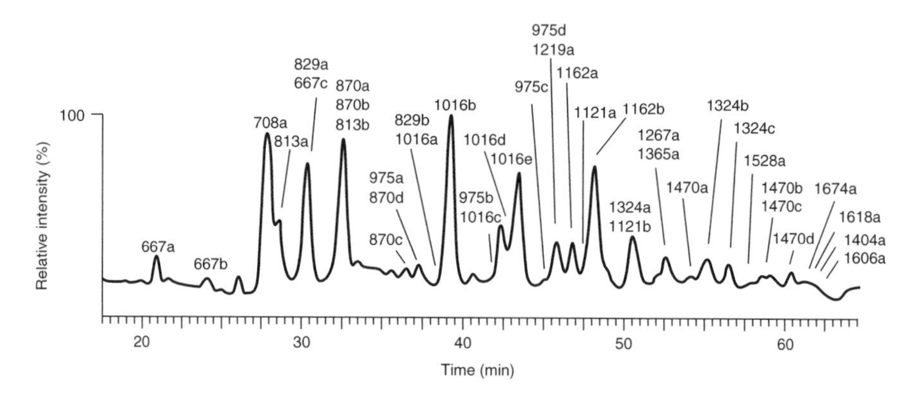


**Figure 5.28** LC–electrospray-MS total ion chromatogram of sulfated oligosaccharides from mucins purified from the porcine large intestine, where the annotations indicate the molecular ions observed from each component. Reprinted with permission from Thomsson, K. A., Karlsson, H. and Hansson, G. C., *Anal. Chem.*, **72**, 4543–4549 (2000). Copyright (2000) American Chemical Society.

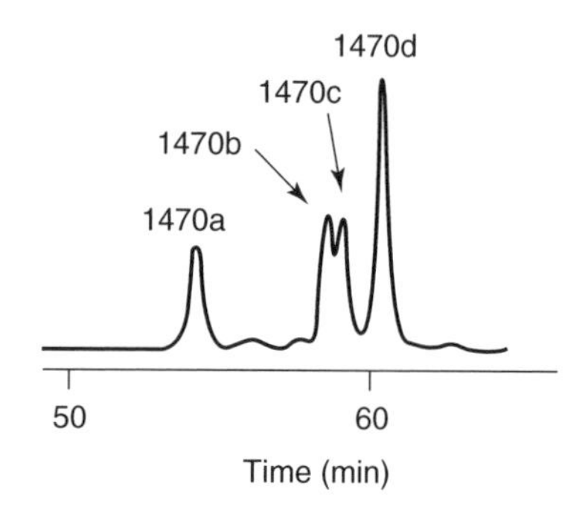


**Figure 5.29** Reconstructed ion chromatogram from $m/z$ 1470 for a retention time-range of 50 to 65 min, showing the presence of four components with a molecular weight of 1471 Da. Reprinted with permission from Thomsson, K. A., Karlsson, H. and Hansson, G. C., *Anal. Chem.*, **72**, 4543–4549 (2000). Copyright (2000) American Chemical Society.

The RIC for $m/z$ 1470 is shown in Figure 5.29, while the MS–MS spectra from each of their $[M-H]^-$ ions is presented in Figure 5.30. A detailed study of these spectra and a comparison with spectra from compounds with known saccharide sequences is required before structural assignments may be made. It is not considered appropriate to carry out such an exercise in a text of this nature, nor to be necessary for the reader to be able to appreciate the value of this example. For this reason, the following paragraph is taken directly from reference [23] and interested readers may, at their leisure, examine Figure 5.30 in conjunction with this:

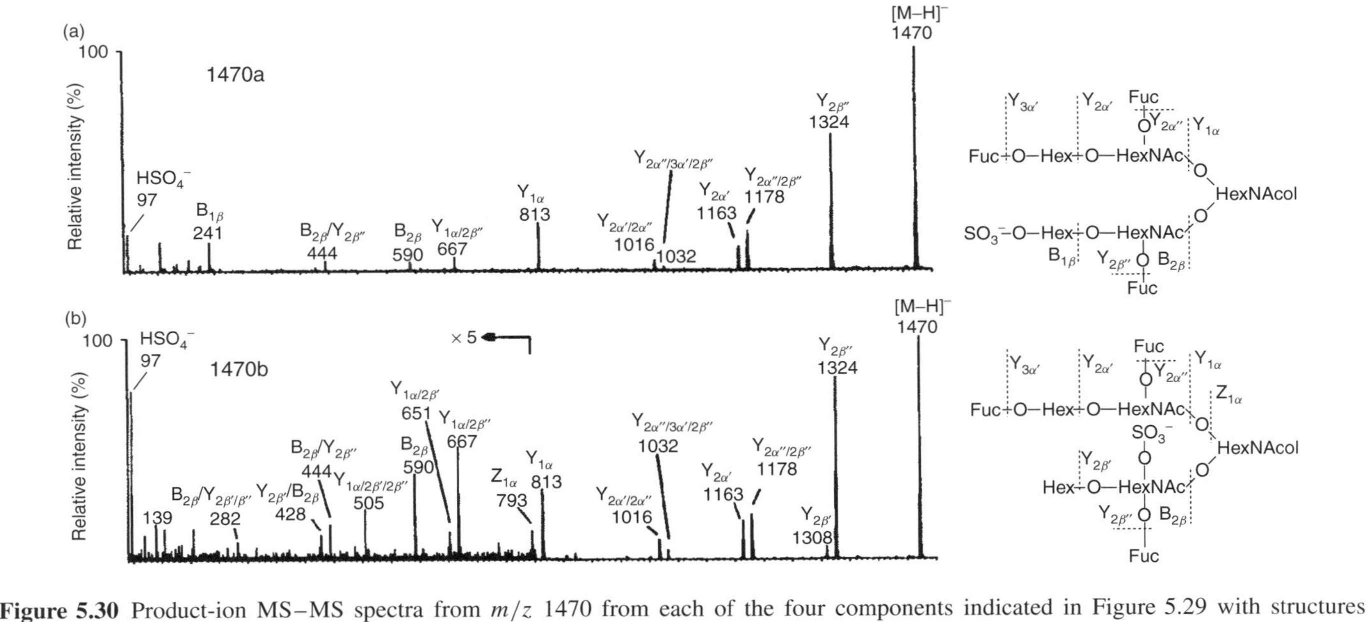


**Figure 5.30** Product-ion MS–MS spectra from $m/z$ 1470 from each of the four components indicated in Figure 5.29 with structures deduced from the fragment ions produced. Reprinted with permission from Thomsson, K. A., Karlsson, H. and Hansson, G. C., *Anal. Chem.*, **72**, 4543–4549 (2000). Copyright (2000) American Chemical Society.

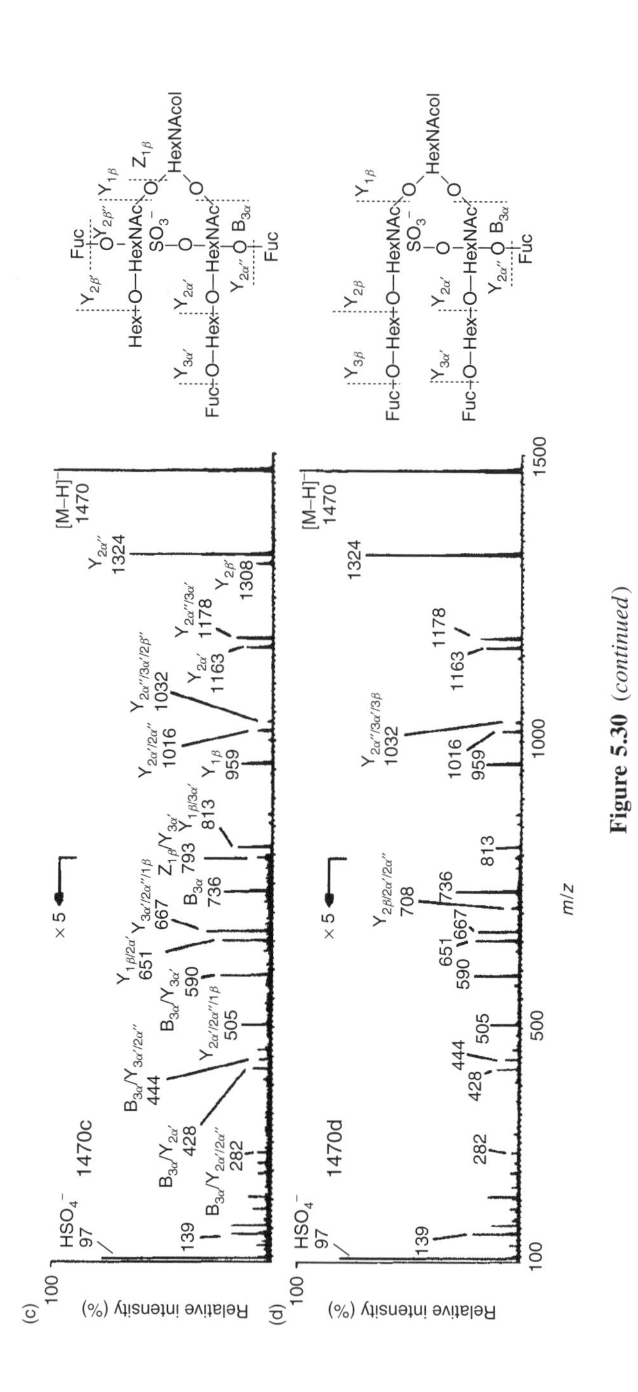


**Figure 5.30** *(continued)*

'The $B_i$ ions indicated that the sulfate-carrying branch of all isomers contained one Fuc, one Hex, one HexNAc ($m/z$ 590 of 1470a and 1470b), and one additional Fuc ($m/z$ 736) for 1470c and 1470d. The 1470a isomer carried the sulfate linked to a Hex residue ($m/z$ 241), whereas the other showed fragment ions at $m/z$ 282, 428 and 651, indicative of sulfate linked to the FucHexNAc unit. The loss of one Hex residue revealed at $m/z$ 1308 indicated terminal Hex in sequence 1470b and 1470c. Only the 1470d compound gave an ion at $m/z$ 708, suggesting that this fragment ion is formed at low abundance when one Fuc residue is attached to the core four HexNAc saccharides, but not when it comes with two Fuc residues (1470a, 1470b and 1470c).'

### 5.3.10 Location of the Position of Attachment of a Glycan on the Polypeptide Backbone of a Glycoprotein

The methodology employed for the study of glycoproteins is somewhat different in that it is not just the saccharide sequence that is required but the position at which it is joined to the peptide backbone.

Diapause is a suspension of development that can occur at the embryonic, larval, pupal or adult stage of a number of insect species. EA4 is a glycoprotein isolated from diapausing eggs of the silkworm *Bombyx mori* which has a major function in terminating the period of diapause through the carbohydrate part of the molecule. The structure of the glycoprotein has therefore been studied in detail [24]. The molecular weight of the glycoprotein was determined, using electrospray ionization, to be 17 336.79 Da. Treatment of EA4 with PNGase F, an enzyme that removes the carbohydrate side-chain, resulted in a reduction in the molecular weight of the protein of 730.58 Da, corresponding to two mannose and two *N*-acetylglucosamine residues.

**SAQ 5.5**

What is the configuration of a Q–ToF mass spectrometer and what are its analytical strengths?

A trypsin digestion of EA4 produced 14 tryptic peptides which were studied by using LC–MS with electrospray ionization and a Q–ToF mass spectrometer. One peptide, designated T3, was, from the electrospray spectrum shown in Figure 5.31, calculated to have a molecular weight of 2037.02 Da, which was reduced after deglycosylation to 1306.65 Da, i.e. by 730.37 Da. corresponding almost exactly with the reduction in molecular weight brought about by deglycosylation of the parent glycoprotein. These molecular weights were rationalized in terms of the sequence from Gly21 to Lys32 shown in Figure 5.32, although the actual position of glycosylation could not be determined by mass spectrometry because the collision energy required to bring about fragmentation of the

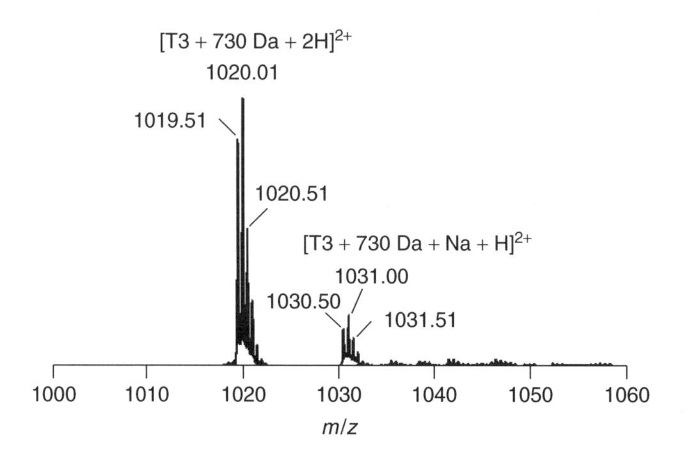


**Figure 5.31** LC–electrospray-MS–MS spectrum of the column eluate at around 22 min in the analysis of the peptide mixture from the tryptic digest of glycoprotein TIME-EA4 from silkworm diapause eggs. Reprinted from *Bioorg. Med. Chem.*, **10**, Kurahashi, T., Miyazaki, A., Murakami, Y., Suwan, S., Franz, T., Isobe, M., Tani, M. and Kai, H., 'Determination of a sugar chain and its linkage site on a glycoprotein TIME-EA4 from silkworm diapause eggs by means of LC–ESI-Q-TOF-MS and MS/MS', 1703–1710, Copyright (2002), with permission from Elsevier Science.

T3

$Gly^{21}-Asn^{22}-Ile^{23}-Thr^{24}-Phe^{25}-Thr^{26}-Gln^{27}-Val^{28}-Gln^{29}-Asp^{30}-Gly^{31}-Lys^{32}$

Monoisotopic mass = 1306.65 Da

T3Y1

$Gly^{21}-Asn^{22}-Ile^{23}-Thr^{24}-Phe^{25}$

Monoisotopic mass = 550.28 Da

T3Y2

$Thr^{26}-Gln^{27}-Val^{28}-Gln^{29}-Asp^{30}-Gly^{31}-Lys^{32}$

Monoisotopic mass = 774.39 Da

**Figure 5.32** Amino acid sequence and monoisotopic masses of the peptides T3, T3Y1 and T3Y2 (see text) obtained during the LC–MS–MS analysis of the peptide mixture from the tryptic digest of glycoprotein TIME-EA4 from silkworm diapause eggs. Reprinted from *Bioorg. Med. Chem.*, **10**, Kurahashi, T., Miyazaki, A., Murakami, Y., Suwan, S., Franz, T., Isobe, M., Tani, M. and Kai, H., 'Determination of a sugar chain and its linkage site on a glycoprotein TIME-EA4 from silkworm diapause eggs by means of LC–ESI-Q-TOF-MS and MS/MS', 1703–1710, Copyright (2002), with permission from Elsevier Science.

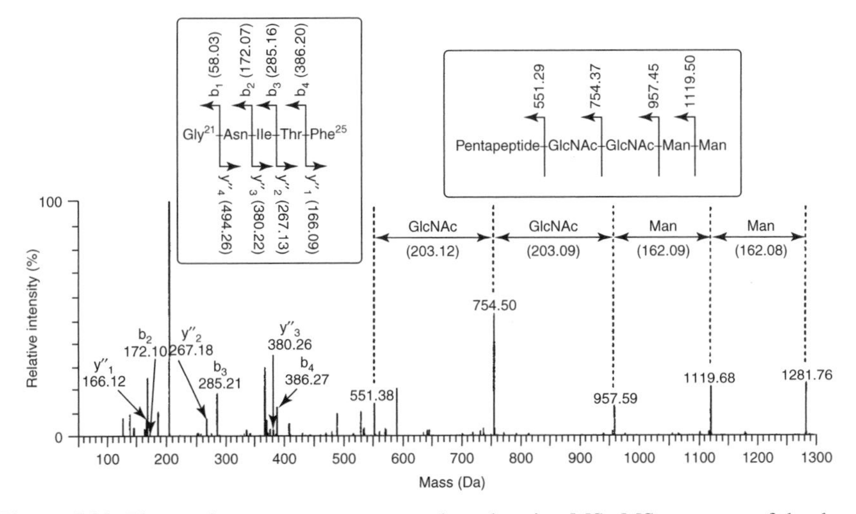


**Figure 5.33** The maximum-entropy-processed product-ion MS–MS spectrum of the doubly charged molecular ion of the glycosylated T3Y1 peptide at $m/z$ 641.31, together with the ions predicted from both the peptide and glycan parts of the molecule. Reprinted from *Bioorg. Med. Chem.*, **10**, Kurahashi, T., Miyazaki, A., Murakami, Y., Suwan, S., Franz, T., Isobe, M., Tani, M. and Kai, H., 'Determination of a sugar chain and its linkage site on a glycoprotein TIME-EA4 from silkworm diapause eggs by means of LC–ESI-Q-TOF-MS and MS/MS', 1703–1710, Copyright (2002), with permission from Elsevier Science.

peptide backbone is much greater than that required to produce fragmentation of the carbohydrate moiety. It has been recognized that there is some resistance to fragmentation via breaking of the bond between the peptide chain and the first saccharide residue when the length of the peptide chain to which the carbohydrate is attached is also reduced and ions generated with this bond intact would provide evidence for the position of glycosylation. In this example, the length of the peptide chain was reduced by further enzymatic digestion of peptide T3, using chymotrypsin, which yielded two smaller fragments, designated T3Y1 and T3Y2, with molecular weights of 1280.62 and 774.39 Da, respectively. By the use of MS–MS, the latter was shown to correspond to residues 26 to 32 of the parent glycoprotein (see Figure 5.32), providing clear evidence that the carbohydrate moiety was attached to the region between residues 21 and 25. The molecular weight of T3Y1 was determined to be 1280.62 Da, with this being consistent with the sequence from residues 21 to 25, with an increase of 730.34 Da as before. The MS–MS spectrum of the doubly charged ion at $m/z$ 641.31 from T3Y1 is shown in Figure 5.33, with the ions present allowing sequencing not only of the carbohydrate moiety but also of the peptide backbone. Closer examination of this spectrum, however, reveals a number of low-intensity ions which allow the position of attachment of the carbohydrate to be determined, as shown in Figure 5.34(a). Ions corresponding to the b2, b3 and b4

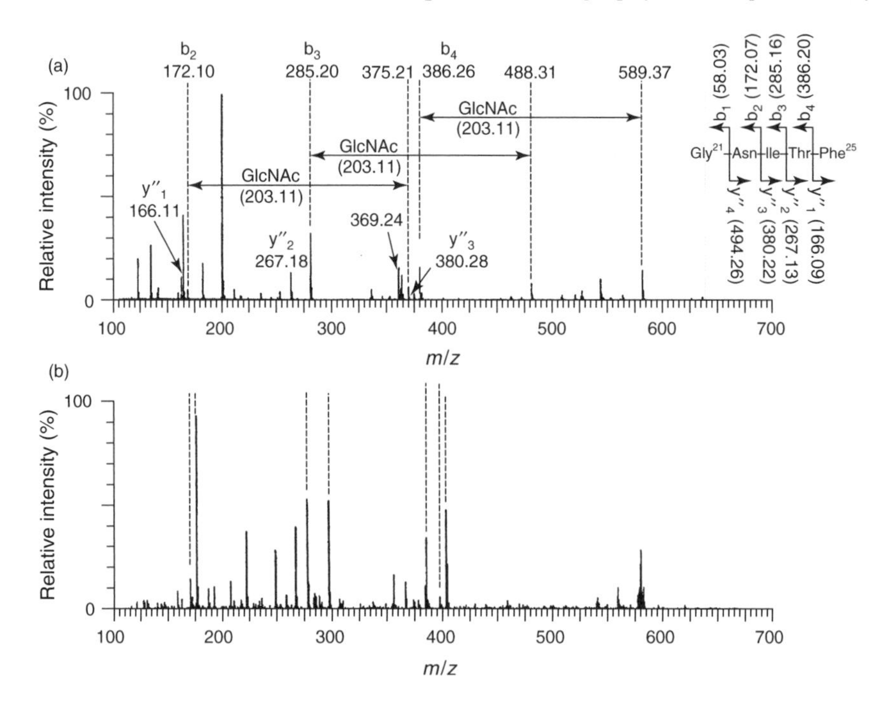


**Figure 5.34** LC–MS–MS spectra from (a) the doubly charged molecular ion of the glycosylated T3Y1 peptide at $m/z$ 641.31, and (b) the singly charged molecular ion of the peptide moiety at $m/z$ 551.40, generated in the mass spectrometer by fragmentation with a high cone-voltage. Reprinted from *Bioorg. Med. Chem.*, **10**, Kurahashi, T., Miyazaki, A., Murakami, Y., Suwan, S., Franz, T., Isobe, M., Tani, M. and Kai, H., 'Determination of a sugar chain and its linkage site on a glycoprotein TIME-EA4 from silkworm diapause eggs by means of LC–ESI-Q-TOF-MS and MS/MS', 1703–1710, Copyright (2002), with permission from Elsevier Science.

fragmentations of the peptide have all increased in $m/z$ by 203 while ions from the $y_1''$, $y_2''$ and $y_3''$ fragmentations are found at the same $m/z$ values as in the deglycosylated species. The presence of these ions indicate that the carbohydrate moiety is attached to Asn22. There is one small ambiguity, however, with that being the presence of an ion at $m/z$ 369.24 which might be considered to arise from the expected $y_1''$ ion with GlcNAc attached, i.e. 166.09 + 203.08. The spectrum shown in Figure 5.34(b) has been generated from the glycopeptide T3Y1 by using a high cone-voltage, which has brought about deglycosylation within the source of the mass spectrometer. While the ions previously rationalized in terms of attachment of the first sugar residue to the peptide backbone are, as expected, no longer observed, the ion at $m/z$ 369 is still present and is therefore clearly not associated with the intact linkage. This piece of information is essential in the unambiguous determination of the position of the carbohydrate moiety as Asn22.

Both of the studies involving glycoproteins have employed Q–ToF mass spectrometers and both sets of authors comment on the sensitivity of this type of instrument, allowing useful data to be obtained from the limited amounts of sample usually available from natural sources.

# 5.4 Molecular Weight Determination of Small (<1000 Da) Molecules

If a 'high'-molecular-weight compound is being studied by LC–MS, the analyst has little choice in the ionization method to use, with atmospheric-pressure chemical ionization (APCI) being wholly inappropriate. However, when 'low'-molecular-weight compounds are involved, both electrospray ionization and APCI are potentially of value.

**SAQ 5.6**

Why is APCI 'wholly inappropriate' for the study of high-molecular-weight materials such as proteins and peptides?

As discussed earlier in Section 4.7, electrospray ionization produces predominantly multiply charged ions, with the range of charges carried by an analyte being related to its structure, in particular the number of possible sites at which a proton may be gained (for positive ionization) or lost (for negative ionization). Low-molecular-weight compounds tend to have fewer potential sites for protonation/deprotonation and therefore their spectra show few (if any) multiply charged ions. Other ionization techniques encountered with LC–MS, as discussed in the remainder of Chapter 4, produce almost exclusively singly charged ions. The spectra of low-molecular-weight compounds produced during LC–MS analyses tend to yield molecular ions, usually $(M + H)^+$ or $(M - H)^-$, and molecular weight determination is therefore not usually difficult, although the presence of adducts of the analyte with species such as $Na^+$, $K^+$, $H_3O^+$, $NH_4{}^+$ and molecules of the mobile phase must always be considered when spectra are being interpreted. The subject does not, therefore, need to be considered in as much detail as it was for biopolymers.

## *5.4.1 The Use of Fast-LC–MS in Combinatorial Chemistry*

Combinatorial chemistry has become of increasing importance, particularly within drug companies, in allowing the very rapid synthesis of large numbers of candidate molecules which require characterization and testing for their effectiveness. An important part of the characterization of any molecule is the determination of its molecular weight and LC–MS has been used extensively for this purpose. With the large number of samples involved, an important feature of any such methodology is the analysis time and fast HPLC methods have been developed. As analysis times decrease, the chromatographic peak widths also decrease and, as discussed above in Section 3.5.2.1, greater performance is required of the mass spectrometer which has to scan more quickly to obtain sufficient mass spectra of an adequate quality.

A method to reduce the overall analysis time by allowing eluates from a number of HPLC columns to be introduced into a mass spectrometer in such a way that spectra from each can be obtained independently has been described [25] and this allows a number of LC–MS determinations to be carried out simultaneously. The hardware is shown in Figure 5.35 and consists of a single pump, the flow from which is split into each of the analytical columns which are served by an autosampler that allows multiple simultaneous or rapidly sequenced injections. Rapid HPLC separations are carried out on 2.1 × 30 mm $C_{18}$ columns with each column eluate passing through a UV detector. The pump initially delivers mobile phase at a flow rate of 8 ml $min^{-1}$, with 2 ml $min^{-1}$ being directed through each individual column. A post-column split delivers 0.2 ml $min^{-1}$ from each into the electrospray source inlet shown in Figure 5.36. This consists of four electrospray probe tips, each of which is connected to a separate HPLC column, and a rotating aperture which allows spray from each tip in turn to be directed into the sampling cone of the electrospray source.

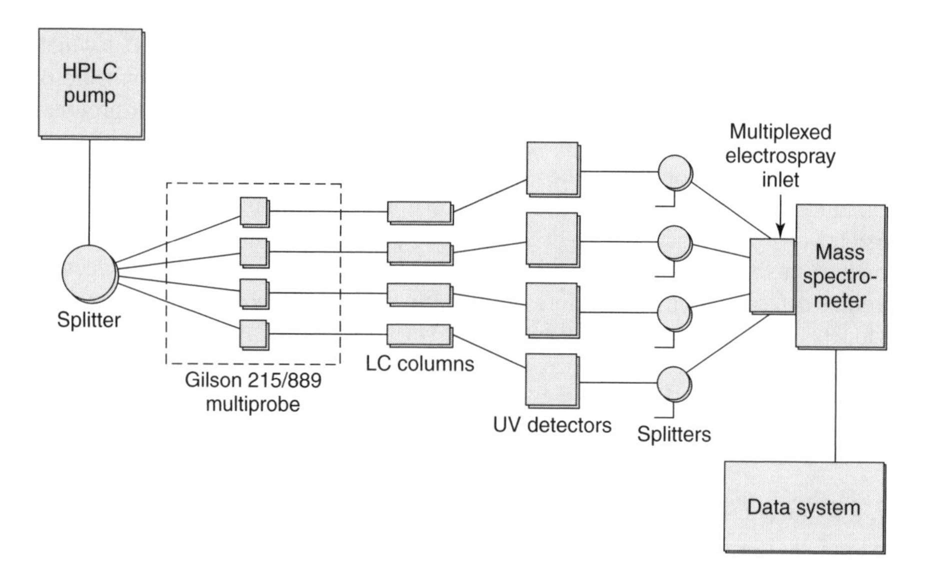


**Figure 5.35** Schematic of a system which allows eluates from four HPLC columns to be introduced simultaneously into a mass spectrometer. From de Biasi, V., Haskins, N., Organ, A., Bateman, R., Giles, K. and Jarvis, S., *Rapid Commun. Mass Spectrom.*, **13**, 1165–1168, Copyright 1999. © John Wiley & Sons Limited. Reproduced with permission.

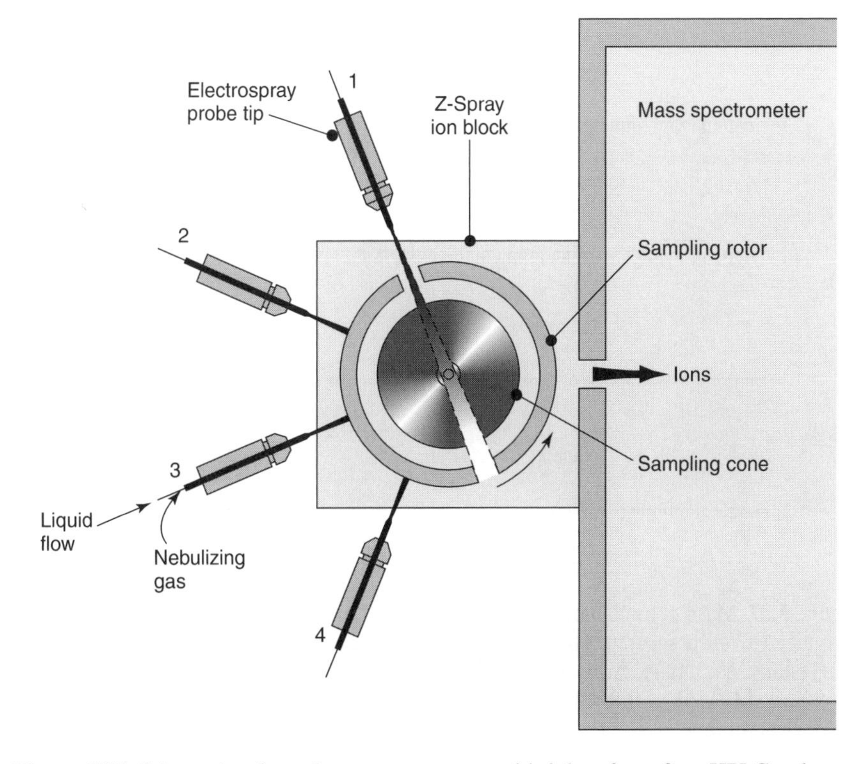


**Figure 5.36** Schematic of an electrospray source with inlets from four HPLC columns. From de Biasi, V., Haskins, N., Organ, A., Bateman, R., Giles, K. and Jarvis, S., *Rapid Commun. Mass Spectrom.*, **13**, 1165–1168, Copyright 1999. © John Wiley & Sons Limited. Reproduced with permission.

In the work described, mass spectra from 200 to 1000 Da were collected from each column for 0.1 s into a separate data file. A period of 0.1 s was required for rotation of the aperture before data could be collected from the next column. A time-of-flight mass analyser (see Section 3.3.4), which is capable of fast scanning without loss of sensitivity, was used for mass spectral data acquisition. The mass chromatograms of the $(M + H)^+$ ions of the four analytes, each of which was injected into a separate column, i.e. furosemide, $m/z$ 331, reserpine, $m/z$ 609, triamterene, $m/z$ 254, and warfarin, $m/z$ 309, are shown in Figure 5.37 and a mass spectrum of warfarin in Figure 5.38. This spectrum is typical of those of low-molecular-weight compounds generated by both APCI and electrospray ionization, showing a base peak corresponding to the molecular weight of the analyte, and no fragmentation. Of importance, in the context of the LC–MS system described, is that no signal is observed from the analyte from the column previously monitored, i.e. triamterene. A system capable of allowing an increase in sample throughput is a valuable asset to the analyst.

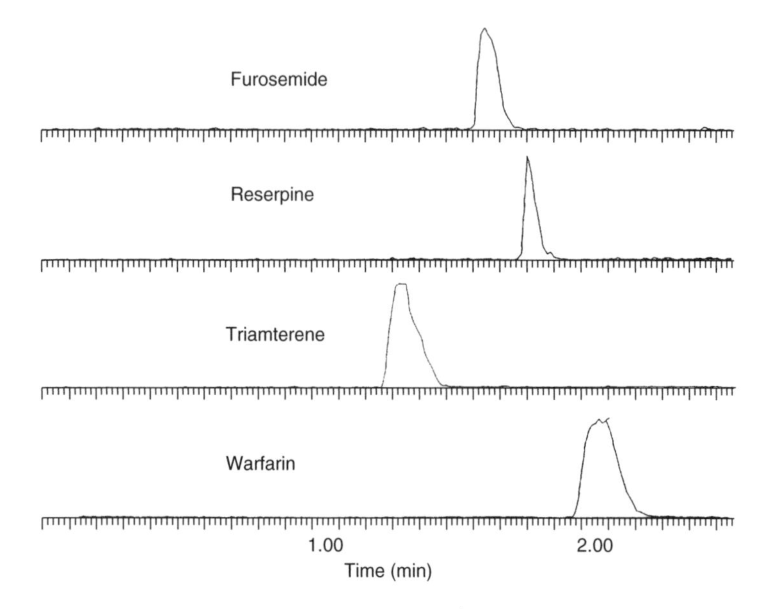


**Figure 5.37** Mass chromatograms of the $(M + H)^+$ ions from four analytes, with each introduced from a separate HPLC column into an electrospray source. From de Biasi, V., Haskins, N., Organ, A., Bateman, R., Giles, K. and Jarvis, S., *Rapid Commun. Mass Spectrom.*, **13**, 1165–1168, Copyright 1999. © John Wiley & Sons Limited. Reproduced with permission.

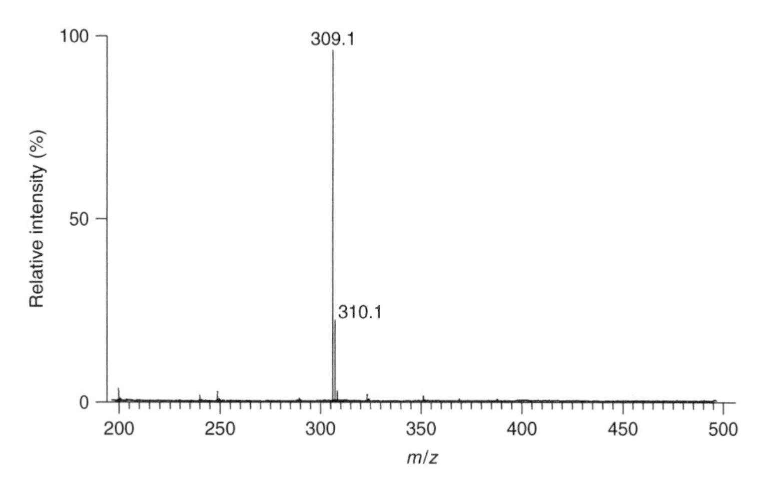


**Figure 5.38** LC–electrospray mass spectrum of warfarin obtained from an LC–MS system which allows eluate from four HPLC columns to be analysed simultaneously. From de Biasi, V., Haskins, N., Organ, A., Bateman, R., Giles, K. and Jarvis, S., *Rapid Commun. Mass Spectrom.*, **13**, 1165–1168, Copyright 1999. © John Wiley & Sons Limited. Reproduced with permission.

# 5.5 Structure Determination of Low-Molecular-Weight Compounds

When 'low'-molecular-weight compounds are involved, both APCI and electrospray ionization are potentially of value and the first task is to decide which of these will give the more useful data.

## *5.5.1 Method Development for Structural Studies*

The major difference between the electrospray spectra of 'low'-molecular-weight compounds (<1000 Da) and those of the 'high'-molecular-weight materials discussed above in Sections 5.2 and 5.3, is that they are much simpler, containing few, if any, multiply charged ions. They therefore resemble in many respects the spectra generated by atmospheric-pressure chemical ionization (APCI), consisting, predominantly, of singly charged ions from molecular species. The relationship between the ions found in a cone-voltage fragmentation spectrum is therefore more easily determined and the technique of greater potential value.

In practice, the value depends on the type of analysis being attempted. Structural studies may require the extent of fragmentation to be maximized, while quantification may require the opposite, i.e. the efficient production of a small number of ions of different $m/z$ ratios, in order to maximize sensitivity. Selectivity may be obtained simply by monitoring of the molecular ions formed. On occasions, when significant background is observed, this may not be adequate and fragmentation of the molecular species may be necessary to provide a number of ions to be monitored. Some compounds may be ionized very effectively under positive-ionization conditions, while others may require the formation of negative ions to allow analysis.

In general terms, electrospray is a more effective ionization method for compounds of higher polarity than is APCI, although 'higher' is a very subjective term and there are a range of compounds for which both techniques are applicable.

A method has been reported for the quantification of five fungicides (shown in Figure 5.39) used to control post-harvest decay in citrus fruits to ensure that unacceptable levels of these are not present in fruit entering the food chain [26]. A survey of the literature showed that previously [27] APCI and electrospray ionization (ESI) had been compared for the analysis of ten pesticides, including two of the five of interest, i.e. carbendazim and thiabendazole, and since it was found that APCI was more sensitive for some of these and had direct flow rate compatibility with the HPLC system being used, APCI was chosen as the basis for method development.

A subsequent comparison of these ionization techniques for the study of another eight pesticides, this time including three of the five of interest [28], i.e. carbendazim, thiabendazole and thiophanate methyl, showed that ESI gave enough sensitivity to allow reliable determination of the pesticides at concentrations below their respective maximum residue levels.

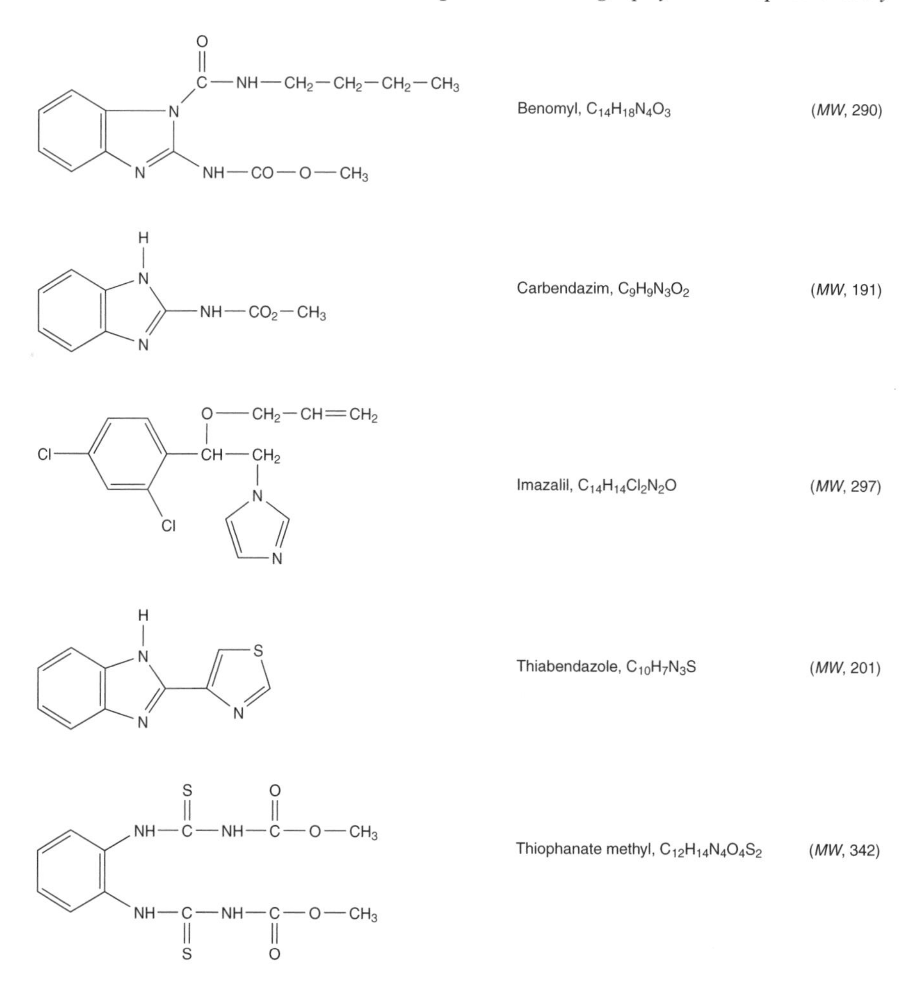


**Figure 5.39** Structures of various fungicides used in the control of post-harvest decay in citrus fruits. Reprinted from *J. Chromatogr., A*, **912**, Fernandez, M., Rodriguez, R., Pico, Y. and Manes, J., 'Liquid chromatographic–mass spectrometric determination of post-harvest fungicides in citrus fruits', 301–310, Copyright (2001), with permission from Elsevier Science.

No APCI or ESI data had been previously reported for two of the five pesticides which were to be determined, i.e. imazalil and benomyl, and therefore although some information was available from the literature it was not possible to make a totally informed decision on the best methodology to employ.

APCI and ESI spectra were obtained by using a range of cone-voltages, with the major ions observed being detailed in Table 5.12. With the exception of benomyl, under APCI conditions all compounds produced abundant molecular

**Table 5.12** Mass spectral fragments and their relative abundances for five fungicides obtained by APCI-MS at different cone-voltages. Reprinted from *J. Chromatogr., A*, **912**, Fernandez, M., Rodriguez, R., Pico, Y. and Manes, J., 'Liquid chromatographic–mass spectrometric determination of post-harvest fungicides in citrus fruits', 301–310, Copyright (2001), with permission from Elsevier Science

| Compound (*MW*) | $m/z$[a] | APCI (V) | | | ES (V) | | |
|---|---|---|---|---|---|---|---|
| | | 20 | 70 | 140 | 20 | 70 | 140 |
| Benomyl (290) | 291 $[M + H]^+$ | — | — | — | 100 | 30 | — |
| | 192 $[M + H–C_4H_9NHCO]^+$ | 100 | 100 | 5 | 15 | 100 | — |
| | 160 $[M + H–C_4H_9NHCO–CH_3O]^+$ | — | 5 | 100 | — | 10 | 100 |
| | 134 $[M + H–C_4H_9NHCO–CH_3OCO]^+$ | 15 | 10 | 25 | — | — | 10 |
| Carbendazim (191) | 192 $[M + H]^+$ | 100 | 100 | — | 100 | 100 | 100 |
| | 160 $[M–CH_3O]^+$ | — | 10 | 100 | 8 | 8 | 30 |
| | 134 $[M + H–CH_3OCO]^+$ | — | — | 25 | — | — | — |
| Imazalil (296) | 297 $[M + H]^+$ | 100 | 100 | 100 | 100 | 100 | 100 |
| Thiabendazole (201) | 202 $[M + H]^+$ | 100 | 100 | 100 | 100 | 100 | 100 |
| | 175 $[M–CN]^+$ | — | — | 20 | — | — | 40 |
| Thiophanate methyl (342) | 343 $[M + H]^+$ | 100 | 100 | — | 100 | 100 | — |
| | 311 $[M–CH_3O]^+$ | — | 5 | 20 | 8 | 12 | — |
| | 268 $[M + H–CH_3NHCOO]^+$ | — | 5 | 10 | — | 5 | — |
| | 226 $[M + H–CH_3OCONHCS]^+$ | — | 5 | 5 | — | 5 | — |
| | 192 $[M + H–CH_3OCONHCS_2]^+$ | — | — | 5 | — | 5 | 10 |
| | 151 $[M + H–CH_3–OCONH–CH_3OCONHCS]^+$ | — | 5 | 100 | — | 5 | 100 |

[a]Tentative ion structure.

ions although their relative intensities are cone-voltage-dependent. These data may be misleading if maximum sensitivity is required since the ESI spectrum of benomyl at a cone-voltage of 20 V shows the molecular ion to be the base peak – this is not the situation at a cone-voltage of 70 V, although the overall sensitivity at this voltage is doubled.

In this case, a cone-voltage of 70 V gave the best sensitivities for all of the analytes being studied but this is not always so and it may be beneficial to be able to change the cone-voltage during the course of an experiment. The spectral data obtained enabled an analytical method, based on selected-ion monitoring, to be proposed, with the ions monitored being $m/z$ 192 from benomyl and carbendazim, $m/z$ 202 from thiabendazole, $m/z$ 343 from thiophanate methyl and $m/z$ 297 from imazalil. Based on a standard series, a detection limit of 0.1 $\mu g\,ml^{-1}$ was obtained for benomyl and 0.01 $\mu g\,ml^{-1}$ for the remaining analytes.

The method was then applied to extracts of orange peel but neither benomyl or thiophanate methyl were recovered as the parent compounds due to their well-known conversion to carbendazim during the extraction process – this is an effect that is not encountered when analysing standards.

While it is true, therefore, that a general method for analysis could be developed and used successfully for a number of applications based on previous knowledge, for those in which optimum performance is required it is advisable to carry out preliminary experiments in which the variables of ionization technique, polarity, cone-voltage, etc., as well as those of sample isolation, are investigated (see Section 5.1 above).

### *5.5.2 The Use of Target-Compound Analysis and LC–MS–MS for the Identification of Drug Metabolites*

A knowledge of the metabolic fate of a particular drug, both in terms of the identification and quantification of the metabolites, is necessary not only to ensure that its use will not cause more problems than the medical condition it is designed to alleviate but, from the drug company's perspective, to facilitate the design of more effective drugs.

Drug metabolism, sometimes known as *biotransformation*, is the route by which drugs are made more water-soluble to facilitate their elimination from the body. This normally occurs in two stages. Phase I metabolism involves the introduction of a polar group, such as an hydroxyl group, into the parent drug structure. These metabolites are often inactive and if they are sufficiently polar may be eliminated from the body directly. Many Phase I metabolites, however, are insufficiently polar to be eliminated and undergo further reaction (conjugation) with an endogenous material, such as glucuronic acid, to form a much more polar compound and thus facilitate elimination. This is termed Phase II metabolism.

The complementary nature of APCI and ESI, APCI for less polar compounds (Phase I metabolites) and ESI for more polar compounds (Phase II metabolites), make them an ideal combination for the characterization of metabolites, particularly with the hardware allowing for relatively simple changing between the two techniques.

While there is a vast range of different drug structures, there are only a relatively small number of chemical reactions, some of which are shown below in Table 5.13 (p. 199), involved in the production of metabolites. Based on the structure of the drug, it is therefore possible to predict the most likely metabolites. Use may then be made of reconstructed ion chromatograms (RICs) of $m/z$ values corresponding to the predicted molecular weights of these metabolites to locate them within the LC–MS data obtained.

This general procedure will only be completely successful, of course, if the correct RICs are examined and it may be that the analyst feels it more appropriate to carry out a manual analysis of the TIC trace. This is not perhaps as daunting as it may at first appear, since although the TIC trace may contain a large amount of information, the ionization techniques employed invariably yield simple spectra and changes in $m/z$ values of the intense ions present are often obvious. The RICs of the ions selected in this way are then constructed for further examination of the spectra involved.

The structures of compounds of interest may then be investigated, if required, by using MS–MS. An analysis of the MS–MS fragmentation of the molecular species from the parent drug may be useful in this respect. Ions at particular $m/z$ values may then be linked to specific structural elements of the analyte and if these are observed in the MS–MS spectra from the metabolites it may be concluded that this part of the structure is intact. If the $m/z$ value of any of these diagnostic fragments has changed by a characteristic number, e.g. an increase of 16 Da, biotransformation of that part of the molecule is suggested.

'Indinavir' is an HIV protease inhibitor whose metabolism has been studied and the LC–MS data generated in this study has been used in the development of an automatic procedure for the study of drugs and their metabolites in *in vitro* biological matrices [29]. The MS–MS spectrum from the electrospray-generated $(M + H)^+$ ion at $m/z$ 614, together with the structural assignments for the major ions observed, are shown in Figure 5.40.

The RICs for $m/z$ 614, indicating unchanged Indinavir, and $m/z$ 630 and 646, i.e. those expected from its mono- and dihydroxymetabolites, are shown in Figure 5.41. This allows the presence of three monohydroxylated and one dihydroxylated metabolite to be demonstrated; the other responses observed were shown by the authors not to be associated with the drug. The MS–MS spectrum of the molecular species of the dihydroxylated compound did not allow its structure to be determined but those from the molecular species from the monohydroxylated compounds allowed the structures shown in Figure 5.42 to be proposed and these 'correlate well with the findings from other laboratories'.

The MS–MS data from metabolite 4 shows a series of ions, i.e. $m/z$ 481, 437 and 380, at $m/z$ values which are 16 greater than those in the MS–MS spectrum

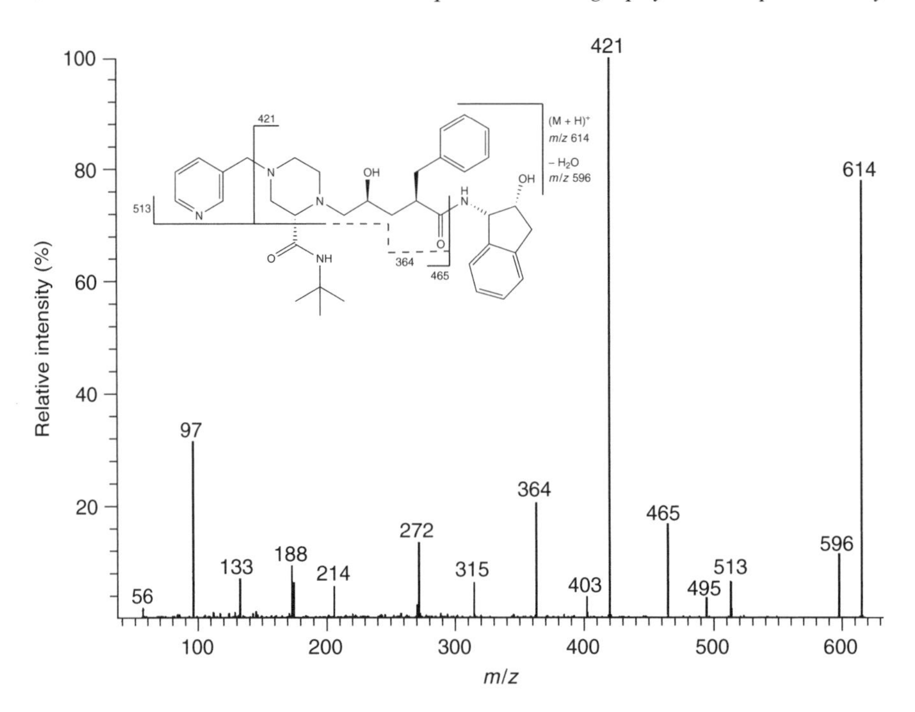


**Figure 5.40** Product-ion spectrum of the $(M + H)^+$ ion ($m/z$ 614) of Indinavir with the proposed origins of the ions observed. Reprinted by permission of Elsevier Science from 'Identification of *in vitro* metabolites of Indinavir by "Intelligent Automated LC–MS/MS" (INTAMS) utilizing triple-quadrupole tandem mass spectrometry', by Yu, X., Cui, D. and Davis, M. R., *Journal of the American Society for Mass Spectrometry*, Vol. 10, pp. 175–183, Copyright 1999 by the American Society for Mass Spectrometry.

from Indinavir. If the metabolite and parent compound fragment in an analogous fashion, the extra oxygen atom cannot therefore be associated with the pyridine or indan rings or the amido side-chain and is thus most likely to be associated with the aromatic ring.

The MS–MS data from metabolite 5 shows a base peak at $m/z$ 437, at an increase of 16 Da over the parent drug, but, in common with Indinavir, ions at $m/z$ 364 and 465. Of most significance is the ion at $m/z$ 465 which indicates that the extra oxygen atom is associated with the indan ring structure.

The MS–MS data from metabolite 6 is not so easy to interpret in isolation in that it shows significant changes to those from the other metabolites, in particular the base peak is now found at $m/z$ 512, in this case 91 Da higher than that in the parent drug. This corresponds to the mass of the pyridyl function attached to the piperazine ring but must include a significant difference in the structure as its route of fragmentation has been affected so dramatically. The structure shown in Figure 5.42 has been proposed and is consistent with the presence of the ion

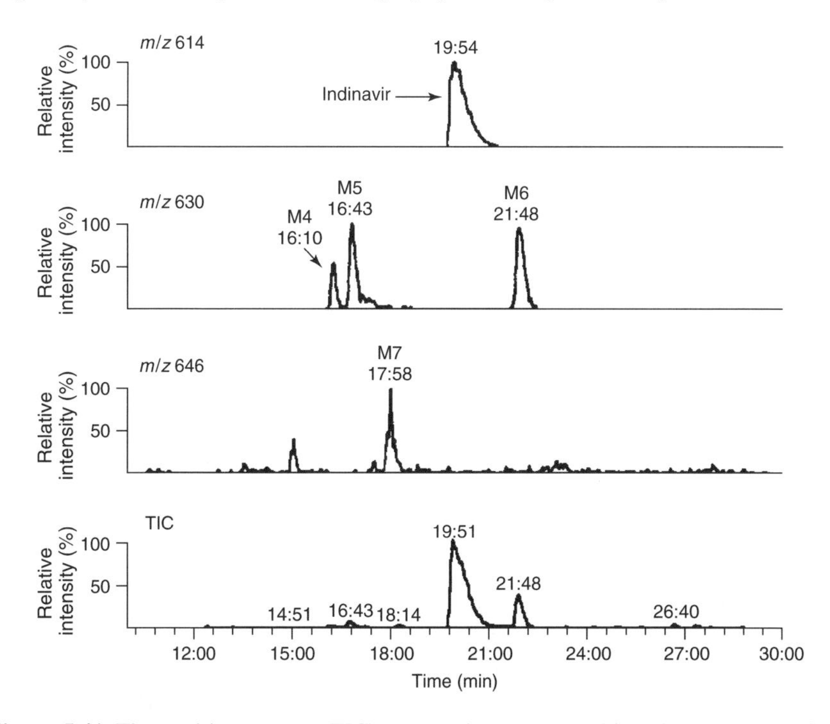


**Figure 5.41** The total-ion-current (TIC) trace and reconstructed ion chromatograms from the predicted pseudomolecular ions of Indinavir ($m/z$ 614) and its mono- ($m/z$ 630) and dihydroxy metabolites ($m/z$ 646), generated from full-scan LC–MS analysis of an incubation of Indinavir with rat liver S9. Reprinted by permission of Elsevier Science from 'Identification of *in vitro* metabolites of Indinavir by "Intelligent Automated LC–MS/MS" (INTAMS) utilizing triple-quadrupole tandem mass spectrometry', by Yu, X., Cui, D. and Davis, M. R., *Journal of the American Society for Mass Spectrometry*, Vol. 10, pp. 175–183, Copyright 1999 by the American Society for Mass Spectrometry.

at $m/z$ 481, as in the mass spectrum of metabolite 4, again indicating that the indan ring is intact and has not been involved in this biotransformation.

Further intense ions, at $m/z$ 483, 523 and 539, were observed when the TIC data were scrutinized manually. The RICs for these ions are shown in Figure 5.43. These masses are not expected from any of the biotransformations given in Table 5.13 and indicate significantly lower molecular weights than the parent drug.

The MS–MS spectrum of the $(M + H)^+$ ion from the parent drug contains an ion at $m/z$ 465, the structure of which is indicated in Figure 5.40. The mass spectrum of metabolite 1 indicates that it has a molecular weight of 482 Da, while the MS–MS spectrum from its $MH^+$ ion contains both an ion at $m/z$ 466 and at $m/z$ 364, also present in that from the $MH^+$ of the parent drug. It is not unreasonable, although not necessarily always correct, to assume that the ion of

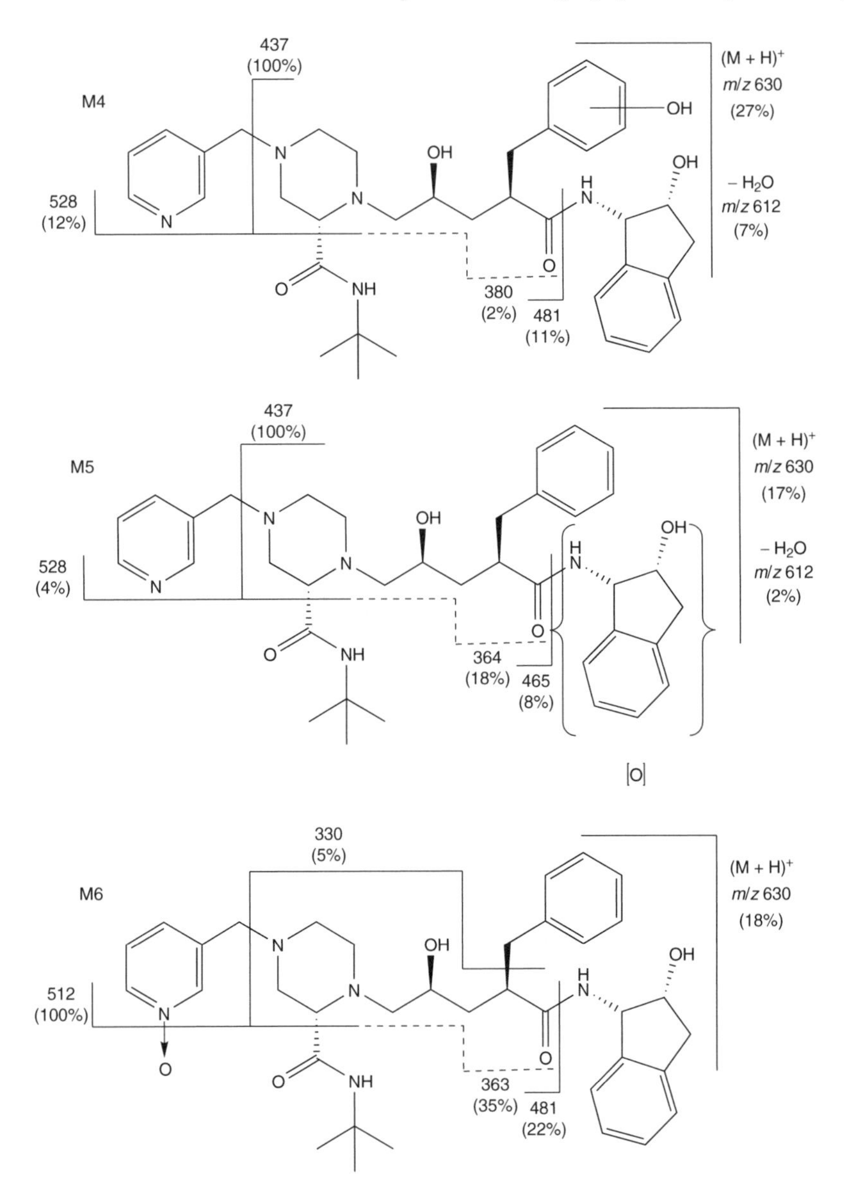


**Figure 5.42** Structures of three monhydroxylated metabolites of Indinavir proposed on the basis of the product-ion scans of their $(M+H)^+$ ions. Reprinted by permission of Elsevier Science from 'Identification of *in vitro* metabolites of Indinavir by "Intelligent Automated LC–MS/MS" (INTAMS) utilizing triple-quadrupole tandem mass spectrometry', by Yu, X., Cui, D. and Davis, M. R., *Journal of the American Society for Mass Spectrometry*, Vol. 10, pp. 175–183, Copyright 1999 by the American Society for Mass Spectrometry.

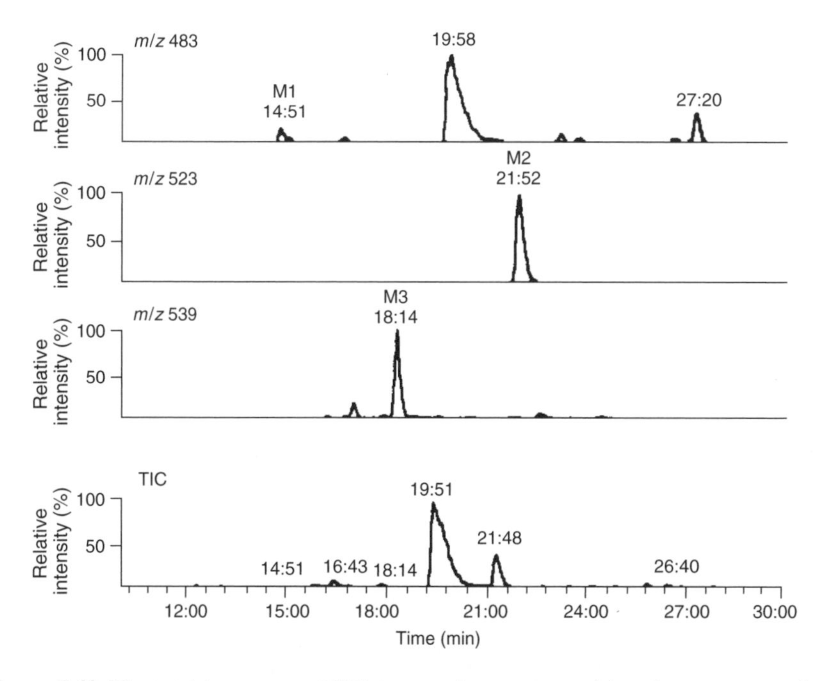


**Figure 5.43** The total-ion-current (TIC) trace and reconstructed ion chromatograms from intense ions at $m/z$ 483, 523 and 539 observed in spectra generated during the full-scan LC–MS analysis of an incubation of Indinavir with rat liver S9. Reprinted by permission of Elsevier Science from 'Identification of *in vitro* metabolites of Indinavir by "Intelligent Automated LC–MS/MS" (INTAMS) utilizing triple-quadrupole tandem mass spectrometry', by Yu, X., Cui, D. and Davis, M. R., *Journal of the American Society for Mass Spectrometry*, Vol. 10, pp. 175–183, Copyright 1999 by the American Society for Mass Spectrometry.

$m/z$ 466 is a protonated version of that at $m/z$ 465 and suggest that metabolite 1 has the structure shown in Figure 5.44.

The electrospray mass spectrum of metabolite 2 indicates it has a molecular weight of 522 Da, while the MS–MS spectrum of the $(M + H)^+$ ion contains an intense ion at $m/z$ 422, 1 Da greater than the base peak of the MS–MS spectrum of the protonated molecular ion of the parent drug. If we assume a similar relationship between these ions as assumed for $m/z$ 465 and $m/z$ 466 above, it is not unreasonable to postulate the structure of metabolite 2 to be that shown in Figure 5.44.

There are many similarities between the MS–MS spectra of the $(M + H)^+$ ions of metabolites 2 and 3, with the molecular weight of the latter being 16 Da greater than the former. Ions occur in both spectra at $m/z$ 374 and 273, thus suggesting that this additional oxygen atom has been incorporated into the indan-ring part of the molecule.

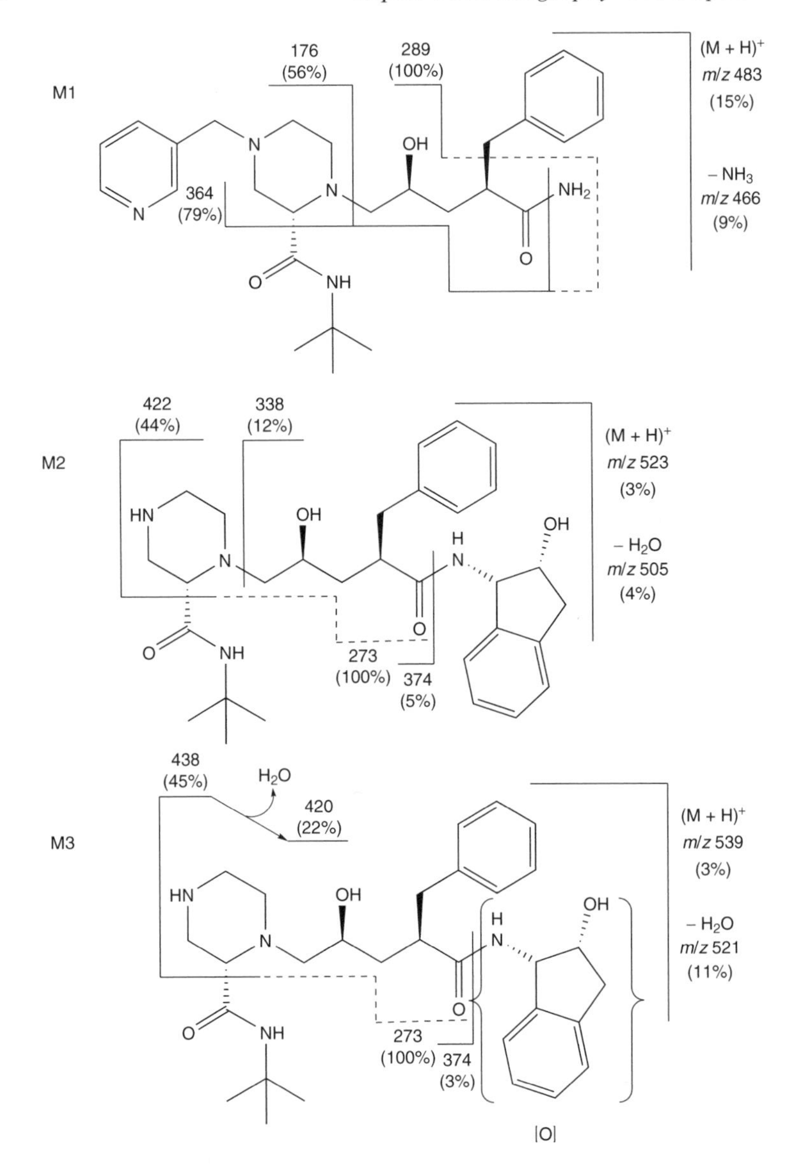


**Figure 5.44** Structures of metabolites of Indinavir proposed on the basis of the product-ion scans of the $(M + H)^+$ ions at $m/z$ 483, 523 and 539 obtained during the LC–MS analysis of an incubation of Indinavir with rat liver S9. Reprinted by permission of Elsevier Science from 'Identification of *in vitro* metabolites of Indinavir by "Intelligent Automated LC–MS/MS" (INTAMS) utilizing triple-quadrupole tandem mass spectrometry', by Yu, X., Cui, D. and Davis, M. R., *Journal of the American Society for Mass Spectrometry*, Vol. 10, pp. 175–183, Copyright 1999 by the American Society for Mass Spectrometry.

**Table 5.13** Mass changes associated with common biotransformation processes

| Mass difference (Da) | Biotransformation process |
|---|---|
| −28 | Demethylation × 2 |
| −16 | Loss of oxygen |
| −14 | Demethylation |
| −2 | Two-electron oxidation |
| +2 | Two-electron reduction |
| +14 | Addition of oxygen and two-electron oxidation |
| +16 | Addition of oxygen atom (hydroxylation) |
| +18 | Hydration |
| +30 | Addition of two oxygen atoms and two-electron oxidation (carboxylic acid formation) |
| +32 | Addition of two oxygen atoms (dihydroxylation) |
| +36 | Hydration × 2 |
| +80 | *O*-sulfate conjugation |
| +192 | *O*-glucuronide conjugation |

### *5.5.3 The Use of High-Accuracy Mass Measurements in Combination with LC–MS for the Structure Determination of Drug Metabolites*

One approach to drug metabolism studies is therefore to predict the molecular weights of possible metabolites of the drug under consideration, to use reconstructed ion chromatograms to locate any components that have the appropriate molecular weights and then use MS–MS to effect fragmentation of the $(M + H)^+$ ions from these metabolites, and then to finally link the $m/z$ values of the ions observed with ions of known structure from the parent drug or from other metabolites whose structures have been elucidated.

In the previous example, it was assumed that the ions of the same nominal $m/z$ ratio that appeared in the MS–MS spectra of the $(M + H)^+$ ions from different molecules had the same structure, and therefore the same atomic composition, and this assumption was borne out by the fact that this allowed the spectra to be rationalized in terms of the known metabolites of Indinavir.

This is not always the case, and the ability to use accurate mass measurements to confirm that certain ions do, or do not, have the same atomic composition would certainly be an advantage. As discussed earlier in Chapter 3, the instruments most widely used for MS–MS studies, i.e. the triple quadrupole and the ion-trap, do not routinely have accurate mass capability for product ions.

An MS–MS instrument only relatively recently made available commercially for LC–MS applications is the Q–ToF system, i.e. the combination of a quadrupole mass analyser for precursor-ion selection and a time-of-flight analyser for product-ion detection. As described earlier in Section 3.4.1.4, this instrument has the

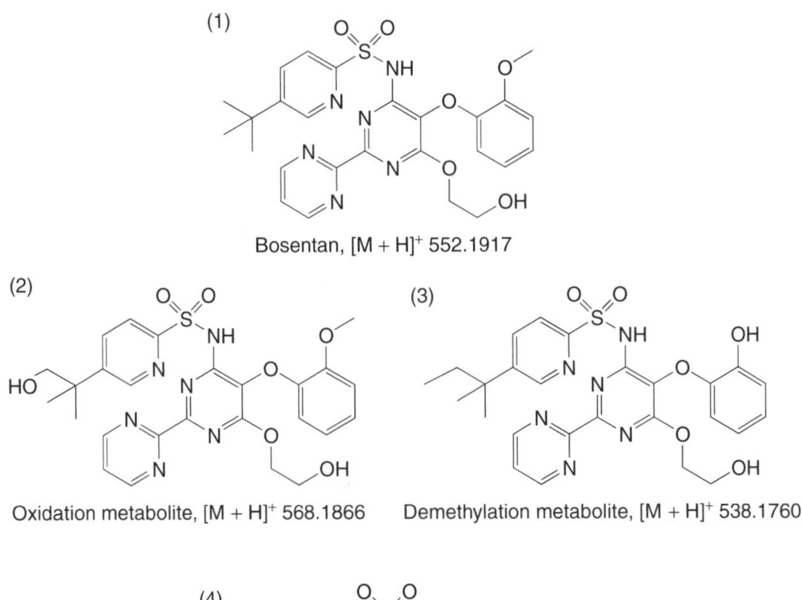


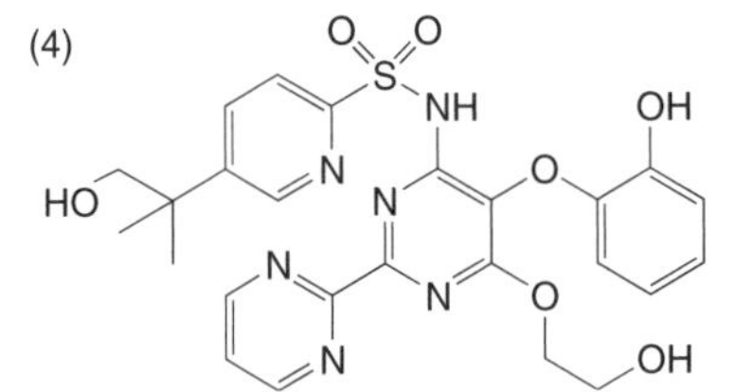


Oxidation and demethylation metabolite, $[M + H]^+$ 554.1709

**Figure 5.45** Structures of (1) Bosentan ($C_{27}H_{29}N_5O_6S$; $[M + H]^+$ 552.1917) and three of its metabolites, formed by (2) oxidation ($C_{27}H_{29}N_5O_7S$; $[M + H]^+$ 568.1866), (3) demethylation ($C_{26}H_{27}N_5O_6S$; $[M + H]^+$ 538.1760), and (4) demethylation–oxidation ($C_{26}H_{27}N_5O_7S$; $[M + M]^+$ 554.1709). Reprinted by permission of Elsevier Science from 'Exact mass measurement of product ions for the structural elucidation of drug metabolites with a tandem quadrupole orthogonal-acceleration time-of-flight mass spectrometer', by Hopfgartner, G., Chernushevich, I. V., Covey, T., Plomley, J. B. and Bonner, R., *Journal of the American Society for Mass Spectrometry*, Vol. 10, pp. 1305–1314, Copyright 1999 by the American Society for Mass Spectrometry.

capability of providing accurate mass measurements on ions that have been fragmented in the first stage of mass spectrometry.

The structure of 'Bosentan' [30] and three of its metabolites are shown in Figure 5.45 and the product-ion spectra from the $[M + H]^+$ ions from these compounds in Figure 5.46. All show an ion at $m/z$ 280 which might be assumed, simplistically, to share the same structure. Their accurate masses, determined by using a Q-ToF instrument, however, show that the ions from compounds (1)

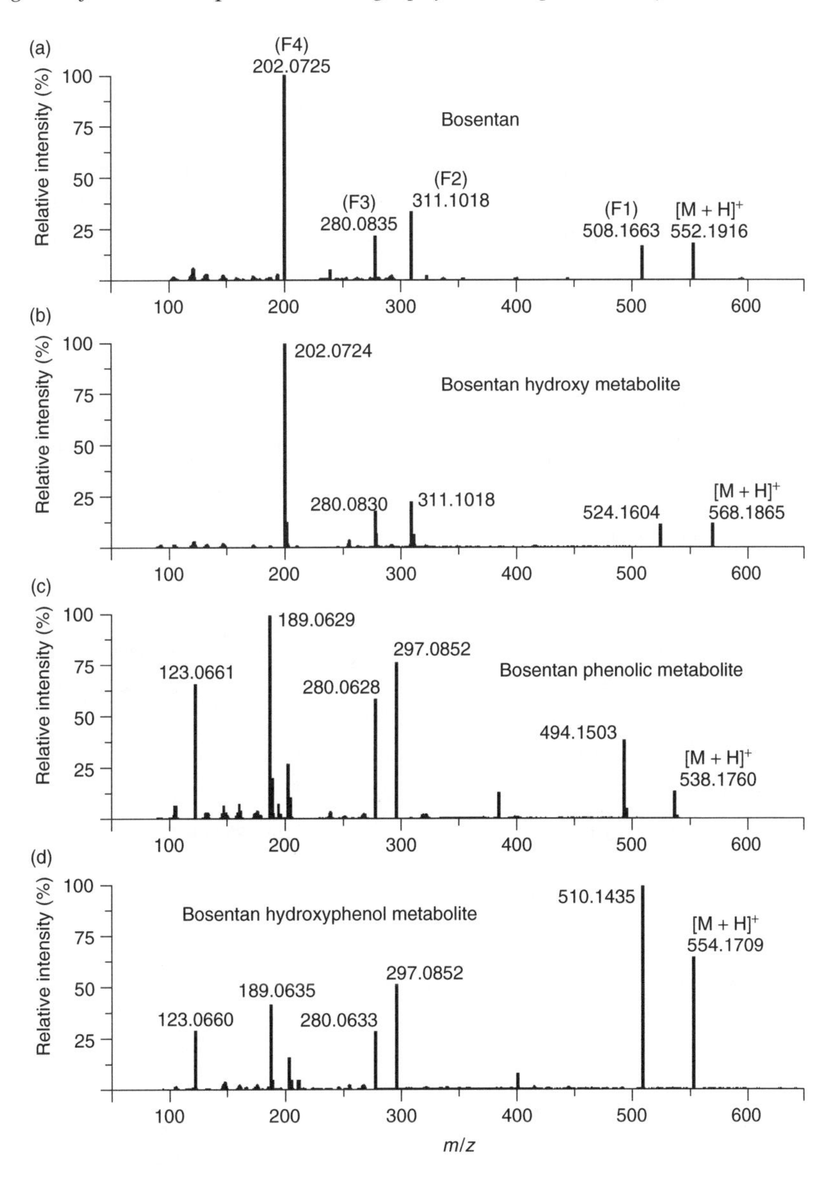


**Figure 5.46** Product-ion spectra of (a) Bosentan (1) $[M + H]^+$ 552.1917, and its (b) hydroxy metabolite (2) $[M + H]^+$ 568.1866, (c) phenol metabolite (3) $[M + H]^+$ 538.1760, and (d) hydroxyphenol metabolite (4) $[M + H]^+$ 554.1790. Reprinted by permission of Elsevier Science from 'Exact mass measurement of product ions for the structural elucidation of drug metabolites with a tandem quadrupole orthogonal-acceleration time-of-flight mass spectrometer', by Hopfgartner, G., Chernushevich, I. V., Covey, T., Plomley, J. B. and Bonner, R., *Journal of the American Society for Mass Spectrometry*, Vol. 10, pp. 1305–1314, Copyright 1999 by the American Society for Mass Spectrometry.

and (2) have masses of 280.0830 and 280.0835 Da, respectively, while those from compounds (3) and (4) have masses of 280.0628 and 280.0633 Da, respectively, a difference of some 20 mDa. This is sufficient to indicate that they do not have the same atomic composition, and therefore structure, and that the fragmentation of the $[M+H]^+$ ions from these two pairs of compounds is different.

Care must be taken in the interpretation of accurate mass data. In this case, the experimentally determined values of 280.0628 and 280.0633 Da have a mass

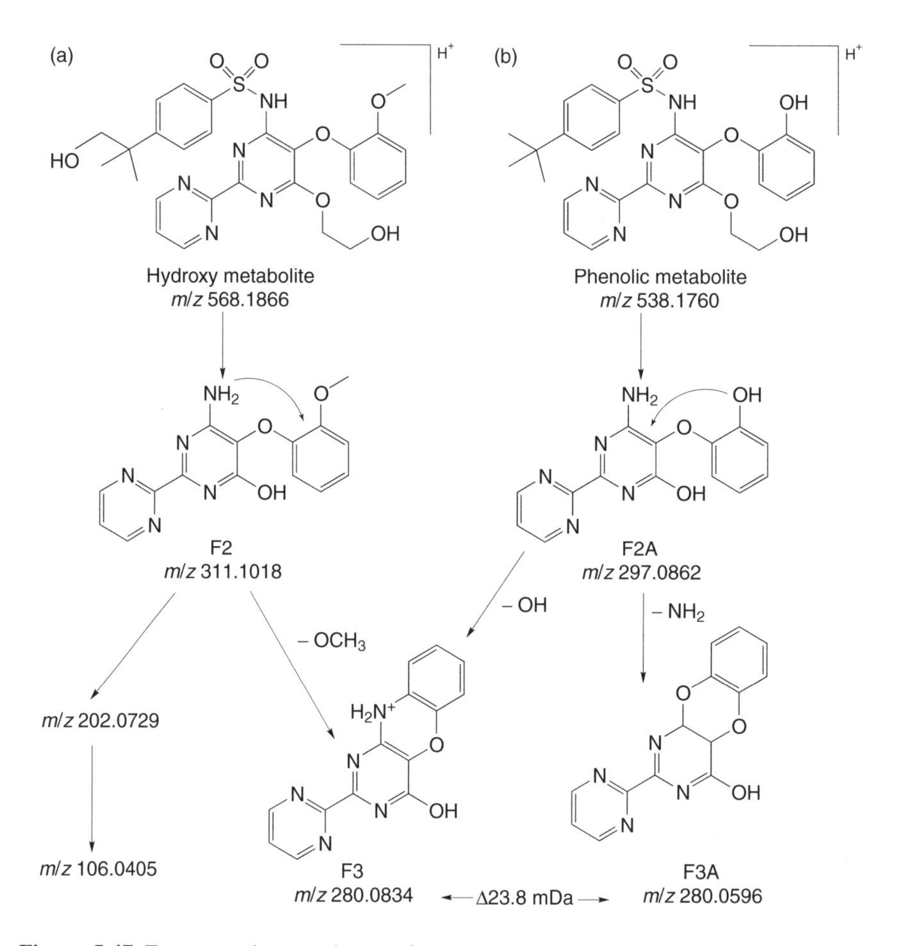


**Figure 5.47** Fragmentations pathways for (a) non-demethylated, and (b) demethylated metabolites of Bosentan. Reprinted by permission of Elsevier Science from 'Exact mass measurement of product ions for the structural elucidation of drug metabolites with a tandem quadrupole orthogonal-acceleration time-of-flight mass spectrometer', by Hopfgartner, G., Chernushevich, I. V., Covey, T., Plomley, J. B. and Bonner, R., *Journal of the American Society for Mass Spectrometry*, Vol. 10, pp. 1305–1314, Copyright 1999 by the American Society for Mass Spectrometry.

error of around 3.5 mDa with respect to the calculated mass for $C_{14}H_8N_4O_3$ of 280.0596 Da, compared to the ions of *m/z* 280.0830 and 280.0835 which are less than 0.5 mDa in error for the calculated mass of 280.0835 Da for the atomic composition $C_{14}H_{10}N_5O_2$. The latter error is more typical of the values that should be obtained and prompts a further consideration of the data obtained from the first two ions. It was suggested that the ion at around *m/z* 280.063 in fact consisted of two unresolved components of *m/z* 280.0596 and 280.0835, thus indicating that fragmentation of the $[M+H]^+$ ion occurred in two distinct ways, as shown in Figure 5.47.

**SAQ 5.7**

Calculate the resolution required to separate ions occurring at 280.0596 and 280.0835 Da.

## 5.5.4 *The Use of Cone-Voltage Fragmentation in Conjunction with High-Accuracy Mass Measurements and LC–MS for Metabolite Identification*

A slightly different use of accurate mass measurement in drug metabolism studies has been reported [31]. In this investigation, the accurate masses of the ions derived from 'Glyburide' (Figure 5.48) and some of its metabolites were used to calculate the difference in mass, and thus the elemental composition, between certain of the ions observed in the spectra of the various compounds.

The mass differences between the $[M+H]^+$ ion from Glyburide and each of the four metabolites studied were shown to correspond to a single oxygen atom with a determined mass difference very close to the 15.9949 Da required by theory, with the results obtained summarized in Table 5.14. The elemental composition of the $MH^+$ ion from each of these metabolites could therefore be assigned as $C_{23}H_{29}ClN_3O_6S$, based on the composition of the equivalent ion from the parent drug being $C_{23}H_{29}ClN_3O_5S$.

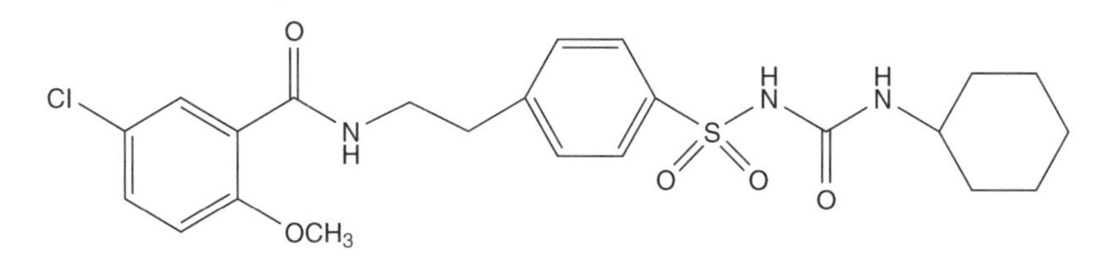


**Figure 5.48** Structure of Glyburide, with the elemental composition of $C_{23}H_{28}Cl\ N_3O_5S$ (exact mass of 493.1438 Da). Reprinted with permission from Zhang, H., Henion, J., Yang, Y. and Spooner, N., *Anal. Chem.*, **72**, 3342–3348 (2000). Copyright (2000) American Chemical Society.

**Table 5.14** Summary of the accurate mass differences (in Da) between the molecular ion of Glyburide and the molecular ions of five of its metabolites.[a] Reprinted with permission from Zhang, H., Henion, J., Yang, Y. and Spooner, N., *Anal. Chem.*, **72**, 3342–3348 (2000). Copyright (2000) American Chemical Society

| Metabolite | Injection | | | Average value | $CV^b$ (%) | $RE^c$ |
|---|---|---|---|---|---|---|
| | 1 | 2 | 3 | | | |
| M1 | 15.9948 | 15.9943 | 15.9974 | 15.9955 | 0.01 | $3.8 \times 10^{-5}$ |
| M2 | 15.9997 | 15.9922 | 15.9944 | 15.9954 | 0.02 | $3.3 \times 10^{-5}$ |
| M3 | 15.9945 | 15.9945 | 15.9962 | 15.9951 | 0.01 | $1.0 \times 10^{-5}$ |
| M4 | 15.9950 | 15.9962 | 15.9932 | 15.9948 | 0.01 | $-6.3 \times 10^{-6}$ |
| M5 | 15.9998 | 15.9945 | 15.9932 | 15.9958 | 0.02 | $5.8 \times 10^{-5}$ |

[a]The theoretical mass of oxygen is 15.9949 Da.
[b]Coefficient of variation.
[c]Relative error.

The use of cone-voltage fragmentation (CVF) to obtain structural information was discussed earlier in Section 4.7.4 and the comment was made there that because this has the potential to bring about fragmentation of all ions in the source of the mass spectrometer the spectra were often difficult to interpret. When the analyte under consideration is of a low molecular weight, its spectrum is likely to consist only of a few ions and CVF is of more practical value.

The structures of the Glyburide metabolites whose elemental compositions had been determined were investigated by using CVF. Separation of the metabolites was achieved by using a 2 mm × 100 mm $C_{18}$ column with gradient elution, employing a mobile phase of aqueous acetonitrile with ammonium acetate at pH 5. A flow rate of 200 $\mu l\,min^{-1}$ was used. The analysis methodology was as previously described, with the locations of the metabolites carried out by using an RIC of $m/z$ 510, i.e. 16 Da above the $[M + H]^+$ of the parent drug. The results of this analysis are shown in Figure 5.49. The mass spectrum of the drug was then obtained by using cone-voltage fragmentation and is shown in Figure 5.50. Of interest are the ions at $m/z$ 532 and $m/z$ 570 which may be attributed to the formation of $[M + K]^+$ and $[M + 2K - H]^+$ ions. It should also be noted that the spectrum is simple, thus facilitating interpretation, and the use of a Q-ToF instrument allows the accurate masses of the fragment ions to be determined. This gives an added confidence in proposing the formation of the potassium adducts and the fragmentation scheme shown in Figure 5.51.

The interpretation of the CVF spectra of the metabolites may be attempted by using a similar methodology to that described previously, i.e. subject to the cautionary note above, looking to see which ions occur at the same $m/z$ values in the spectra of the parent drug and its metabolites and which show significant mass variations. The CVF spectrum from one of the metabolites is shown in Figure 5.52. In addition to the potassium adducts noted in the spectrum from

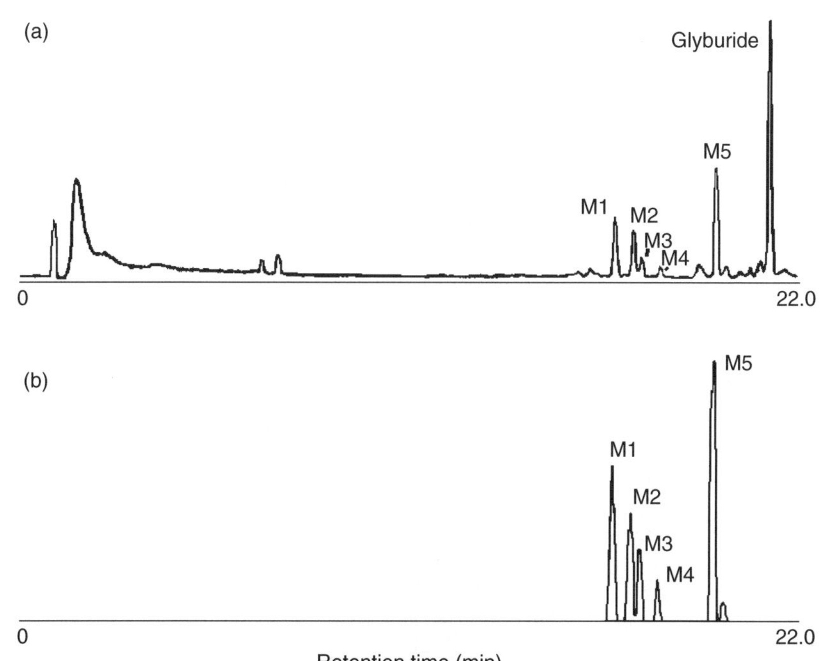


**Figure 5.49** (a) Total-ion-current trace, and (b) the reconstructed ion chromatogram of $m/z$ 510.2 ± 0.5 (mono-oxygenated metabolites) from LC–MS analysis of human microsomal incubation of Glyburide. Reprinted with permission from Zhang, H., Henion, J., Yang, Y. and Spooner, N., *Anal. Chem.*, **72**, 3342–3348 (2000). Copyright (2000) American Chemical Society.

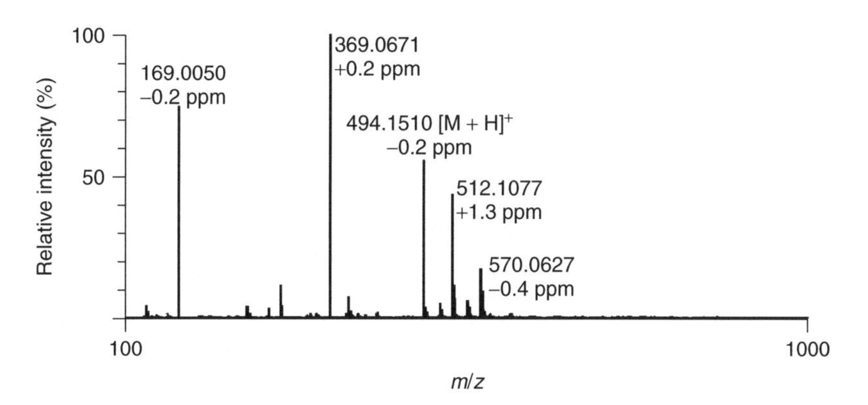


**Figure 5.50** Accurate-mass cone-voltage spectrum of Glyburide determined from *in vitro* incubation samples. Reprinted with permission from Zhang, H., Henion, J., Yang, Y. and Spooner, N., *Anal. Chem.*, **72**, 3342–3348 (2000). Copyright (2000) American Chemical Society.

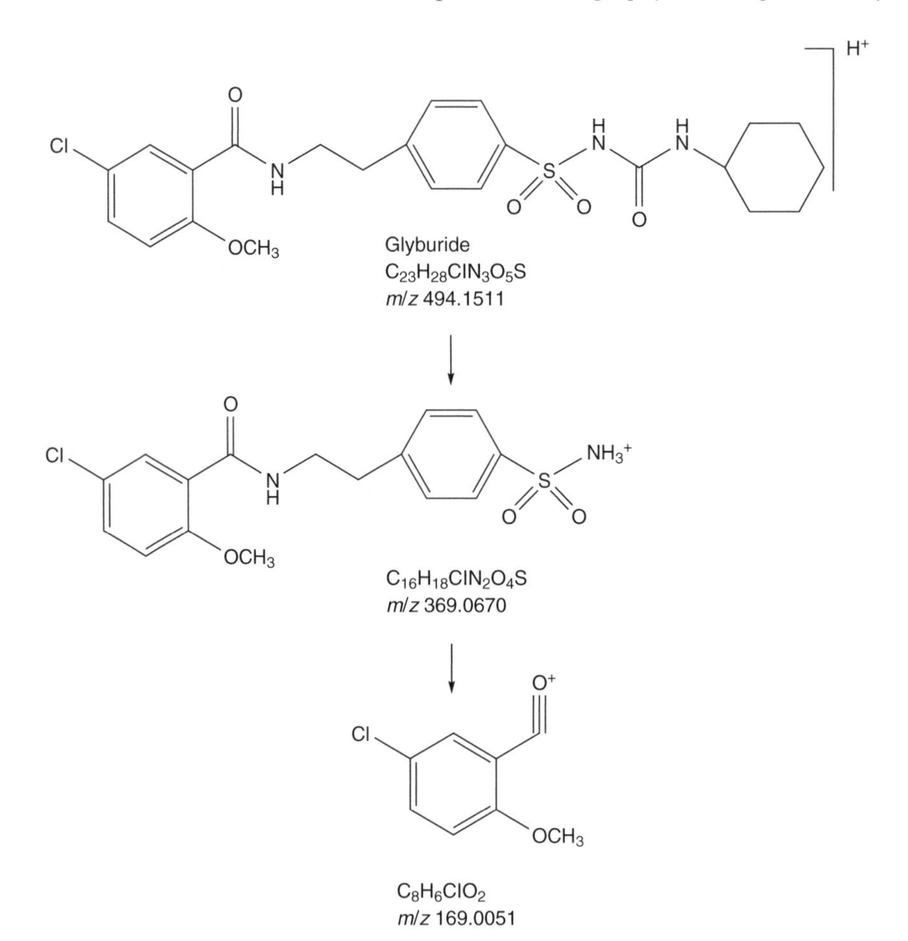


**Figure 5.51** Proposed fragmentation scheme for Glyburide. Reprinted with permission from Zhang, H., Henion, J., Yang, Y. and Spooner, N., *Anal. Chem.*, **72**, 3342–3348 (2000). Copyright (2000) American Chemical Society.

the parent drug, this spectrum contains ions at $m/z$ 169.0045 – also observed in the spectrum of the parent drug – $m/z$ 385.0624 and $m/z$ 367.0504 (with $m/z$ 369.0671 being observed in the spectrum of the parent drug). Only one theoretical elemental composition matching the experimentally determined value within 9 ppm was found for the ions of $m/z$ 169.0045 and 385.0624 – these being $C_8H_6ClO_2$ and $C_{16}H_{18}N_2O_5SCl$, respectively. The ion of $m/z$ 367.0504 yielded two possible atomic compositions, i.e. $C_{16}H_{16}N_2O_4SCl$ and $C_{19}H_{12}N_2O_4Cl$, with the former being chosen on the basis of the relative abundance of the ion at $m/z$ 369 being consistent with the presence of both chlorine and sulfur. The ion of $m/z$ 385 was considered to arise from elimination of cyclohexylisocyanate and that at

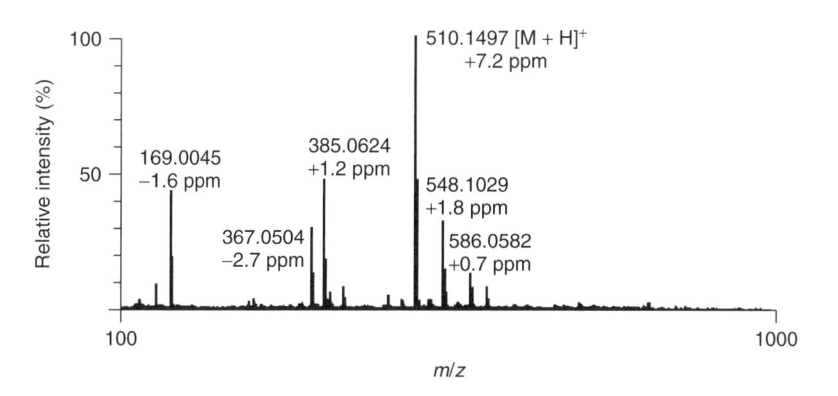


**Figure 5.52** Accurate-mass cone-voltage spectrum of the ethylhydroxy metabolite of Glyburide, formed *in vitro* by human liver microsomes. Reprinted with permission from Zhang, H., Henion, J., Yang, Y. and Spooner, N., *Anal. Chem.*, **72**, 3342–3348 (2000). Copyright (2000) American Chemical Society.

$m/z$ 367 from elimination of both cyclohexylisocyanate and water. It was then argued that the elimination of the isocyanate indicated that metabolism had not occurred on the cyclohexyl ring and from the elimination of water that metabolism had not occurred on either of the aromatic rings – the ion at $m/z$ 169 being evidence for the presence of the unchanged chlorine-containing aromatic ring. It was proposed that the structure of the metabolite was as shown in Figure 5.53, which also shows a rationale for the CVF fragmentation observed.

### 5.5.5 The Use of LC–MS$^n$ for the Identification of Drug Metabolites

The examples chosen above have illustrated how CVF or MS–MS may be used to generate useful structural information but these do not always provide sufficient detail to allow an unequivocal structural assignment. There may still be instances where it might be necessary to probe fragmentation pathways further. This can be accomplished by combining MS–MS with CVF, i.e. use CVF to effect fragmentation of an ion of interest and then study one of the product ions so formed by using conventional MS–MS. This may be considered to be 'MS–MS–MS'.

One of the features of an ion-trap is that ion selection is carried out in *time* rather than *space*. In this type of instrument, MS–MS data are generated by ionizing the analyte of interest in the normal way but then, instead of causing ions of all $m/z$ values to become unstable and reach the detector, ions other than those being studied by MS–MS are ejected from the trap. The selected ion is then caused to fragment, in the trap, and the ions so generated are made unstable in order to generate the MS–MS spectrum. The procedure may then be

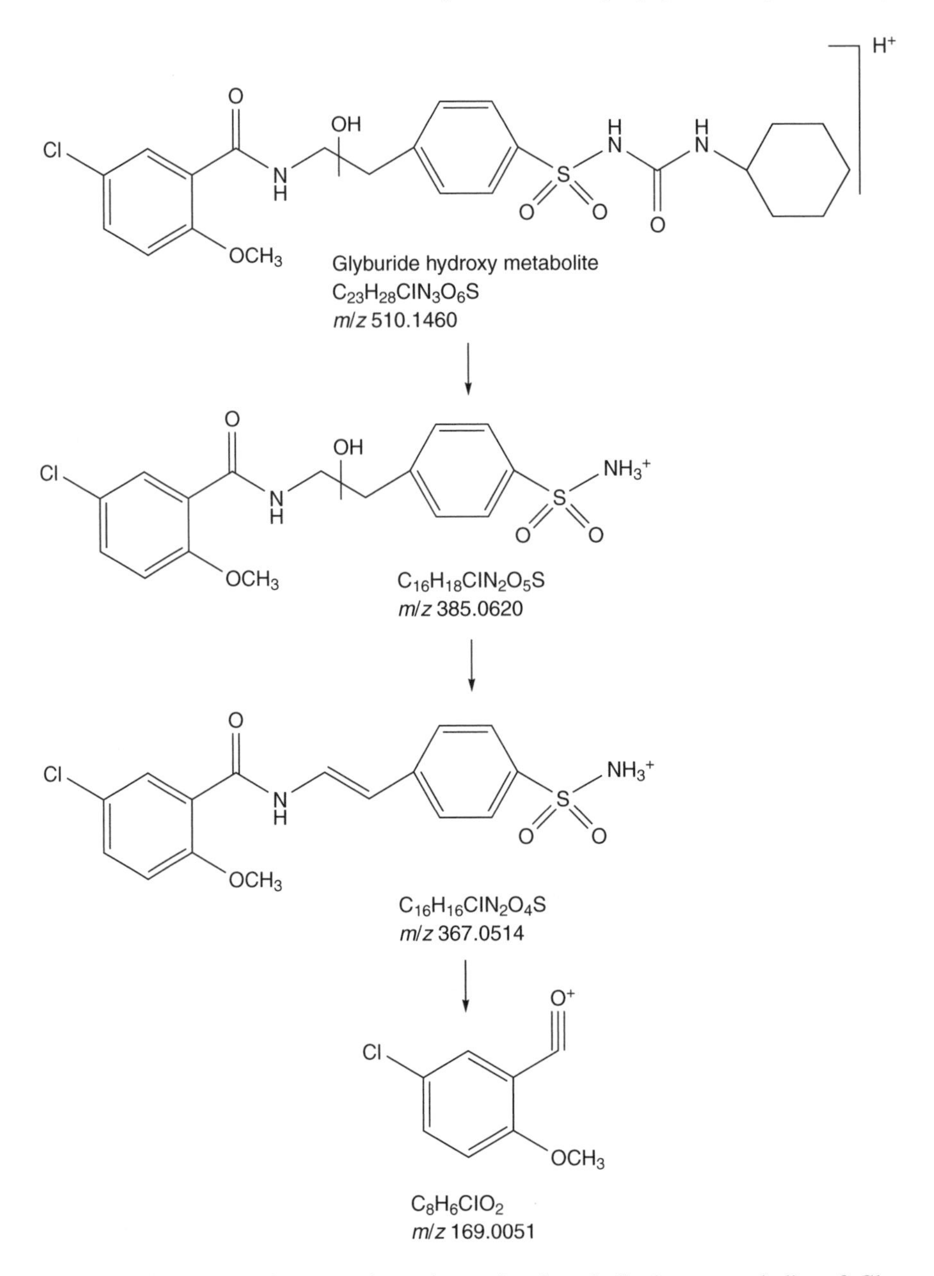


**Figure 5.53** Proposed fragmentation scheme for the ethylhydroxy metabolite of Glyburide. Reprinted with permission from Zhang, H., Henion, J., Yang, Y. and Spooner, N., *Anal. Chem.*, **72**, 3342–3348 (2000). Copyright (2000) American Chemical Society.

repeated, i.e. a single ion produced during the first stage of MS–MS may then be selected – ions of all other $m/z$ values being ejected from the trap – and this ion then caused to fragment. An MS–MS–MS (or $MS^3$) spectrum may then be generated. This process may again be repeated by using one of the ions generated using $MS^3$ to give an $MS^4$ spectrum, etc. In this way, more detailed information on the structure of the ions involved may be obtained. These techniques are often referred to as $MS^n$ to indicate multiple stages of MS–MS.

The metabolism of the chiral drug 'Praziquantel', the structure of which is shown in Figure 5.54, was investigated by using multiple stages of MS–MS. The drug is administered as a racemate but the activity is associated mainly with the R- (−) enantiomer. Biotransformation yields five metabolites which were studied by using a combination of capillary electrophoresis and LC–MS [32], with the LC–MS studies involving separation of the metabolites produced during the incubation of Praziquantel with rat liver microsomes.

The mass spectrometry employed electrospray ionization and each metabolite gave an $[M + H]^+$ ion which was then used as a precursor ion for a product-ion MS–MS scan. For subsequent $MS^n$ experiments, the base peak of the previous MS–MS experiment was chosen under computer control and this allowed all analytes to be studied in a single chromatographic separation.

The protonated molecular ion of metabolite M2 occurred at $m/z$ 329, an increase of 16 Da over the parent drug, thus suggesting that an hydroxyl group had been formed. The product-ion MS–MS spectrum of this ion gave a base peak at $m/z$ 311 which was then used as the precursor ion for $MS^3$, while $MS^4$ was carried out on $m/z$ 283 and $MS^5$ on $m/z$ 173. The resulting spectra are shown in Figure 5.55. A detailed analysis of these spectra allowed the authors to suggest, on the basis of an ion appearing in the $MS^4$ spectrum, that initial hydroxylation was likely to have occurred on positions 6 or 7 of the parent compound.

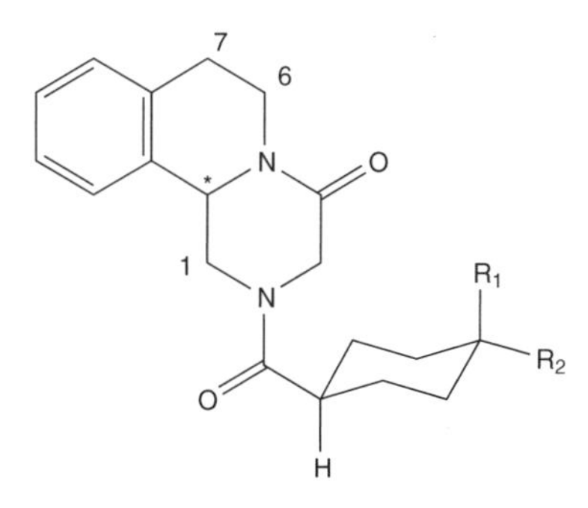


| Compound | $R_1$ | $R_2$ |
|---|---|---|
| Praziquantel (1) | H | H |
| *trans*-4-Hydroxypraziquantel (2) | H | OH |
| *cis*-4-Hydroxypraziquantel (3) | OH | H |

**Figure 5.54** Structures of Praziquantel and its metabolites, *cis*- and *trans*-4-hydroxypraziquantel. Reprinted from *J. Chromatogr., B*, **708**, Lerch, C. and Blaschke, G., 'Investigation of the stereoselective metabolism of Praziquantel after incubation with rat liver microsomes by capillary electrophoresis and liquid chromatography–mass spectrometry', 267–275, Copyright (1998), with permission from Elsevier Science.

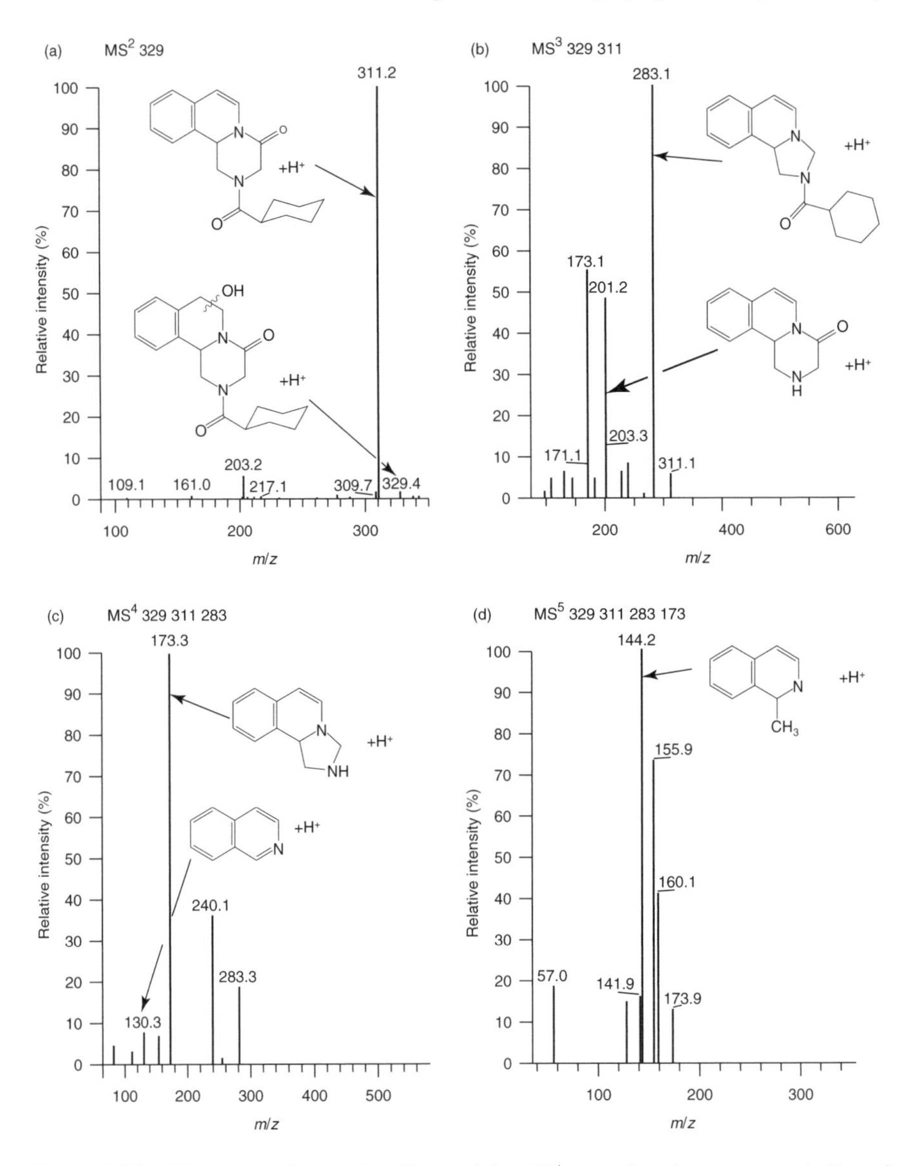


**Figure 5.55** MS$^n$ spectra from $m/z$ 329, the $(M + H)^+$ ion of an hydroxy metabolite of Praziquantel, and the base peaks in subsequent product-ion spectra. Reprinted from *J. Chromatogr., B*, **708**, Lerch, C. and Blaschke, G., 'Investigation of the stereoselective metabolism of Praziquantel after incubation with rat liver microsomes by capillary electrophoresis and liquid chromatography–mass spectrometry', 267–275, Copyright (1998), with permission from Elsevier Science.

An important feature of this is that the mass spectrometer had sufficient sensitivity to obtain three levels of MS–MS spectra during the elution of an HPLC peak and hence yield useful analytical information.

The above selected examples demonstrate the importance of APCI and electrospray in the field of drug metabolism studies but these ionization techniques have been used extensively for structural studies involving environmental pollutants, such as pesticides, and for impurity profiling where all materials present in pharmaceutical formulations at a level greater than 0.1% must be characterized fully. The general procedures employed do not vary, however, as in many cases the structures of the compounds under investigation are not totally unknown. In the study of drug metabolism, it is possible to make certain predictions and this is also usually equally possible in impurity profiling when reagents and experimental conditions are known.

The general procedure is to use reconstructed ion chromatograms at appropriate $m/z$ values in an attempt to locate compounds of interest and then look at the mass spectrum of the unknown to determine its molecular weight. MS–MS can then be employed to obtain spectra from this and related compounds to find ions that are common to both and which may therefore contain common structural features. Having the same $m/z$ value does not necessarily mean the ions are identical and further MS–MS data or the elemental composition may be needed. If these data do not allow unequivocal structure identification, then further $MS^n$ information may be required.

In addition, synthesis of reference compounds and a comparison of HPLC retention characteristics and mass spectral data is often required.

## 5.6 Quantitation

The development of a quantitative method involving LC–MS is, in principle, no different from developing a quantitative method using any other analytical technique; the intensity of signal from the analyte(s) of interest in the 'unknown' sample is compared with that from known amounts of the analyte. The task of the analyst is to decide how this is best achieved knowing the resources available and the purpose for which the results are required.

### *5.6.1 Requirements of a Quantitative Method Involving LC–MS*

The attributes required of a method usually include good sensitivity, low limits of detection, and selectivity. It must be recognized that while low limits of detection will usually require good sensitivity, the latter, in itself, does not guarantee low limits of detection since these are often determined by the levels of interfering materials present and not the absolute sensitivity of the technique. Low limits of detection allow the analyte to be determined at levels at or below those considered to be harmful or prescribed by legislation or at which it is found in a particular matrix. Good selectivity allows the measured signal to be assigned, with certainty, to the analyte of interest rather than any interfering compounds which may be present. Without this certainty, any interpretation of the analytical data may be questioned. The selectivity of the method may often be used to advantage to simplify sample work-up prior to analysis.

The issue of selectivity is one that is often difficult to address. Initial method development is invariably carried out by using standards made up with pure solvents, i.e. free from any matrix effects. It is often only when 'real' samples are analysed that the true extent of interference becomes apparent and the value of the method can be properly assessed. An added complication is that 'interferences', by their very nature, are not constant and a number of samples may have a combination of interferences that defy analysis by a method that is otherwise successful on a routine basis (another example of Murphy's law!).

Other features of an analytical method that should be borne in mind are its linear range, which should be as large as possible to allow samples containing a wide range of analyte concentrations to be analysed without further manipulation, and its precision and accuracy. Method development and validation require all of these parameters to be studied and assessed quantitatively.

Quantitative methodology employing mass spectrometry usually involves selected-ion monitoring (see Section 3.5.2.1) or selected-decomposition monitoring (see Section 3.4.2.4) in which a small number of ions or decompositions of ions specific to the compound(s) of interest are monitored. It is the role of the analyst to choose these ions/decompositions, in association with chromatographic performance, to provide sensitivity and selectivity such that when incorporated into a method the required analyses may be carried out with adequate precision and accuracy.

How then are these ions/decompositions chosen? Before considering this we must define, very carefully, the requirements of the analysis to be carried out. Is a single compound to be determined or are a number of compounds of interest? If a single compound is involved, its mass spectrum and MS–MS spectra can be obtained and scrutinized for any appropriate ions or decompositions. If the requirement is to determine a number of analytes, their chromatographic properties need to be considered. If they are well separated, different ions/decompositions can be monitored for discrete time-periods as each compound elutes, thus obtaining the maximum sensitivity for each analyte. If the analytes are not well separated, this approach may not be possible and it may then be necessary to monitor a number of ions/decompositions for the complete duration of the analysis. If this is the case, the analyst should attempt to find the smallest number of ions/decompositions that give adequate performance for all of the analytes (remember the more ions/decompositions monitored, then the lower the overall sensitivity will be).

The specificity of the ions/decompositions must be considered. Both electrospray ionization and APCI are soft ionization techniques and the resulting mass spectrum of an analyte is likely to consist of a single ion. The analyst must consider whether the presence of this single ion, together with the retention time of the analyte, is sufficiently specific. Since this ion is a molecular species, it is certainly more discriminating than a single ion from an electron ionization (EI) mass spectrum. Selected-decomposition monitoring is inherently more specific in that two ions, a precursor (often a molecular species) and a product ion with a known relative intensity, as well as a retention time, are involved.

### 5.6.2 Quantitative Standardization

Adequate precision and accuracy are only likely to be achieved if some standardization procedure is employed and the nature of this, internal or external standards or the method of standard additions, needs to be chosen carefully. If internal standardization procedures are adopted then appropriate compound(s) must be chosen and their effect on the chromatographic and mass spectrometry methods assessed. The ideal internal standard is an isotopically labelled analogue of the analyte but, although there are a number of commercial companies who produce a range of such molecules, these are not always readily available. An analytical laboratory is then faced with the choice of carrying out the synthesis of the internal standard themselves or choosing a less appropriate alternative with implications on the accuracy and precision of the method to be developed.

### 5.6.3 Matrix Effects in LC–MS

In many cases when methods involve internal or external standards, the solutions used to construct the calibration graph are made up in pure solvents and the signal intensities obtained will not reflect any interaction of the analyte and internal standard with the matrix found in 'unknown' samples or the effect that the matrix may have on the performance of the mass spectrometer. One way of overcoming this is to make up the calibration standards in solutions thought to reflect the matrix in which the samples are found. The major limitation of this is that the composition of the matrix may well vary widely and there can be no guarantee that the matrix effects found in the sample to be determined are identical to those in the calibration standards.

It is well known that electrospray ionization (EI) suffers from suppression effects when polar/ionic compounds other than the analyte(s) of interest, such as those originating from the sample matrix, are present, with this phenomenon being attributed to competitive ionization of all of the appropriate species present [33]. Matrix effects can, therefore, be considerable and these have two distinct implications for quantitative procedures, as follows:

(i) a loss of absolute sensitivity may render some or all of the analytes of interest undetectable;

(ii) the accuracy and precision of the determination may well be affected.

One approach to the problem of matrix effects is to prevent the matrix materials reaching the electrospray source by carrying out some form of clean-up prior to analysis and/or to employ chromatographic separation. It is not always possible, however, to develop a simple procedure for sample clean-up and since this approach involves further work-up with the associated increase in analysis time and potential for sample loss it is therefore not ideal.

The extent of matrix effects on the analysis of three pesticides, i.e. G-fenozide, hydroxyfenozide and methoxyfenozide (Figure 5.56), using LC–MS with electrospray ionization, has been investigated [34, 35]. A matrix extract was prepared from wheat forage material and then spiked with the analytes prior to LC–MS–MS analysis. Separation was carried out by using a 25 cm × 3 mm 5μ $C_{18}$ column with gradient elution at a flow rate of 0.5 ml $min^{-1}$, with a post-column split being employed to allow approximately 50 μl $min^{-1}$ of the column eluate to enter the electrospray source. The initial mobile phase was water:acetonitrile (80/20), decreasing to 50% water at 6 min, 30% water at 10 min and 5% water at 16.5 min, returning to 80% water at 20 min. The column eluate for the first 8 min was sent to waste to minimize possible contamination of the electrospray interface, with the retention time of the first eluting analyte, i.e. G-fenozide, being 9 min.

It was found (Table 5.15) that increasing the sample injection volume brought about a significant increase in the amount of signal suppression observed for all of the analytes studied. The analytes were monitored in negative-ion mode using the decomposition from the $[M - 1]^-$ ion to $m/z$ 149 for hydroxy- and methoxyfenozide and the $[M - 1]^-$ ion to $m/z$ 353 for G-fenozide, with the differences in the signals obtained when 50 μl volumes of 0.02 μg $ml^{-1}$ solutions of the analyte made up in acetonitrile:water (1/1) and in wheat forage extract being shown in Figure 5.57. In each case, the signal from the sample dissolved in the matrix showed a decrease of approximately 90%. The origin of this reduction was probed further by post-column infusion of analyte into the HPLC eluate after

**Table 5.15** Relative signal responses[a] from various injection volumes for the LC–MS–MS analysis of a wheat forage matrix sample. Reprinted from *J. Chromatogr., A*, **907**, Choi, B. K., Hercules, D. M. and Gusev, A. I., 'Effect of liquid chromatography separation of complex matrices on liquid chromatography–tandem mass spectrometry signal suppression', 337–342, Copyright (2001), with permission from Elsevier Science

| Analyte | Injection volume (μl) | | | |
|---|---|---|---|---|
| | 5 | 10 | 50 | 100 |
| Methoxyfenozide | 88.8 | 91.3 | 57.5 | 55.8 |
| Hydroxyfenozide | 91.8 | 92.5 | 58.5 | 44.8 |
| G-fenozide | 63.6 | 65.2 | 30.1 | 27.2 |

[a] Signal response is expressed as a percentage of that obtained from standard samples; 100% is indicative of no signal suppression.

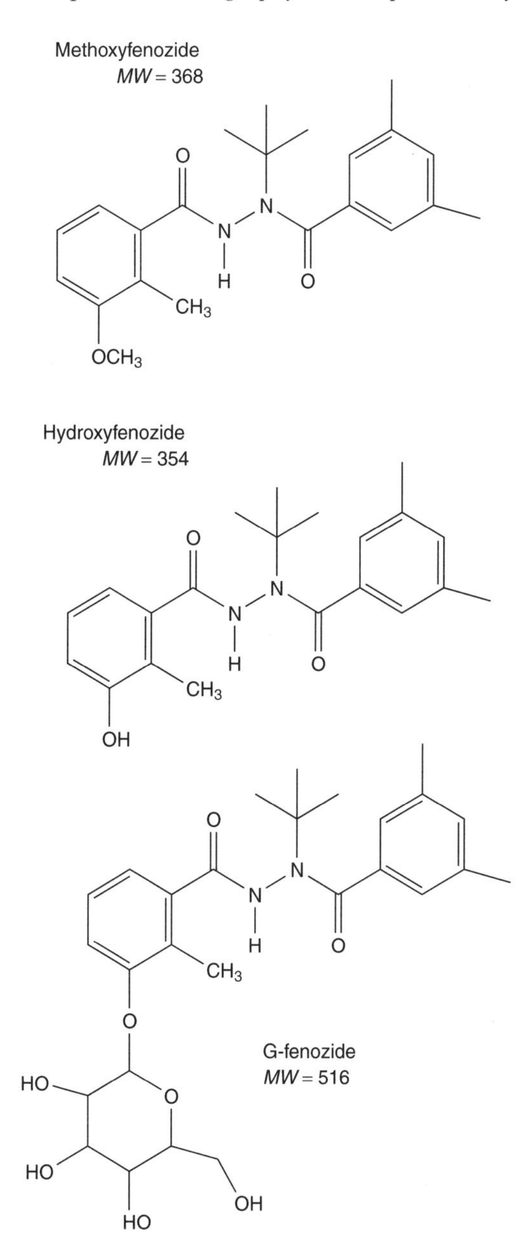


**Figure 5.56** Structures of the three analytes pesticides used in an investigation of the matrix effects observed in LC–MS–MS. Reprinted from *J. Chromatogr., A*, **907**, Choi, B. K., Hercules, D. M. and Gusev, A. I., 'Effect of liquid chromatography separation of complex matrices on liquid chromatography–tandem mass spectrometry signal suppression', 337–342, Copyright (2001), with permission from Elsevier Science.

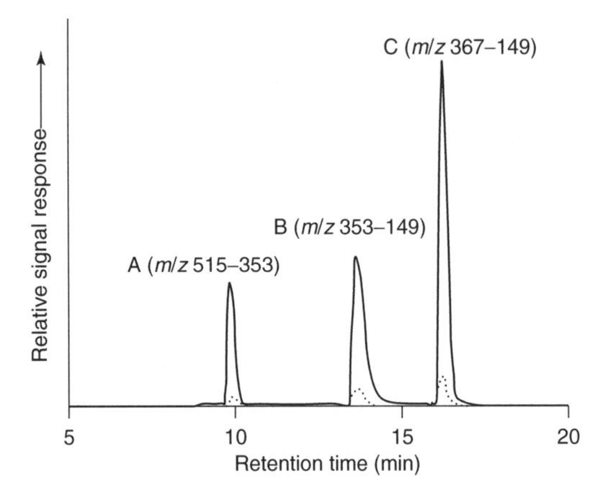


**Figure 5.57** Reconstructed ion chromatograms from the negative-ion LC–MS–MS analysis of (A) G-fenozide, (B) hydroxyfenozide, and (C) methoxyfenozide (for structures, see Figure 5.56) acquired from standard (——) and matrix (· · ·) samples. Reprinted from *J. Chromatogr., A*, **907**, Choi, B. K., Hercules, D. M. and Gusev, A. I., 'Effect of liquid chromatography separation of complex matrices on liquid chromatography–tandem mass spectrometry signal suppression', 337–342, Copyright (2001), with permission from Elsevier Science.

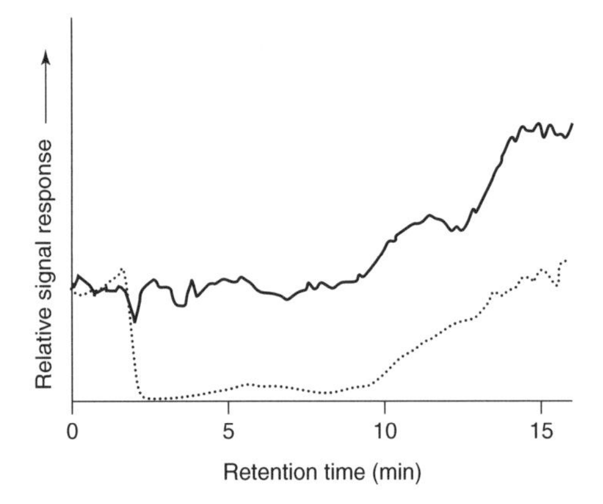


**Figure 5.58** Reconstructed LC–MS–MS ion chromatograms for selected-reaction monitoring of methoxyfenozide using the $m/z$ 367 to $m/z$ 149 transition from the continual post-column infusion of a standard solution of analyte during the HPLC analysis of a blank wash (acetonitrile:water) (——) and a matrix extract (· · ·). Reprinted from *J. Chromatogr., A*, **907**, Choi, B. K., Hercules, D. M. and Gusev, A. I., 'Effect of liquid chromatography separation of complex matrices on liquid chromatography–tandem mass spectrometry signal suppression', 337–342, Copyright (2001), with permission from Elsevier Science.

50 μl injections of a wash of acetonitrile:water (1/1) and of the matrix preparation. Examination of the results presented in Figure 5.58 indicates that no suppression is observed until ca. 2 min into the analysis. The degree of suppression then remains virtually constant until ca. 9 min into the analysis, after which it gradually decreases until after ca. 15 min the signal from the injection of matrix material is around 60% of that from the injection of acetonitrile:water. This behaviour leads the authors to suggest that the suppression, rather than being due to co-elution of analyte and interference, is due to overloading of the HPLC column which results in a continuous elution of matrix materials and continuous suppression. This explanation was supported by reduced suppression effects being observed when the matrix extract was submitted to off-line two-dimensional HPLC clean-up. It was also found that the suppression effects could be altered by the incorporation of different buffers into the mobile phase, as shown in Table 5.16. If this course of action is to be contemplated, the effect of these on the HPLC separation must also be considered.

This work [34, 35] reinforces the view that chromatographic separation itself is not always sufficient to effect adequate clean-up to remove all matrix effects but that these will be reduced if the minimum possible injection volume is used.

**Table 5.16** LC–MS–MS signal responses[a] obtained from wheat forage matrix samples using various mobile-phase additives (injection volumes of 50 μl). From Choi, B. K., Hercules, D. M. and Gusev, A. I., 'LC–MS/MS signal suppression effects in the analysis of pesticides in complex environmental matrices', *Fresenius' J. Anal. Chem.*, **369**, 370–377, Table 2, 2001. © Springer-Verlag GmbH & Co. KG. Reproduced with permission

| Analyte | Ion mode | Acetonitrile/ water | Additive | | | |
|---|---|---|---|---|---|---|
| | | | Formic acid[b] | Ammonium hydroxide[b] | Ammonium acetate[c] | Ammonium formate[c] |
| Methoxy-fenozide | Negative | 9.85 | 77.8 | 67.2 | 49.8 | 73.0 |
| Hydroxy-fenozide | Negative | 12.50 | 77.9 | 70.9 | 44.6 | 81.6 |
| G-fenozide | Negative | 5.30 | 55.2 | 59.1 | 32.0 | 49.3 |
| Methoxy-fenozide | Positive | 5.82 | 63.7 | 58.9 | 97.7 | 87.4 |
| Hydroxy-fenozide | Positive | 3.35 | 68.7 | 39.5 | 89.5 | 86.7 |
| Hydroxy-fenbucon-azole | Positive | 7.25 | 70.1 | 56.4 | 77.3 | 73.7 |

[a]Signal response is expressed as a percentage of that obtained from standard samples; 100% is indicative of no signal suppression.
[b]Concentration of 0.01 vol%.
[c]Concentration of 1 mM.

### *5.6.4 The Method of Standard Additions to Overcome Matrix Effects*

A method for the analysis of four diarrhetic shellfish poisoning toxins, namely okadaic acid (OA), dinophysistoxin-1 (DTX1), pectenotoxin-6 (PTX6) and yessotoxin (YTX) (Figure 5.59), in scallops using LC–electrospray-MS has been reported [36]. The intensity of the molecular species, $[M - 2Na + H]^-$ for YTX and $[M - H]^-$ for OA, DTX1 and PTX6, were monitored after injection in either pure methanol or a solution of extract from poison-free scallop into a 150 × 2.1 mm, 5 μm ODS2 column. Elution was accomplished with a linear gradient of a mobile phase composed of methanol and 1 mM ammonium acetate in water at a flow rate of 200 $\mu l\,min^{-1}$. Good linear calibration graphs over the complete concentration range studied, up to 500 $\mu g\,l^{-1}$, were obtained for both sets of data with $R^2$ values between 0.9991 and 0.9999[†] The intensity of the signal for the analyte in the scallop extract was lower in each case, as shown in Figure 5.60.

Four methods of overcoming suppression effects were discussed in this paper, as follows:

(i) Complete removal of co-eluting substances by sample clean-up – time consuming and difficult, if not impossible, if the matrix is complex.

(ii) The use of *internal standards* – this would improve the precision but, if the calibration standards were made up in pure solvents, would not necessarily improve the accuracy. In this particular example, appropriate internal standards were not, in any case, available.

(iii) Make the calibration standards up in a matrix extract rather than in a pure solvent. The problem with this approach is that the composition of such an extract cannot be guaranteed to be identical to that in which the analytes to be determined are found and the extent to which matrix effects are being corrected cannot be certain. Figure 5.61 shows the quantitative results obtained when three different matrix extracts spiked with 200 $\mu g\,l^{-1}$ of each of the four toxins were analysed and illustrates the variability that may be encountered if this methodology is employed.

---

**Figure 5.59** Molecular structures of the diarrhetic shellfish poisons: (a) pectenotoxin-6 (PTX6); (b) okadaic acid (OA); (c) dinophysistoxin-1 (DTX1); (d) yessotoxin (YTX). Reprinted from *J. Chromatogr., A*, **943**, 'Matrix effect and correction by standard addition in quantitative liquid chromatographic–mass spectrometric analysis of diarrhetic shellfish poisoning toxins', Ito, S. and Tsukada, K., 39–46, Copyright (2002), with permission from Elsevier Science.

[†] $R$ (see Figure 5.60) is known as the *correlation coefficient*, and provides a measure of the quality of calibration. In fact, $R^2$ (the *coefficient of determination*) is used because it is more sensitive to changes. This varies between −1 and +1, with values very close to −1 and +1 pointing to a very tight 'fit' of the calibration curve.

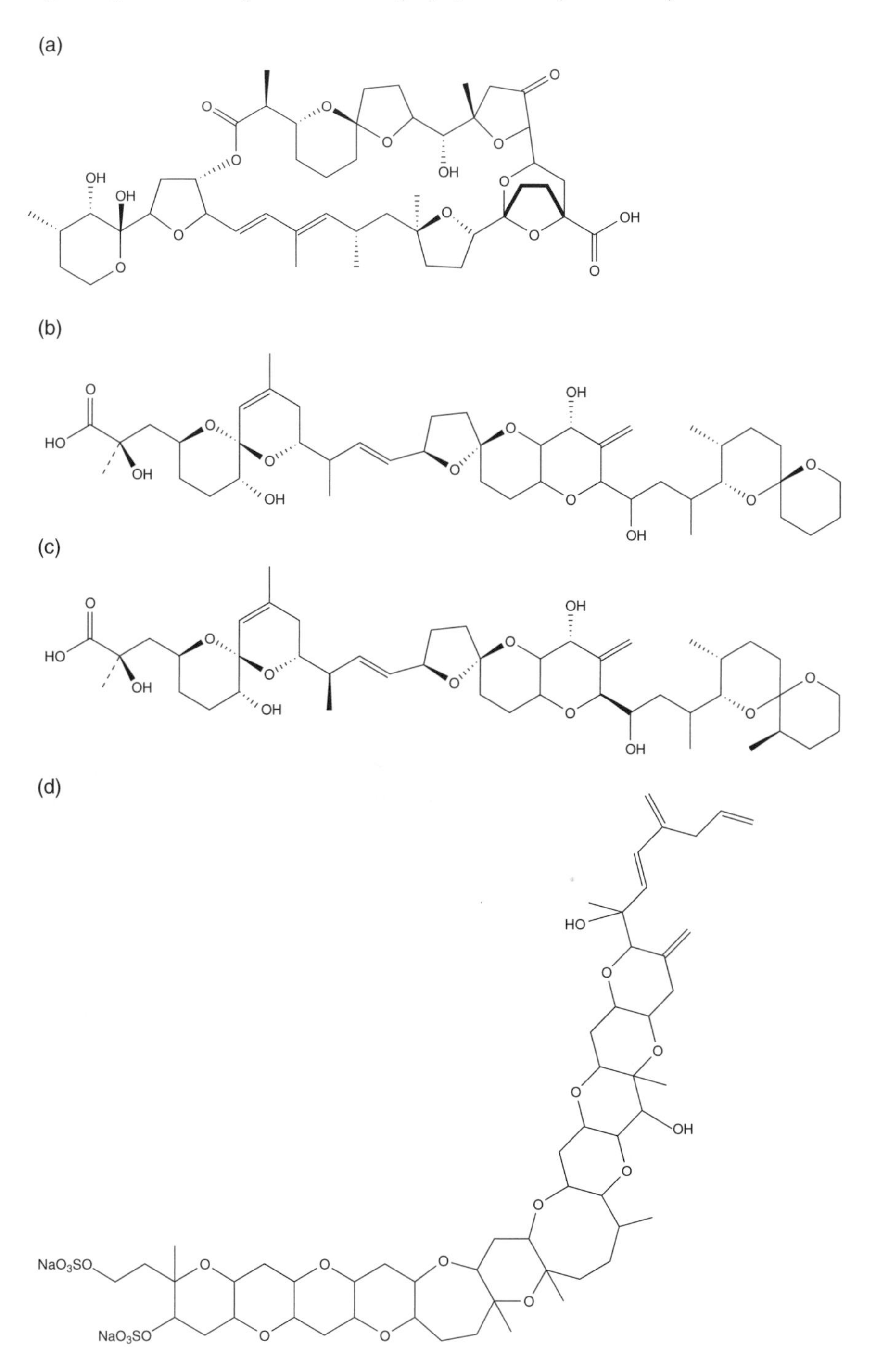
(a)
(b)
(c)
(d)
OH
O
HO
NaO3SO
NaO3SO

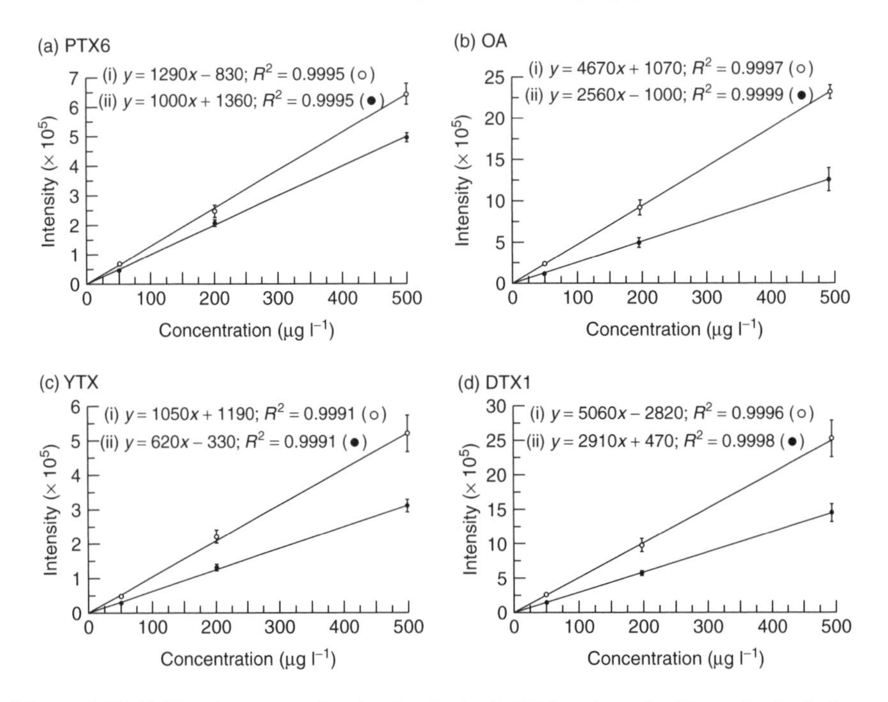


**Figure 5.60** Calibration curves for the diarrhetic shellfish poisons in (i) standard solutions in methanol (○), and (ii) standard solutions in poison-free scallop extract solutions (●): (a) pectenotoxin-6; (b) okadaic acid; (c) yessotoxin; (d) dinophysistoxin-1. Reprinted from *J. Chromatogr., A*, **943**, 'Matrix effect and correction by standard addition in quantitative liquid chromatographic–mass spectrometric analysis of diarrhetic shellfish poisoning toxins', Ito, S. and Tsukada, K., 39–46, Copyright (2002), with permission from Elsevier Science.

(iv) Calibration based on the method of *standard additions*. This involves determination of the 'unknown', followed by addition of a known amount(s) of each of the analytes of interest to this sample. The increase in signal can then be related to the amount of analyte added and the amount originally present in the sample calculated. The main disadvantage of this approach is that further LC–MS analyses must be carried out – one if a single-point calibration is to be used, and more if further additions are made.

A matrix extract was prepared from poison-free scallop and spiked at the level of 200 ng $g^{-1}$ of scallop hepatopancreas. The toxins were determined by using LC–MS with calibration employing external standards prepared in methanol. The matrix extract was then spiked further with 300 ng $g^{-1}$ of each of the toxins and redetermined. The results obtained for each analyte are summarized in Table 5.17 and show that, when using the external calibration method, the values obtained range from 138 to 170 ng $g^{-1}$, a reduction in accuracy of between 15

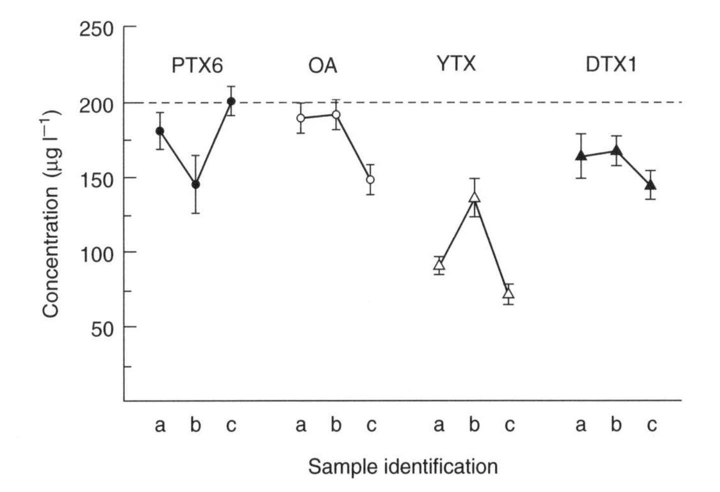


**Figure 5.61** Quantitative results for four diarrhetic shellfish poisons obtained when identical amounts of analyte were spiked into three different poison-free scallop extract solutions. Reprinted from *J. Chromatogr., A*, **943**, 'Matrix effect and correction by standard addition in quantitative liquid chromatographic–mass spectrometric analysis of diarrhetic shellfish poisoning toxins', Ito, S. and Tsukada, K., 39–46, Copyright (2002), with permission from Elsevier Science.

**Table 5.17** Quantitative results obtained for the determination of four diarrhetic shellfish poisons (DSPs) using external standards and the method of standard additions. Reprinted from *J. Chromatogr., A*, **943**, 'Matrix effect and correction by standard addition in quantitative liquid chromatographic–mass spectrometric analysis of diarrhetic shellfish poisoning toxins', Ito, S. and Tsukada, K., 39–46, Copyright (2002), with permission from Elsevier Science

| Calibration method | Amount ± SD(ng g$^{-1}$)[a] | | | |
|---|---|---|---|---|
| | PTX6 | OA | YTX | DTX1 |
| Model sample (theoretical) | 200 | 200 | 200 | 200 |
| External standard method[b] | 170 ± 8 | 134 ± 14 | 135 ± 8 | 138 ± 6 |
| Standard addition method[c] | 197 ± 9 | 213 ± 20 | 215 ± 12 | 214 ± 10 |

[a] SD, standard deviation for $n = 6$; ng g$^{-1}$, hepatopancrea extract.
[b] The calibration curves were prepared by using DSP standard solutions in methanol.
[c] The calculation was carried out by adding the DSP standard solution to the model sample.

and 33%. In contrast, the method of standard additions gave values between 197 and 215 ng g$^{-1}$ with detection limits of between 30 and 60 pg of injected analyte. The overall method performance is summarized in Table 5.18, and clearly illustrates the effectiveness of this approach.

**Table 5.18** Intra-day accuracy and precision data for the determination of four diarrhetic shellfish poisons using LC–MS and the method of standard additions. Reprinted from *J. Chromatogr., A*, **943**, 'Matrix effect and correction by standard addition in quantitative liquid chromatographic–mass spectrometric analysis of diarrhetic shellfish poisoning toxins', Ito, S. and Tsukada, K., 39–46, Copyright (2002), with permission from Elsevier Science

| Parameter | Compound | | | |
|---|---|---|---|---|
| | PTX6 | OA | YTX | DTX1 |
| Concentration ($\mu g\ l^{-1}$) | 200 | 200 | 200 | 200 |
| Mean $\pm$ SD[a] ($\mu g\ l^{-1}$) | 197 $\pm$ 9 | 213 $\pm$ 20 | 215 $\pm$ 12 | 214 $\pm$ 10 |
| Relative standard deviation (RSD) (%) | 4.4 | 9.4 | 5.6 | 4.5 |
| Bias (%) | −1.5 | 6.5 | 7.5 | 7.0 |

[a] SD, standard deviation for $n = 6$.

### 5.6.5 The Quantitative Determination of DNA Oxidation Products

A method has been reported for the quantification of the DNA oxidation products, 8-hydroxy-2′-deoxyguanosine (8-OH-dG), 8-hydroxy-2′-deoxyadenosine (8-OH-dA), 5-hydroxymethyl-2-deoxyuridine (HMDU), thymidine glycol (TG) and 2-hydroxy-2′-deoxyadenosine (2-OH-dA) [37]. The HPLC system employed consisted of a 2.0 × 250 mm $C_{18}$ column and gradient elution from water:methanol, (94:6) to (10:90) over 28 min, at a flow rate of 200 $\mu l\,min^{-1}$.

Each of the compounds studied gave intense ions from molecular species with both positive and negative electrospray ionization. The product-ion scan from 8-OH-dG is shown in Figure 5.62 and shows, in positive-ion mode, a loss of 116 Da from the sugar moiety, and in negative-ion mode a loss of 90 Da. The rationale for this was confirmed by studies of the product-ion spectrum from an isotopically labelled compound, with the proposed fragmentation pathway being shown in Figure 5.63. Both positively and negatively charged molecular species showed decompositions of appropriate intensity to be monitored but for TG the decomposition from $m/z$ 275 to $m/z$ 116 in the negative-ion mode was 100 times more sensitive than the equivalent decomposition in the positive-ion mode and the former mode was therefore chosen for method development. The decompositions monitored for each compound are shown in Table 5.19.

It may seem obvious but it is always worthwhile checking that the ions/decompositions selected for monitoring are 'sensible' with respect to the structure of the analyte(s) involved rather than waste time in developing a method based on ions from a background component.

The results obtained from the LC–MS analysis of a mixture of standards containing ca. 4 $ng\,\mu l^{-1}$ are shown in Figure 5.64. Note that because it does not contain a chromophore, TG is not observed in the UV trace and that because

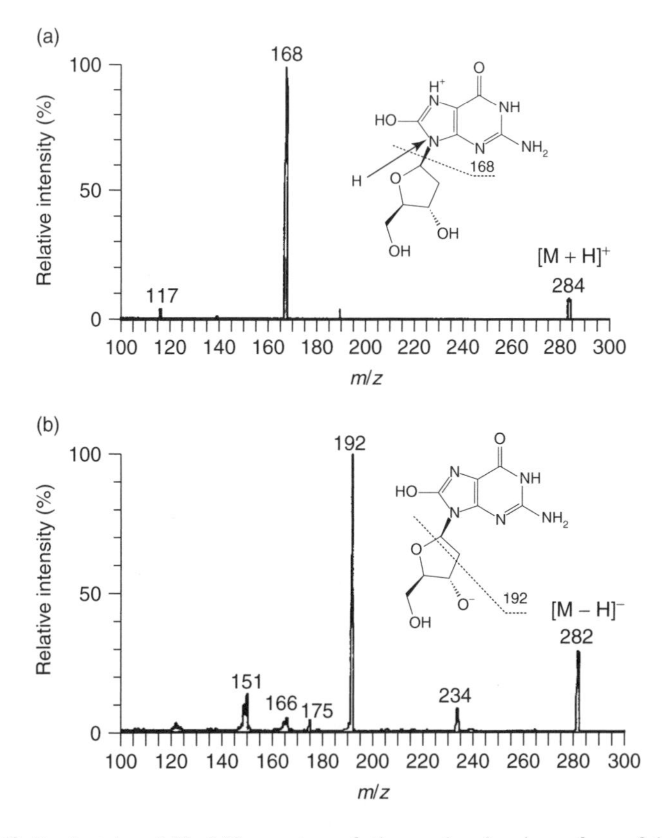


**Figure 5.62** Product-ion MS–MS spectra of the molecular ions from 8-hydroxy-2′-deoxyguanosine, obtained by (a) positive, and (b) negative ionization. Reprinted by permission of Elsevier Science from 'Comparison of negative- and positive-ion electrospray tandem mass spectrometry for the liquid chromatography–tandem mass spectrometry analysis of oxidized deoxynucleosides', by Hua, Y., Wainhaus, S. B., Yang, Y., Shen, L., Xiong, Y., Xu, X., Zhang, F., Bolton, J. L. and van Breemen, R. B., *Journal of the American Society for Mass Spectrometry*, Vol. 12, pp. 80–87, Copyright 2000 by the American Society for Mass Spectrometry.

the decompositions are characteristic of the oxidation products the four deoxynucleotides present do not interfere with the analysis.

This method was applied to the determination of these oxidized nucleosides in salmon testes using $[^{13}C_{10},^{15}N_5]$-8-hydroxy-2′-deoxyguanosine (L-8-OH-dG) as the internal standard (Figure 5.65). Four of the oxidized products were below the limits of detection of the method, while the concentration of 8-OH-dG was determined to be 0.93 ppb.

The need for choosing appropriate ions/decompositions to provide adequate sensitivity and selectivity has been stressed and the selectivity provided by mass

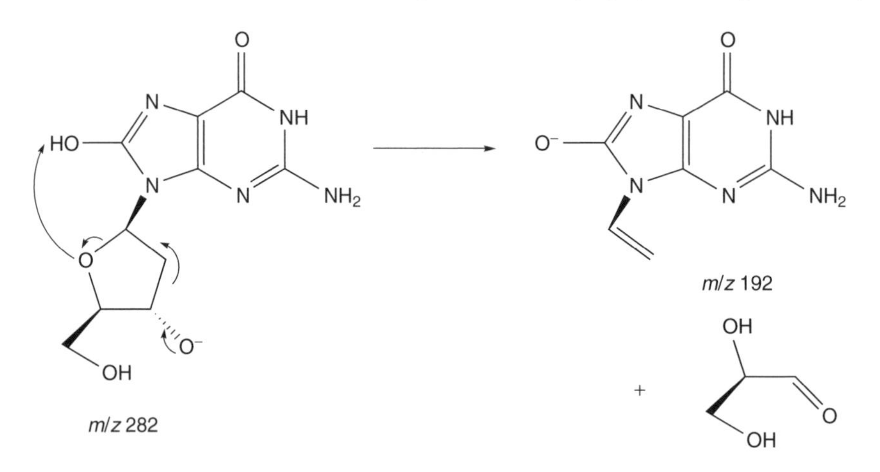


**Figure 5.63** Proposed fragmentation pathway of the molecular ion from 8-hydroxy-2′-deoxyguanosine generated by negative ionization. Reprinted by permission of Elsevier Science from 'Comparison of negative- and positive-ion electrospray tandem mass spectrometry for the liquid chromatography–tandem mass spectrometry analysis of oxidized deoxynucleosides', by Hua, Y., Wainhaus, S. B., Yang, Y., Shen, L., Xiong, Y., Xu, X., Zhang, F., Bolton, J. L. and van Breemen, R. B., *Journal of the American Society for Mass Spectrometry*, Vol. 12, pp. 80–87, Copyright 2000 by the American Society for Mass Spectrometry.

**Table 5.19** Analytical data obtained for the LC–MS determination of DNA oxidation products

| Compound | HPLC retention time (min) | Decomposition monitored in SRM[a] |
|---|---|---|
| TG | 8.8 | 275 → 116 |
| HMDU | 13.3 | 257 → 141 (M − 116) |
| 2-OH-dA | 15.4 | 266 → 150 (M − 116) |
| 8-OH-dG | 20.6 | 282 → 192 (M − 90) |
| 8-OH-dA | 28.0 | 266 → 176 (M − 90) |

[a] Selected-reaction monitoring.

spectra generated by soft ionization techniques put forward as a subject for thought. This example method clearly demonstrates that the incorporation of MS–MS methodology can provide adequate selectivity.

## 5.6.6 The Use of MS–MS for Quantitative Determinations

The advent of the time-of-flight analyser as part of an MS–MS instrument has already been noted. Its value has been compared with selected-decomposition monitoring (SDM) for the quantitative determination of Idoxifene (Figure 5.66),

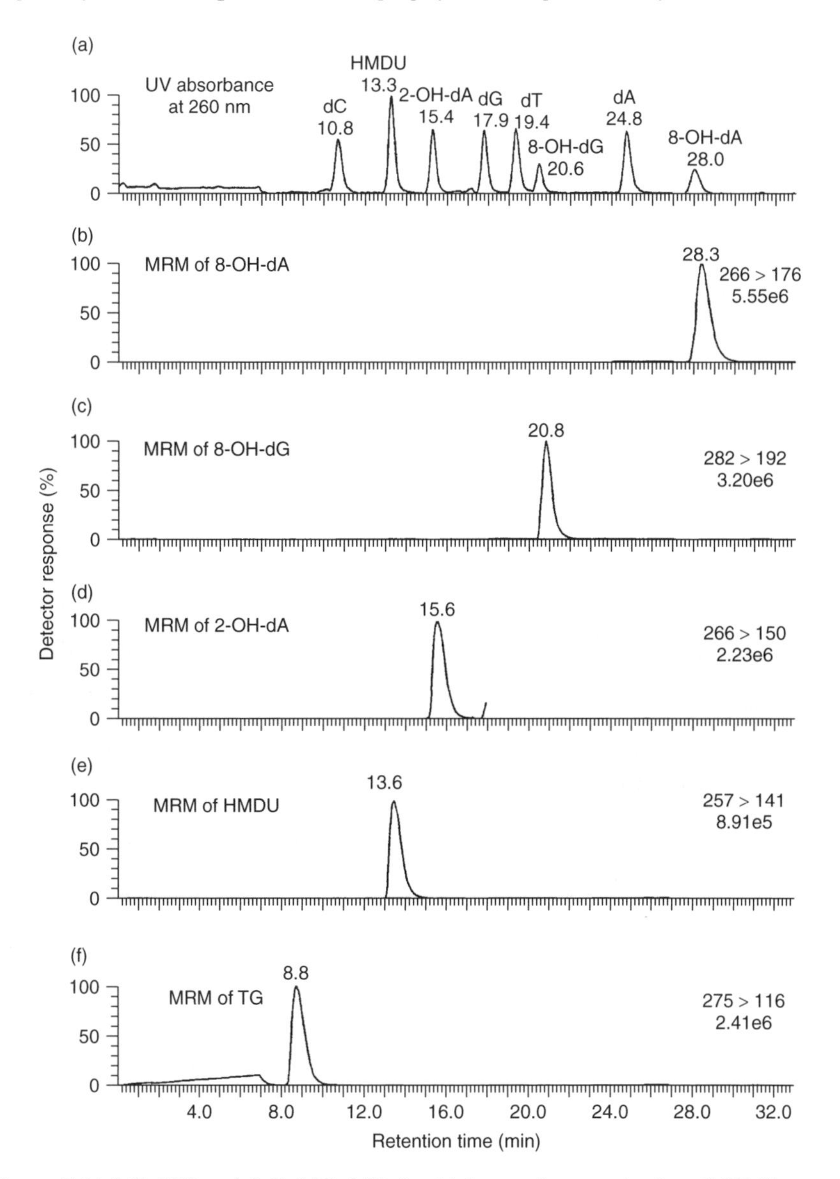


**Figure 5.64** LC–UV and LC–MS–MS (multiple-reaction monitoring (MRM)) traces from the analysis of a synthetic mixture of four native and five oxidized deoxynucleosides (for nomenclature, see text). Reprinted by permission of Elsevier Science from 'Comparison of negative- and positive-ion electrospray tandem mass spectrometry for the liquid chromatography–tandem mass spectrometry analysis of oxidized deoxynucleosides', by Hua, Y., Wainhaus, S. B., Yang, Y., Shen, L., Xiong, Y., Xu, X., Zhang, F., Bolton, J. L. and van Breemen, R. B., *Journal of the American Society for Mass Spectrometry*, Vol. 12, pp. 80–87, Copyright 2000 by the American Society for Mass Spectrometry.

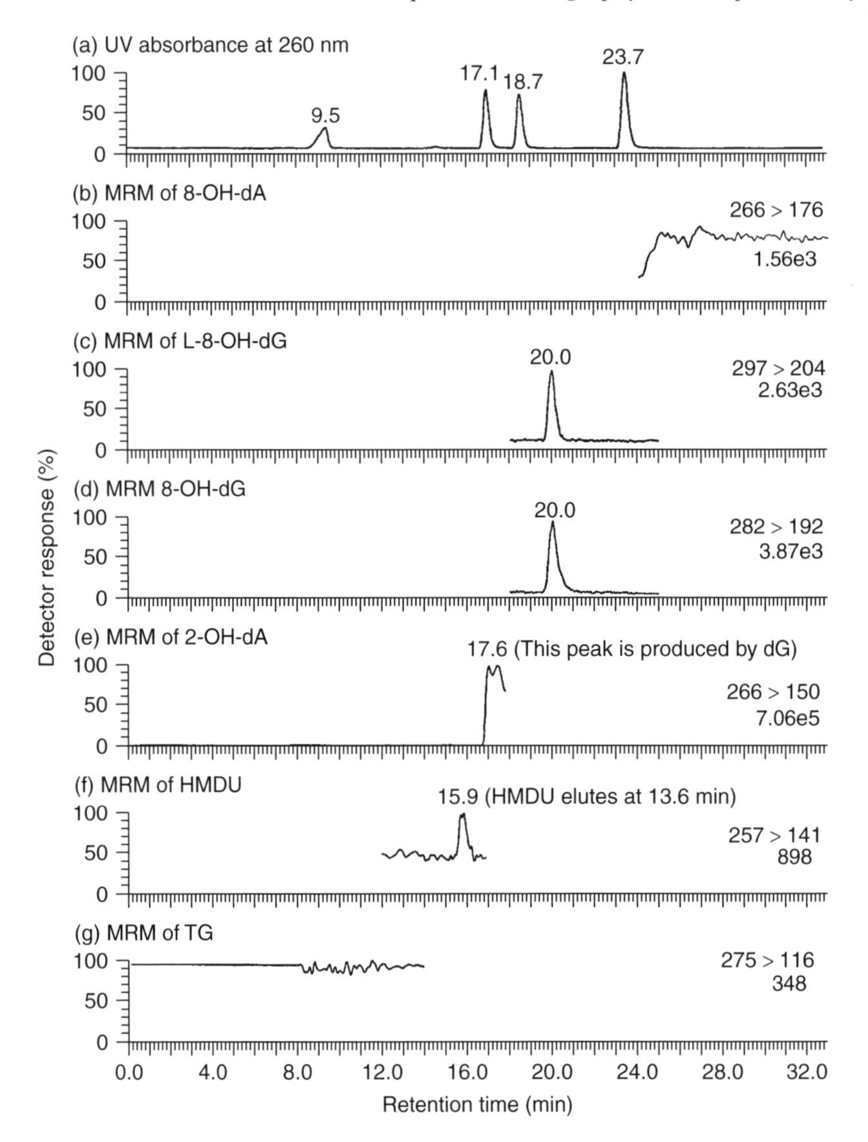


**Figure 5.65** LC–UV and LC–MS–MS (multiple-reaction monitoring (MRM)) traces from the analysis of a enzymatically digested solution of 100 μg salmon testes DNA (for nomenclature, see text). Reprinted by permission of Elsevier Science from 'Comparison of negative- and positive-ion electrospray tandem mass spectrometry for the liquid chromatography–tandem mass spectrometry analysis of oxidized deoxynucleosides', by Hua, Y., Wainhaus, S. B., Yang, Y., Shen, L., Xiong, Y., Xu, X., Zhang, F., Bolton, J. L. and van Breemen, R. B., *Journal of the American Society for Mass Spectrometry*, Vol. 12, pp. 80–87, Copyright 2000 by the American Society for Mass Spectrometry.

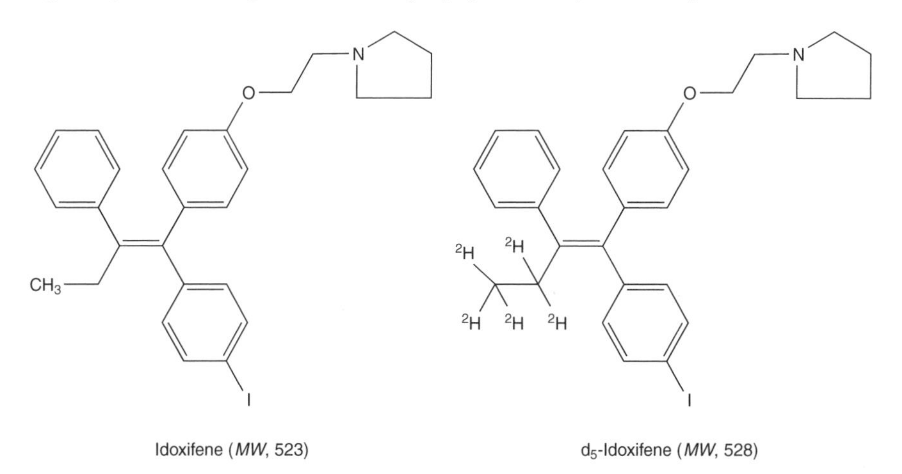


**Figure 5.66** Molecular structures of Idoxifene and its deutrated internal standard $d_5$-Idoxifene. Reprinted from *J. Chromatogr., B*, **757**, 'Comparison between liquid chromatography–time-of-flight mass spectrometry and selected-reaction monitoring liquid chromatography–mass spectrometry for quantitative determination of Idoxifene in human plasma', Zhang, H. and Henion, J., 151–159, Copyright (2001), with permission from Elsevier Science.

a selective oestrogen receptor modulator, in human plasma using $d_5$-Idoxifene as the internal standard [38].

The HPLC system used consisted of a 30 × 2 mm Luna CN column with linear gradient elution employing two mobile phases A and B (A, 90% $H_2O$:10% acetonitrile; B, 10% $H_2O$:90% acetonitrile) with both phases containing 5 mM ammonium acetate and 0.2% formic acid. The linear gradient commenced with 50:50 A:B increasing to 100% B after 1 min of the analysis; this composition was maintained for 1 min before returning to 50:50 A:B after 4 min. Positive-ion ionspray (pneumatically assisted electrospray) was used to obtain mass spectra, with the spectrometer operating at a resolution of 5000.

The way in which the ToF mass spectrometer is operated is quite different from the more conventional quadrupole and ion-trap instruments. The source is pulsed at 20 000 Hz, i.e. 20 000 spectra are obtained per second, over a mass range of 300 to 600 Da. While it would be possible to store and manipulate this number of spectra, the implications in terms of the computer storage requirement and the time required for data processing would be severe. For this reason, a number of spectra are added so that effectively a full spectrum is acquired, in this case, every 0.3 s. The number of spectra added can be modified in light of the analytical requirements, i.e. for fast chromatography where chromatographic peak widths are reduced an increased number of spectra may be obtained, without sacrificing sensitivity, by combining fewer spectra. Quantitation was effected by comparison of the peak areas of the RICs from the $[M + H]^+$ ions from Idoxifene ($m/z$

524.144 ± 0.01) and the internal standard, those obtained from a blank and from a sample containing 5 ng ml$^{-1}$ Idoxifen, the limit of quantitation (LOQ) for this ToF method, with the LOQ being defined by these authors as the lowest concentration on the calibration curve where the accuracy and precision are within 20% of the known value, are shown in Figure 5.67. Even at high resolution, a signal is obtained from the background and, although this does not interfere with the determination shown, the authors did comment that 'adequate chromatographic resolution of the targeted compound from metabolites or endogenous compounds is more important with LC–ToF-MS than with LC–MS–MS.'

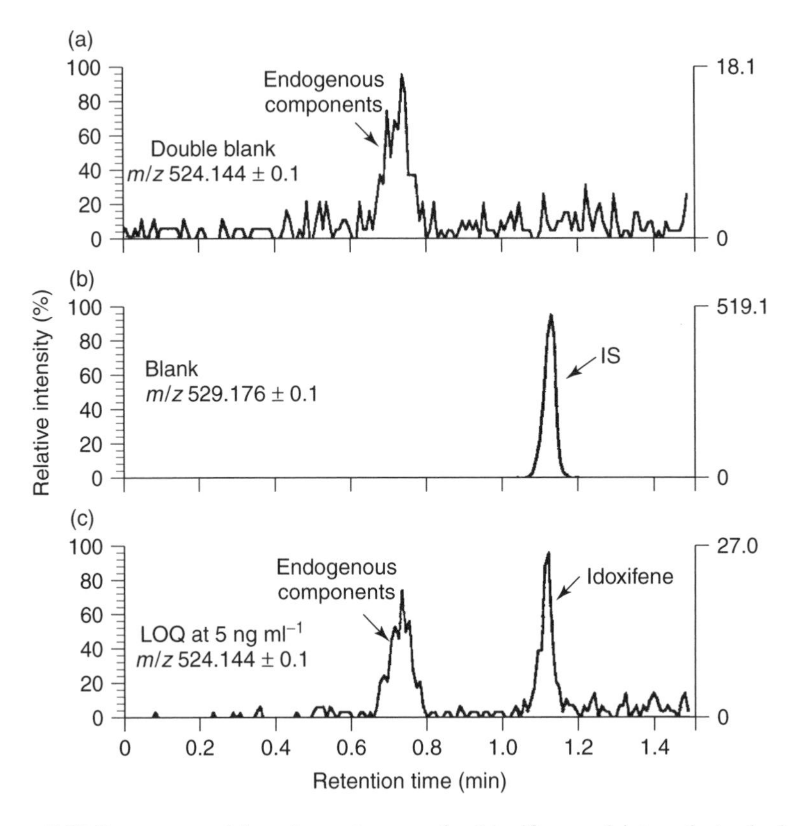


**Figure 5.67** Reconstructed ion chromatograms for Idoxifene and internal standard ($d_5$-Idoxifene) using LC–ToF-MS for (a) double-blank human plasma extract, (b) extract of blank human plasma containing internal standard (IS), and (c) control-blank human plasma spiked with Idoxifene at 5 g ml$^{-1}$, the LOQ of the method. Reprinted from *J. Chromatogr., B*, **757**, 'Comparison between liquid chromatography–time-of-flight mass spectrometry and selected-reaction monitoring liquid chromatography–mass spectrometry for quantitative determination of Idoxifene in human plasma', Zhang, H. and Henion, J., 151–159, Copyright (2001), with permission from Elsevier Science.

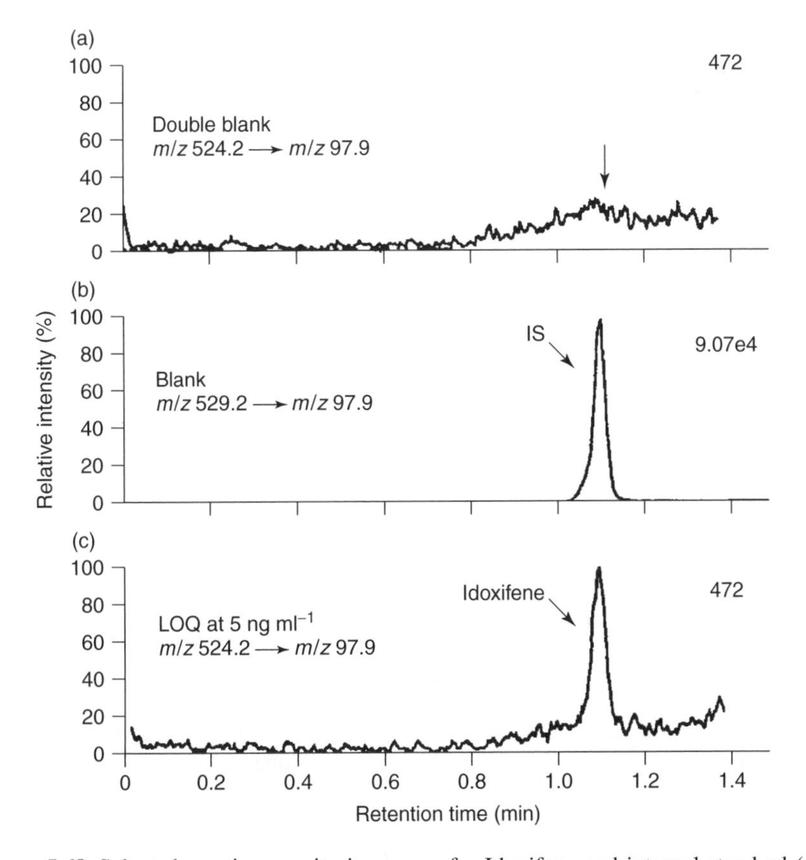


**Figure 5.68** Selected-reaction monitoring traces for Idoxifene and internal standard ($d_5$-Idoxifene) using LC–MS–MS with a triple quadrupole for (a) double-blank human plasma extract, (b) extract of blank human plasma containing internal standard (IS), and (c) control-blank human plasma spiked with Idoxifene at 0.5 ng ml$^{-1}$, the LOQ of the method. Reprinted from *J. Chromatogr., B*, **757**, 'Comparison between liquid chromatography–time-of-flight mass spectrometry and selected-reaction monitoring liquid chromatography–mass spectrometry for quantitative determination of Idoxifene in human plasma', Zhang, H. and Henion, J., 151–159, Copyright (2001), with permission from Elsevier Science.

SDM was carried out by monitoring the decomposition of the $[M + H]^+$ ion of Idoxifen at $m/z$ 524.2 to the ion at $m/z$ 97.9 and the corresponding decomposition of $m/z$ 529.2 to $m/z$ 97.9 for the internal standard. Data obtained from a blank and a sample spiked at the LOQ, for this methodology 0.5 ng ml$^{-1}$, are shown in Figure 5.68. An obvious difference is the reduction in background in the MS–MS method. The calibration graphs for both methods are shown in Figure 5.69, while Tables 5.20 and 5.21 show the accuracy and precision obtained

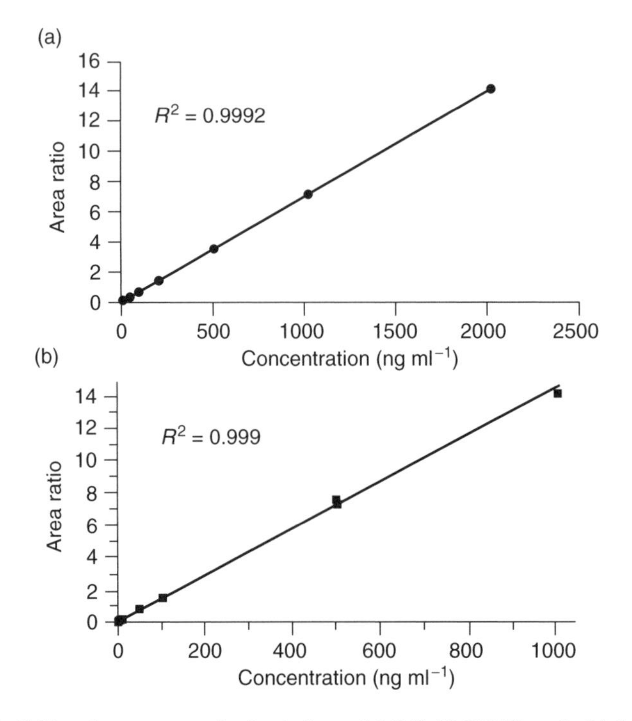


**Figure 5.69** Calibration curves obtained from (a) LC–ToF-MS and (b) LC–MS–MS using selected-reaction monitoring for Idoxifene in human plasma, fortified from 5 to 2000 ng ml$^{-1}$ for LC–ToF-MS and 0.5 to 1000 ng ml$^{-1}$ for LC–MS–MS with a triple quadrupole: $R^2$ is the correlation coefficient, a measure of the quality of calibration (see p. 218). Reprinted from *J. Chromatogr., B*, **757**, 'Comparison between liquid chromatography–time-of-flight mass spectrometry and selected-reaction monitoring liquid chromatography–mass spectrometry for quantitative determination of Idoxifene in human plasma', Zhang, H. and Henion, J., 151–159, Copyright (2001), with permission from Elsevier Science.

for the determination of quality control (QC) samples at 'low', 'medium' and 'high' levels. The two methods are compared in Table 5.22 and, with the exception of the LOQ, both show a similar performance. Whether or not this increase in the LOQ is of real importance depends upon whether the higher level provided by the ToF instrument is inadequate for the samples encountered. Limitations in the linear range of the ToF mass analyser for quantitative applications have been previously identified and said [39] to be associated with the very high acquisition rates employed. Improvements in this area are being made but those evaluating this type of instrument for quantitative applications should ensure that they have adequate performance.

**Table 5.20** Intra-assay precision and accuracy of the LC–ToF-MS determination of Idoxifene. Reprinted from *J. Chromatogr., B*, **757**, 'Comparison between liquid chromatography–time-of-flight mass spectrometry and selected-reaction monitoring liquid chromatography–mass spectrometry for quantitative determination of Idoxifene in human plasma', Zhang, H. and Henion, J., 151–159, Copyright (2001), with permission from Elsevier Science

| Replicate number[a] | Low, 15 ng $ml^{-1}$ | Middle, 1000 ng $ml^{-1}$ | High, 1500 ng $ml^{-1}$ |
|---|---|---|---|
| 1 | 14.20 | 955.8 | 1562 |
| 2 | 14.33 | 935.1 | 1501 |
| 3 | 14.69 | 944.4 | 1450 |
| 4 | 14.63 | 1097 | 1393 |
| 5 | 16.14 | 1095 | 1486 |
| 6 | 15.07 | 1070 | 1451 |
| Mean | 14.84 | 1016 | 1474 |
| RSD (%)[b] | 4.3 | 7.1 | 3.5 |
| % Deviation[c] | −1.0 | 1.6 | −1.7 |

[a] $n = 6$.
[b] RSD, relative standard deviation = (standard deviation/mean) × 100.
[c] % Deviation = [(mean − nominal)/nominal] × 100.

**Table 5.21** Intra-assay precision and accuracy of the LC–MS–MS determination of Idoxifene using a triple quadrupole mass spectrometer. Reprinted from *J. Chromatogr., B*, **757**, 'Comparison between liquid chromatography–time-of-flight mass spectrometry and selected-reaction monitoring liquid chromatography–mass spectrometry for quantitative determination of Idoxifene in human plasma', Zhang, H. and Henion, J., 151–159, Copyright (2001), with permission from Elsevier Science

| Replicate number[a] | Low, 1.5 ng $ml^{-1}$ | Middle, 500 ng $ml^{-1}$ | High, 800 ng $ml^{-1}$ |
|---|---|---|---|
| 1 | 1.513 | 503.3 | 761.7 |
| 2 | 1.509 | 511.3 | 766.5 |
| 3 | 1.548 | 524.5 | 800.5 |
| 4 | 1.569 | 508.0 | 779.8 |
| 5 | 1.520 | 514.8 | 780.6 |
| 6 | 1.462 | 518.4 | 794.6 |
| Mean | 1.535 | 511.8 | 777.1 |
| RSD (%)[b] | 1.9 | 1.8 | 2.2 |
| % Deviation[c] | 2.3 | 2.4 | −2.9 |

[a] $n = 6$.
[b] RSD, relative standard deviation = (standard deviation/mean) × 100.
[c] % Deviation = [(mean − nominal)/nominal] × 100.

**Table 5.22** Comparison of method performance for LC–ToF-MS and LC–MS–MS determination of Idoxifene

| Parameter | LC–ToF-MS | LC–MS–MS |
|---|---|---|
| Limit of quantitation | 5 $ng\,ml^{-1}$ | 0.5 $ng\,ml^{-1}$ |
| Precision | 5.1% | 8.1% |
| Accuracy | −5.7% | −2.0% |
| Linear range | 5–2000 $ng\,ml^{-1}$ | 0.5–1000 $ng\,ml^{-1}$ [a] |

[a]The upper limit to the linear range was 5000 $ng\,ml^{-1}$, but at concentrations >1000 $ng\,ml^{-1}$ 'carry-over' from the autosampler was observed. This could be reduced by extensive washing.

# Summary

In this chapter, a number of applications of LC–MS have been described. The examples have been chosen to illustrate the variety of types of molecules for which LC–MS is appropriate and the wide range of analytical information that can be obtained when using the mass spectrometer as a detector.

Method development is important. LC–MS performance, probably more than any other technique involving organic mass spectrometry, is dependent upon a range of experimental parameters, the relationship between which is often complex. While it is possible (but not always so) that conditions may be chosen fairly readily to allow the analysis of simple mixtures to be carried out successfully, the widely variable ionization efficiency of compounds with differing structures often makes obtaining optimum performance for the study of all components of a complex mixture difficult. In such cases, the use of experimental design should be seriously considered.

The analyses that may be carried out can be conveniently classified as *qualitative*, the determination of the molecular weights and/or structures of both high- and low-molecular-weight materials, and *quantitative*, their precise and accurate quantitation.

Within each of these classifications the methodology employed is essentially simple. Both APCI and electrospray are soft ionization techniques and therefore molecular weight information is usually readily available. Structural information may be generated by the use of either cone-voltage (in-source) fragmentation (CVF) and, if an appropriate instrument is available, through MS–MS. Electrospray spectra of high-molecular-weight materials invariably contain a number of ions, with each corresponding to different charge states of the intact molecule and, although the way in which these molecules fragment is well understood, the complexity of the spectra often makes it difficult to extract useful structural information. The use of enzyme digestion to cleave a biomolecule in a predictable way to produce smaller molecules which give mass spectra from which structural information can be more readily obtained, is often required and is now well accepted methodology. While the CVF or MS–MS spectra of smaller molecules, i.e. <1000 Da, usually contain fewer ions, understanding their structural significance may require the study of equivalent spectra from related compounds of known structure and a knowledge of the atomic compositions of the ions being studied.

The fact that APCI and electrospray are soft ionization techniques is often advantageous because the molecular ion alone, in conjunction with HPLC separation, often provides adequate selectivity and sensitivity to allow an analytical method to be developed. Again, method development is important, particularly when more than one analyte is to be determined, when the effect of experimental parameters, such as pH, flow rate, etc., is not likely to be the same for each. Electrospray, in particular, is susceptible to matrix effects and the method of standard additions is often required to provide adequate accuracy and precision.

# References

1. Science Direct®, [http://www.sciencedirect.com], Elsevier Science BV, 2002.
2. Miller, J. N. and Miller, J. C., *Statistics and Chemometrics for Analytical Chemistry*, 4th Edn, Pearson Educational, Harlow, UK, 2000.
3. Asperger, A., Efer, J., Koal, T. and Engewald, W., *J. Chromatogr., A*, **937**, 65–72 (2001).
4. Naidong, W., Chen, Y., Shou, W. and Jiang, X., *J. Pharm. Biomed. Anal.*, **26**, 753–767 (2001).
5. Seto, C., Bateman, K. P. and Gunter, B., *J. Am. Soc. Mass Spectrom.*, **13**, 2–9 (2002).
6. Pamme, N., Steinbach, K., Ensinger, W. J. and Schmidt, T. C., *J. Chromatogr., A*, **943**, 47–54 (2001).
7. McAtee, C. P., Zhang, Y., Yarbough, P. O., Fuerst, T. R., Stone, K. L., Samander, S. and Williams, K. R., *J. Chromatogr., B*, **685**, 91–104 (1996).
8. Wan, H. Z., Kaneshiro, S., Frenz, J. and Cacia, J., *J. Chromatogr., A*, **913**, 437–446 (2001).
9. Lehninger, A. L., Nelson, D. L. and Cox, M. M., *Principles of Biochemistry*, 3rd Edn, Worth Publishers, New York, 2000.
10. Chapman, J. R., (Ed.), *Protein and Peptide Analysis by Mass Spectrometry*, Methods in Molecular Biology, Vol. 61, Humana Press, Totowa, NJ, 1996.
11. Turula, V. E., Bishop, R. T., Ricker, R. D. and de Haseth, J. A., *J. Chromatogr., A*, **763**, 91–103 (1997).
12. Klarskov, K., Leys, D., Backers, K., Costa, H. S., Santos, H., Guisez, Y. and Van Beeumen, J. J., *Biochim. Biophys. Acta*, **1412**, 47–55 (1999).
13. Poutanen, M., Salusjarvi, L., Ruohonen, L., Penttila, M. and Kalkkinen, N., *Rapid Commun. Mass Spectrom.*, **15**, 1685–1692 (2001).
14. [http://prowl.rockefeller.edu/cgi-bin/ProFound].
15. [http://pepsea.protana.com].
16. [http://prospector.ucsf.edu].
17. Bonomo, R. A., Liu, J., Chen, Y., Ng, L., Hujer, A. M. and Anderson, V. E., *Biochim. Biophys. Acta*, **1547**, 196–205 (2001).
18. Huddleston, M. J., Annan, R. S., Bean, M. F. and Carr, S. A., *J. Am. Soc. Mass Spectrom.*, **4**, 710–717 (1993).
19. Leonil, J., Gagnaire, V., Molle, D., Pezennec, S. and Bouhallab, S., *J. Chromatogr., A*, **881**, 1–21 (2000).
20. Huddleston, M. J., Bean, M. F. and Carr, S. A., *Anal. Chem.*, **65**, 877–884 (1993).
21. Molle, D., Morgan, F., Bouhallab, S. and Leonil, J., *Anal. Biochem.*, **259**, 152–161 (1998).
22. Karlsson, N. G., Karlsson, H. and Hansson, G. C., *Glycoconj. J.*, **12**, 69–76 (1995).
23. Thomsson, K. A., Karlsson, H. and Hansson, G. C., *Anal. Chem.*, **72**, 4543–4549 (2000).
24. Kurahashi, T., Miyazaki, A., Murakami, Y., Suwan, S., Franz, T., Isobe, M., Tani, M. and Kai, H., *Bioorg. Med. Chem.*, **10**, 1703–1710 (2002).

25. de Biasi, V., Haskins, N., Organ, A., Bateman, R., Giles, K. and Jarvis, S., *Rapid Commun. Mass Spectrom.*, **13**, 1165–1168 (1999).
26. Fernandez, M., Rodriguez, R., Pico, Y. and Manes, J., *J. Chromatogr., A*, **912**, 301–310 (2001).
27. Barnes, K. A., Fussell, R. J., Startin, J. R., Pegg, M. K., Thorpe, S. A. and Reynolds, S. L., *Rapid Commun. Mass Spectrom.*, **11**, 117–123 (1997).
28. Lacassie, E., Dreyfuss, M. F., Daguet, J. L., Vignaud, M., Marquet, P. and Lachâtre, G., *J. Chromatogr., A*, **830**, 135–143 (1999).
29. Yu, X., Cui, D. and Davis, M. R., *J. Am. Soc. Mass Spectrom.*, **10**, 175–183 (1999).
30. Hopfgartner, G., Chernushevich, I. V., Covey, T., Plomley, J. B. and Bonner, R., *J. Am. Soc. Mass Spectrom.*, **10**, 1305–1314 (1999).
31. Zhang, H., Henion, J., Yang, Y. and Spooner, N., *Anal. Chem.*, **72**, 3342–3348 (2000).
32. Lerch, C. and Blaschke, G., *J. Chromatogr., B*, **708**, 267–275 (1998).
33. Constantopoulos, T. L., Jackson, G. S. and Enke, C. G., *J. Am. Soc. Mass Spectrom.*, **10**, 625–634 (1999).
34. Choi, B. K., Hercules, D. M. and Gusev, A. I., *J. Chromatogr., A*, **907**, 337–342 (2001).
35. Choi, B. K., Hercules, D. M. and Gusev, A. I., *Fresenius' J. Anal. Chem.*, **369**, 370–377 (2001).
36. Ito, S. and Tsukada, K., *J. Chromatogr., A*, **943**, 39–46 (2002).
37. Hua, Y., Wainhaus, S. B., Yang, Y., Shen, L., Xiong, Y., Xu, X., Zhang, F., Bolton, J. L. and van Breemen, R. B., *J. Am. Soc. Mass Spectrom.*, **12**, 80–87 (2000).
38. Zhang, H. and Henion, J., *J. Chromatogr., B*, **757**, 151–159 (2001).
39. Zhang, H. W., Heinig, K. and Henion, J., *J. Mass Spectrom.*, **35**, 423–431 (2000).

# Responses to Self-Assessment Questions

## Chapter 2

### *Response 2.1*

There are two methods that are commonly employed to overcome this situation, i.e. (a) derivatization of the analyte to introduce a chromophore, and (b) the use of indirect UV detection.

A number of derivatizing agents are in use, such as benzoyl chloride, which contains a benzene ring capable of absorbing UV radiation. Derivatization may be undertaken pre- or post-column. If the former, it is the derivative that undergoes the chromatographic process and the retention characteristics measured will be that of the derivative. If the derivatization takes place post-column, the retention characteristic is that of the parent compound. Pre-column derivatization has the advantage that the experimental conditions, in particular the reaction time, can be more closely controlled to optimize precision and accuracy and will have no effect on the chromatography. It is also possible to remove excess derivatization reagent to minimize any effect it may have on degradation of the column or detector. Post-column derivatization may allow a greater degree of automation, although *well-plate technology*[†] is now being utilized for pre-column derivatization, but if the reaction does not take place very quickly extra peak broadening may be introduced. In addition, relatively large amounts of derivatizing agent will enter

[†] The analytical scientist often has to deal with large numbers of samples and one of the current interests is the development of methodology which allows these to be studied rapidly. One way of achieving this is to use a plate that can accommodate a number of samples, e.g. 96, each of which is confined within an individual hole ('well') in the plate. With appropriate hardware, experimentation may be carried out simultaneously in each of these wells, thus dramatically increasing throughput of samples.

the detector and this may be undesirable. Whether pre- or post-column derivatization is undertaken, the reproducibility and extent of derivatization should be determined in separate experiments.

The alternative is to add a UV-absorbing material to the mobile phase. If a compound elutes from the HPLC column that does not absorb UV radiation, the detector response will decrease. An additive should be chosen which has significant absorption in order that it may be added at low concentration and thus have minimal effect on the chromatographic separation. It is also important that reaction between the analyte(s) and additive does not occur.

## *Response 2.2*

By using equation (2.6), the following figures should be obtained:

*System A*

$$\alpha = 4.89/4.83 = 1.01$$

$$R = 0.25 \times [(1.01 - 1)/1.01] \times [4.89/(1 + 4.89)] \times \sqrt{3500}$$

$$= 0.25 \times 0.099 \times 0.83 \times 59.16 = 1.21$$

*System B*

$$\alpha = 3.32/2.32 = 1.43$$

$$R = 0.25 \times [(1.43 - 1)/1.43] \times [3.32/(1 + 3.32)] \times \sqrt{3500}$$

$$= 0.25 \times 0.30 \times 0.77 \times 59.16 = 3.42$$

## *Response 2.3*

If the signal is symmetrical and well separated from others in the chromatographic trace, peak heights and peak areas should give equal precision and accuracy. In real life, when columns have been in use for some time and the mixture being analysed contains a number of components and 'chemical noise', these ideal conditions are less likely to be encountered and considerable thought has to be given to the subject. If well-resolved asymmetrical peaks are obtained, then peak areas are likely to give greater precision and accuracy. If unresolved components are of interest, an important consideration is how the peak is defined in terms of its base line and retention-time limits. A number of methods are available for this and experimentation is often required to ascertain how the best accuracy and precision may be obtained.

## *Response 2.4*

At first sight, the result from analysis (a), i.e. 1.48 mg ml$^{-1}$, clearly indicates that the level of pollutant is above that allowed by law. Examination of the precision of the measurement, however, indicates that the correct result, assuming that the method has been shown to have adequate accuracy, lies between 0.48 and 2.48 $mg\,ml^{-1}$. It is possible, therefore, that the sample analysed contains less than the level for prosecution and on the basis of this result prosecution should not be undertaken. In case (b), the level found was the same but the precision of measurement, again assuming adequate accuracy has been proven, indicates that the true level lies between 1.38 and 1.58 $mg\,ml^{-1}$, with this complete range being above the level at which prosecution should be undertaken. In this situation, prosecution would be recommended.

---

# Chapter 3

## *Response 3.1*

1 mol $\equiv$ 22.4 l of vapour at STP. The volumes can therefore be calculated as follows:

(a) methanol, RMM = 32, and so 1 g = 0.031 mol $\equiv$ 700 $cm^3$ at STP.

(b) acetonitrile, RMM = 41, and so 1 g = 0.024 mol $\equiv$ 546 $cm^3$ at STP.

(c) water, RMM = 18, and so 1 g = 0.056 mol $\equiv$ 1244 $cm^3$ at STP.

To obtain the equivalent volume at $10^{-6}$ torr, we must use the ideal gas equation and multiply the values above by $7.6 \times 10^8$ (1 atm = 760 torr), thus giving values of (a) $5.32 \times 10^{11}$ $cm^3$, (b) $4.15 \times 10^{11}$ $cm^3$, and (c) $9.45 \times 10^{11}$ $cm^3$.

## *Response 3.2*

This is either an $(M + 1)^+$ or an $(M + 18)^+$ ion, depending upon the relative proton affinities of the analyte and ammonia. The molecular weight is therefore either 221 or 204 Da. Since the molecule contains an even number of nitrogen atoms, it must have an even molecular weight. The molecular species must therefore be $(M + 18)^+$ and so the molecular weight is 204 Da.

## *Response 3.3*

In effect, the composition of the mobile phase, and thus the selectivity of the chromatographic system, has been changed. As mentioned in the text, dynamic FAB operates effectively with lower concentrations of matrix than static FAB and although its effect may be minimal it should always be considered. Post-column addition of matrix overcomes potential problems of this nature.

## *Response 3.4*

As discussed in Section 3.2.1, an electron ionization (EI) spectrum arises from a number of competing and consecutive fragmentation reactions of the molecular ion and any ions arising from such fragmentation. An ion at any particular $m/z$ value may therefore originate from a number of precursors but not necessarily directly from the molecular ion itself. A product-ion MS–MS spectrum of a molecular ion will show **only** those any ions which arise by direct fragmentation of that ion and will not show any ions which arise by subsequent fragmentation of the first generation of product ions.

### *Response 3.5*

Although not annotated, there is an ion at $m/z$ 199, with the ion at $m/z$ 201 being around 33% of its intensity. The most likely explanation is therefore that these two ions arise from a species containing a single chlorine atom, with $m/z$ 199 being from the $^{35}Cl$ isotope, and $m/z$ 201 from the $^{37}Cl$ isotope. There is a difference of 22 Da between $m/z$ 221 and $m/z$ 199 and the most likely explanation therefore is that $m/z$ 199 is the $[M+H]^+$ adduct and $m/z$ 221 the $[M+Na]^+$ adduct of the same molecule. A careful inspection of the spectrum shows that there is an ion at $m/z$ 223 of approximately 30% of the intensity of $m/z$ 221, thus indicating the presence of a single chlorine atom in this molecule.

---

# Chapter 4

### *Response 4.1*

By using equation (4.6), we may define the following:

$$m_1 = 1060.71; \; m_2 = 998.25$$

and $n_1$ is given by:

$$(998.25 - 1)/(1060.71 - 998.25) = 997.25/62.46 = 15.97$$

By using equation (4.3), we then obtain the molecular weight of horse heart myoglobin as 16 955.36.

### *Response 4.2*

The accurate mass for $C_{35}H_{48}N_8O_{11}S$ is:

$$(35 \times 12.0000) + (48 \times 1.0078) + (8 \times 14.0031)$$
$$+ (11 \times 15.9949) + 31.9721 = 788.3152$$

The single $^{13}C$ satellite occurs at $m/z$ $(788.3152 - 12.0000 + 13.0034) = 789.3186$.

The resolution is therefore given by:

$$788.3162/(789.3186 - 788.3162) = 788.3162/1.0034 = 786$$

Similarly, the accurate mass for $C_{284}H_{432}N_{84}O_{79}S_7$ is:

$$(284 \times 12.0000) + (432 \times 1.0078) + (84 \times 14.0031)$$
$$+ (79 \times 15.9949) + (7 \times 31.9721) = 6507.0318$$

The single $^{13}C$ satellite occurs at $m/z$ $(6507.0318 - 12.0000 + 13.0034) = 6508.0352$.

The resolution is therefore given by:

$$6507.0318/(6508.0352 - 6507.0318) = 6507.0318/1.0034 = 6485$$

## *Response 4.3*

The accurate masses for $C_{284}H_{432}N_{84}O_{79}S_7$ have been calculated above as 6507.0318 and 6508.0352. The $m/z$ values for 5, 7 and 10 charges and the corresponding resolutions are therefore as follows:

5 – 1301.4064 and 1301.6070, and $R = 1301.4064/0.2006 = 6488$
7 – 929.5760 and 929.7193, and $R = 929.5760/0.1433 = 6487$
10 – 650.7032 and 650.8035, and $R = 650.7032/0.1003 = 6488$

These figures clearly demonstrate that the number of charges on the ion does not affect the resolution required to separate the ions in the molecular ion region.

# Chapter 5

## *Response 5.1*

APCI is likely to be the best technique for non-polar to moderately polar compounds, while electrospray ionization is more suitable for moderately polar to ionic compounds.

## *Response 5.2*

The spectrometer is behaving as a concentration-sensitive detector as the signal intensity remains constant as the flow rate increases. If it were mass-sensitive, the detector response would increase.

## *Response 5.3*

The mass spectrometry molecular weight is based on the mass of the more/most abundant isotope of each element, and for $C_{284}H_{432}N_{84}O_{79}S_7$ is therefore:

$$(284 \times 12.0000) + (432 \times 1.0078) + (84 \times 14.0031)$$
$$+ (79 \times 15.9949) + (7 \times 31.9721) = 6507.0318$$

## Response 5.4

A reconstructed ion chromatogram is a plot showing the variation in intensity of an ion of a particular $m/z$ ratio as a function of analysis time, while the total-ion-current trace shows the variation in the intensity of all ions being produced as a function of analysis time. Simplistically, the TIC will show an increase as a compound elutes from an HPLC column and is ionized. If an ion with a particular $m/z$ value is found to be diagnostic of a compound or series of compounds of interest, then an RIC of this $m/z$ will show where its intensity increases and, therefore, where a compound of interest may have eluted. The mass spectrum at this point can then be examined for further confirmation that it is of significance.

## Response 5.5

A Q–ToF mass spectrometer is one which consists of a quadrupole mass analyser and a time-of-flight mass analyser in series. Between these is a region in which fragmentation of ions passed by the first quadrupole may be effected. The instrument is thus capable of generating product-ion spectra and carrying out selected-decomposition monitoring. Its main analytical advantages are that it is capable of fast scanning and of determining the masses of product ions at high resolution. The fast-scanning capability allows product-ion spectra to be produced from all ions in a mass spectrum and computer processing of these data effectively allows precursor-ion data to be obtained. The high-resolution capability gives added confidence in interpretation in that the atomic compositions of product ions may be determined.

## Response 5.6

In APCI, droplets are generated by a combination of heat and a nebulizing gas. While the analytes are embedded in a droplet, and thus protected to some extent from the heat, many thermally labile materials are decomposed. In addition, ionization occurs mainly by ion–molecule reactions and yields predominantly singly charged ions. If, therefore, compounds do not undergo thermal degradation, a mass spectrometer with extended mass range would be required to detect any ions formed.

## Response 5.7

Using the procedure adopted earlier for SAQs 4.2 and 4.3, the resolution is given by:

$$280.0596/0.0238 = 11767$$

# Bibliography

There is a wealth of information available on HPLC, MS and LC–MS in a variety of formats, including books, papers/articles in journals and, increasingly, on websites devoted to the subjects. A number of general references to the techniques appear in the appropriate chapters of this book and these, together with a number of other sources of information, are presented below. There are undoubtedly others but the following represent a good starting place for the reader if further information is required.

## Books

### *High Performance Liquid Chromatography*


Lindsay, S., *High Performance Liquid Chromatography*, 2nd Edn, ACOL Series, Wiley, Chichester, UK, 1992.

Meyer, V. R., *Practical High Performance Liquid Chromatography*, Wiley, Chichester, UK, 1994.

Robards, K., Haddad, P. R. and Jackson, P. E., *Principles and Practice of Modern Chromatographic Methods*, Academic Press, London, 1994.


### *Mass Spectrometry*


Ashcroft, A. E., *Ionization Methods in Organic Mass Spectrometry*, RSC Analytical Spectroscopy Monograph, The Royal Society of Chemistry, Cambridge, UK, 1997.

Barker, J., *Mass Spectrometry*, 2nd Edn, ACOL Series, Wiley, Chichester, UK, 1999.

Busch, K. L., Glish, G. L. and McLuckey, S. A., *Mass Spectrometry/Mass Spectrometry: Techniques and Applications of Tandem Mass Spectrometry*, VCH, New York, 1988.

Chapman, J. R., *Practical Organic Mass Spectrometry*, 2nd Edn, Wiley, Chichester, UK, 1993.

De Hoffmann, E., Charette, J. and Stroobant, V., *Mass Spectrometry – Principles and Applications*, Wiley, Chichester, UK, 1996.

McLafferty, F. W. (Ed.), *Tandem Mass Spectrometry*, Wiley, New York, 1983.

McLafferty, F. W. and Turecek, F., *Interpretation of Mass Spectra*, 4th Edn, University Science Books, Mill Valley, CA, 1993.

Pramanik, B. N., Ganguly, A. K. and Gross, M. L., *Applied Electrospray Mass Spectrometry*, Global View Publishing, Pittsburgh, PA, 2002.

Russell, D. H. (Ed.), *Experimental Mass Spectrometry*, Plenum Press, New York, 1989.

Siuzdak, G., *Mass Spectrometry for Biotechnology*, Academic Press, San Diego, CA, 1996.

Sparkman, O. D., *Mass Spectrometry Desk Reference*, Global View Publishing, Pittsburgh, PA, 2000.

Throck Watson, J., *Introduction to Mass Spectrometry*, 3rd. Edn, Lippincott, Williams and Wilkins, Philadelphia, PA, 1997.

## Liquid Chromatography–Mass Spectrometry

Niessen, W. M. A., *Liquid Chromatography–Mass Spectrometry*, 2nd Edn, Chromatographic Science Series, No. 79, Marcel Dekker, New York, 1999.

Willoughby, R., Sheehan, E. and Mitrovich, S., *A Global View of LC/MS – How to Solve Your Most Challenging Analytical Problems*, 2nd Edn, Global View Publishing, Pittsburgh, PA, 2002

Yergey, A. L., Edmonds, C. G., Lewis, I. A. S. and Vestal, M. L., *Liquid Chromatography/Mass Spectrometry Techniques and Applications*, Plenum Press, New York, 1990.

## General Analytical Chemistry

The following texts all contain sections describing HPLC, MS, LC–MS and quantitative analysis.

Harris, D. C., *Quantitative Chemical Analysis*, 6th Edn, W. H. Freeman, New York, 2002.

Kellner, R., Mermet, J.-M., Otto, M. and Widmer, H. M., *Analytical Chemistry*, Wiley-VCH, Weinheim, Germany, 1998.

Rubinson, K. A. and Rubinson, J. F., *Contemporary Instrumental Analysis*, Prentice Hall, Englewood Cliffs, NJ, 1999.

Skoog, D. A., West, D. M. and Holler, F. J., *Fundamentals of Analytical Chemistry*, 7th Edn, Saunders College Publishing, Fort Worth, TX, 1996.

Skoog, D. A., West, D. M., Holler, F. J. and Crouch, S. R., *Analytical Chemistry: An Introduction*, 7th Edn, Saunders College Publishing, Fort Worth, TX, 2000.

# Websites[†]

## *High Performance Liquid Chromatography*

*Basic HPLC*
http://ntri.tamuk.edu/hplc/hplc.html

*HPLC*
http://www.netaccess.on.ca/~dbc/cic_hamilton/liquid.html

*HPLC Book on the Web*
http://hplc.chem.shu.edu/NEW/HPLC_Book/index.html

*HPLC Links*
http://osoon.ut.ee/~lulla/HPLC/HPLC.html

*HPLC Tutorial*
http://www.geocities.com/CapeCanaveral/8775/program.html

*Introduction to Capillary Chromatography*
http://www.ionsource.com/tutorial/capillary/captoc.htm

*Users Guide to HPLC*
http://www.pharm.uky.edu/ASRG/HPLC/hplcmytry.html

## *Mass Spectrometry*

*A History of Mass Spectrometry*
http://masspec.scripps.edu/information/history/

*John Wiley Spectroscopy Resources* (including MS)
http://www.spectroscopynow.com/Spy/basehtml/SpyH/

*Mass Spectrometry and Biotechnology Resources*
http://www.ionsource.com/

*MS Glossary*
http://chemed.chem.purdue.edu/analyticalreview/mass_spec/msglossary.htm

*MS Links to Information Sources*
http://mslinks.com/

*MS Tutorial*
http://www-methods.ch.cam.ac.uk/meth/ms/theory/

[†] As of July 2002. The material displayed is not endorsed by the author or the publisher.

*What is Mass Spectrometry?* – ASMS
http://www.asms.org/whatisms/page_index.html

The websites of manufacturers of LC–MS equipment are often a valuable source of applications data and educational material, as well as providing details of their own products:

*Agilent*
http://www.chem.agilent.com/

*Analytica of Branford*
http://www.aob.com/

*Bruker*
http://www.bruker.co.uk/

*Jeol*
http://www.jeol.com/ms/ms.html

*MDS-Sciex*
http://www.mdssciex.com/index.html

*Micromass*
http://www.micromass.co.uk

*Shimadzu*
http://www.ssi.shimadzu.com/

*ThermoFinnigan*
http://www.thermo.com/eThermo/CDA/BU_Home/BU_Homepage/0,1285,113,00.html

### *Liquid Chromatography–Mass Spectrometry*

*LC and LC–MS*
http://www.forumsci.co.il/HPLC/index.html

*LC–MS Home Page*
http://www.lcms.com/lcms_top.htm

## Meetings

In addition to general scientific meetings which contain papers involving LC–MS, the Annual Montreux Meeting, held since 1980, is a meeting devoted to LC–MS, CE–MS and MS–MS which deals with technical developments in on-line aspects, theoretical considerations, and applications of the techniques in environmental, clinical, industrial and pharmaceutical analysis, and other fields. Recent advances and applications are presented in plenary lectures, in addition to contributed oral and poster presentations [http://www.rtpnet.org/~lcms99/].

# Journals

In addition to the peer review journals which publish developments and applications of LC–MS, the following, which is available on free subscription, contains general articles and is highly recommended:

*LC–GC Magazine*
http://www.lcgcmag.com/lcgc/

# LC–MS Database

This database consists of specially commissioned abstracts covering methods (including instrumental details and conditions), assays, applications and techniques in selected areas. All abstracts are written by specialists and are NOT simply copies of the author abstracts. For further details, contact:

HD Science
LC–MS Applications Database
HD Science Limited
16 Petworth Avenue, Toton
Nottingham NG9 6JF, UK
[http://www.hdscience.com/]

# Glossary of Terms

This section contains a glossary of terms, all of which are used in the text. It is not intended to be exhaustive, but to explain briefly those terms which often cause difficulties or may be confusing to the inexperienced reader.

**Accuracy** The closeness of a result to its true value.

**Accurate mass** The $m/z$ ratio of an ion determined to high accuracy to enable the elemental composition of the ion to be determined.

**Adduct ion** An ion arising from the combination of two species, e.g. the molecular species observed in a positive-ion APCI spectrum is usually an adduct of the analyte molecule with a species such as $H^+$ , $Na^+$ or $NH_4{}^+$.

**Aerospray** An atmospheric-pressure ionization technique in which droplets are formed from a liquid stream by a combination of heat and a nebulizing gas and ions are formed by ion evaporation rather than ion–molecule reactions.

**Affinity chromatography** A form of chromatography in which separation is achieved by utilizing highly specific biochemical interactions, such as steric- or charge-related conditions, between the analyte and a molecule immobilized on a column. It is different from most forms of chromatography in that analytes do not continuously elute from the column – only those that interact with the stationary phase are retained and thus separated from other components of the mixture under investigation. These immobilized materials are eluted from the column after all other materials have been removed.

**Atmospheric-pressure chemical ionization (APCI)** An ionization method in which a liquid stream is passed through a heated capillary and a concentric flow of a nebulizing gas. Ions are formed by ion–molecule reactions between the analyte and species derived from the HPLC mobile phase.

**Atmospheric-pressure ionization (API)** A general term used for all forms of ionization that take place at atmospheric pressure.

**Background-subtracted spectrum** A mass spectrum from which ions arising from species other than the analyte have been removed by computer manipulation.

**Base peak** The most intense ion in a mass spectrum. The intensity of other ions in the spectrum are reported as a percentage of the intensity of the base peak.

**Biotransformation** An alternative term for drug metabolism.

**Buffer** An electrolyte added to the HPLC mobile phase.

**Capacity factor** The parameter used in HPLC to measure the retention of an analyte.

**Capillary column** This term refers to a chromatographic column of 'small' diameter and is used in both gas and high performance liquid chromatography. In HPLC, the term is usually applied to columns with internal diameters of between 0.1 and 2 mm. The term *microbore* column is often used synonymously to describe these columns but is more correctly applied to columns with internal diameters of 1 or 2 mm.

**Charge-residue mechanism** One of the two mechanisms used to account for the production of ions by electrospray ionization.

**Chemical ionization** An ionization method used to maximize the production of intact molecular species. Used for volatile, thermally stable analytes.

**Chemical noise** Signals from species other than the analyte present in the system or sample that cannot be resolved from that of the analyte.

**Chromatographic selectivity** The degree to which compounds are separated on a particular chromatographic system.

**Chromatography** General term for a number of methods used to separate the individual components of a mixture.

**Collision energy** The energy of the collision between an ion and a gas molecule which may be used to vary the amount of fragmentation observed.

**Collision-induced dissociation** Fragmentation of an ion by collision with a gas molecule.

**Concentration-sensitive detector** A detector for which the intensity of response is proportional to the concentration of analyte reaching it.

**Cone-voltage fragmentation** Fragmentation of ions, commonly produced by APCI or electrospray ionization, effected by the application of a voltage within the source of the mass spectrometer.

**Constant-neutral-loss scan** An MS–MS scan in which ions containing a particular structural feature may be identified.

**Corona discharge** Occurs when the field at the tip of the electrode is sufficiently high to ionize the gas surrounding it but insufficiently high to cause a spark. An integral part of the APCI interface.

**Coulombic explosion** The process by which a droplet disintegrates into a number of smaller droplets which occurs when the repulsive forces between charges on the surface of a droplet are greater than the cohesive force of surface tension.

**Diode-array UV detector** A UV detector which monitors all wavelengths simultaneously and therefore allows a complete UV spectrum to be obtained instantaneously. The alternative, a dispersive UV detector, monitors one wavelength at a time and thus requires a considerable amount of time to record a complete spectrum.

**Discharge electrode** An electrode used to generate a corona discharge.

**Double-focusing mass spectrometer** A mass spectrometer consisting of electrostatic and magnetic sector analysers capable of achieving high mass spectral resolution.

**Drug metabolism** The process by which drugs are transformed in the body to a form that is more readily eliminated.

**Dynamic range (of a detector)** The range over which the addition of further analyte brings about an increase, however small, in detector response.

**Edman degradation** A method of amino acid sequencing in proteins in which successive *N*-terminal amino acids are removed from the polypeptide chain and identified.

**Electrohydrodynamic ionization** A process in which a high voltage is used to generate droplets from which ions are desorbed under conditions of high vacuum.

**Electron ionization** An ionization method employed in mass spectrometry in which analytes, in the vapour phase, are bombarded with high-energy electrons.

**Electrospray** The process whereby a liquid stream is broken up into droplets by the action of a high potential.

**Electrospray ionization** The production of ions from droplets produced by the electrospray process.

**Electrostatic analyser (ESA)** An energy-focusing device used in a double-focusing mass spectrometer to increase mass spectral resolution.

**Energy-sudden ionization technique** One in which energy is provided to a thermally labile molecule so rapidly that it is desorbed and ionized before decomposition takes place.

**Enzyme digestion** The treating of a protein with a proteolytic enzyme to form a number of smaller peptides which may then be sequenced.

**Experimental design** A number of formal procedures whereby the effect of experimental variables on the outcome of an experiment may be assessed. These may be used to assess the optimum conditions for an experiment and to maximize the accuracy and precision obtained.

**External standard** A method of relating the intensity of a signal from an analyte measured in an 'unknown' to the amount of analyte present. This method consists of running a series of standards containing known amounts of the analyte independently from the samples to be determined.

**Factor** An experimental variable that has (or may have) an effect on the outcome of an experiment, e.g. temperature, concentration of reactants, presence of a catalyst, etc.

**Factorial design** One method of experimental design that allows interactions between factors to be investigated, i.e. whether changing one experimental variable changes the optimum value of another.

**Fast-atom bombardment** An ionization method used for involatile and thermally labile materials. In this technique, the sample is dissolved in a matrix material and bombarded with a high-energy atom or ion beam.

**Field desorption** An ionization method in which sample is deposited on a wire to which a high voltage is applied.

**Flow programming** Varying the HPLC flow rate during the course of a separation.

**Forward-geometry double-focusing mass spectrometer** A double-focusing mass spectrometer in which the electrostatic analyser precedes the magnetic analyser.

**Four-sector mass spectrometer** A mass spectrometer used for MS–MS studies consisting of two double-focusing mass spectrometers in series.

**Fragmentor voltage** Another term for cone-voltage fragmentation.

**General detector** A (chromatographic) detector which responds to all compounds reaching it.

**Glycoprotein** A protein containing sugar molecules attached to its polypeptide chain.

**Glycosylation** The incorporation of a sugar molecule into a protein.

**Gradient elution** The changing of HPLC mobile phase composition during the course of an analysis.

**High-resolution mass spectrometer** A mass spectrometer capable of high resolution and measuring $m/z$ ratios with high accuracy to enable the atomic composition of an ion to be determined.

**Hybrid mass spectrometer** An MS–MS instrument combining magnetic sector and quadrupole mass analysers.

**Hybrid technique** The combination of two or more analytical techniques.

**Hyphenated technique** The combination of two analytical techniques.

**Injector** A common term for the method of sample introduction into a chromatographic system.

**In-source fragmentation** (*see* Cone-voltage fragmentation)

**Interface** The hardware employed to link two analytical techniques. The primary purpose of an interface is to ensure that the operational requirements of each of the techniques are not compromised by the other.

**Interference** A species other than the analyte of interest which gives a detector response.

**Internal standard** A method of relating the intensity of signal from an analyte measured in an 'unknown' to the amount of analyte present. In this approach, a known amount of an internal standard is added to both calibration and 'unknown' samples and the ratio of signal intensities of the analyte and internal standard in each is calculated. This method of standardization improves both accuracy and precision.

**Ion evaporation** One of the two mechanisms used to account for the production of ions by electrospray ionization.

**Ion–molecule reaction** The reaction between an ion and a neutral molecule which leads to the production of an adduct ion.

**Ion-pairing reagent** A compound that forms a complex with an ionic compound to allow its analysis using HPLC.

**Ionspray®** Pneumatically assisted electrospray – a process in which nebulizing gas is used in conjunction with a high voltage to form droplets from a liquid stream.

**Ion-trap** A low-resolution mass analyser.

**Isocratic elution** The use of a mobile phase of constant composition during the course of an analysis.

**LC–MS–MS** The combination of HPLC with MS–MS.

**Library searching** The use of a computer to compare a mass spectrum to be identified with large numbers of reference spectra.

**Limit of detection** The smallest quantity of an analyte that can be detected reliably.

**Limit of quantitation** The smallest quantity of an analyte that can be determined with accuracy and precision.

**Linear range** The range of concentrations over which the analytical signal is directly proportional to the amount of analyte present.

**Linked scanning** A series of techniques in which the electrostatic analyser voltage and magnetic field strength of a double-focusing mass spectrometer are scanned to obtain MS–MS spectra.

**Low-resolution mass spectrometer** A spectrometer which is capable of measuring the $m/z$ ratio of an ion to the nearest integer value.

**Magnetic sector** A low-resolution mass analyser in which the variation of a magnetic field is used to bring ions of different $m/z$ ratios to a detector.

**Mass-analysed ion kinetic energy spectrometry (MIKES)** A form of MS–MS product-ion scan that may be carried out on a reverse-geometry double-focusing mass spectrometer.

**Mass chromatogram** (*see* Reconstructed ion chromatogram)

**Mass-flow-sensitive detector** A detector for which the intensity of response is proportional to the amount of analyte reaching it.

**Mass-sensitive detector** (*see* Mass-flow-sensitive detector)

**Matrix-assisted laser desorption ionization (MALDI)** A method used for the ionization of high-molecular-weight compounds. In this approach, the analyte is crystallized with a solid matrix and then bombarded with a laser of a frequency which is absorbed by the matrix material.

**Matrix effects** An increase or decrease in signal intensity from an analyte due to the presence of any other materials in the sample in which it is being determined.

**Matrix material** A material used in fast-atom bombardment and matrix-assisted laser desorption ionization to transfer energy to an analyte molecule to bring about its ionization.

**Maximum entropy** A computer algorithm used to predict the theoretical signal from which that observed in a spectrum has been derived. Used in conjunction with electrospray ionization to enhance the quality of the spectra obtained.

**McLafferty rearrangement** A molecular rearrangement that occurs under certain ionization conditions which results in the production of characteristic ions in the mass spectrum of the analyte from which it has been generated.

**Megaflow electrospray** An electrospray system capable of producing droplets directly from HPLC flow rates of the order of 1 ml $min^{-1}$ (true electrospray is most efficient at flow rates of the order of 10 μl $min^{-1}$).

**Microbore column** (*see* Capillary column)

**Mobile phase** That part of a chromatographic system which causes the analyte to move from the point of injection to the detector – in HPLC, this is a liquid.

**Molecular ion** The ion in the mass spectrum corresponding to the unfragmented molecule under investigation.

**Monoisotopic molecular weight** The molecular weight of an analyte, calculated by using the masses of the more/most abundant isotopes of each of the elements present.

**MS–MS** A number of techniques in which two stages of mass spectrometry are used in series to probe the relationship between ions formed from an analyte – also known as tandem mass spectrometry.

**$MS^n$** An extension of MS–MS in which more than two stages of mass spectrometry are used to probe the relationship between ions formed from an analyte.

**Multiple-ion detection** (*see* Selected-ion monitoring)

**Multiply charged ion** An ion with more than one charge. The electrospray spectra from compounds of high molecular weight contain exclusively multiply charged ions.

**Murphy's Law** 'If something can go wrong it will do so *and* at the most inconvenient time!'[†]

**Nanoflow electrospray** A form of electrospray ionization, carried out at flow rates of the order of nl $min^{-1}$.

**Negative ionization** The production of negative ions of analytical significance from the analyte of interest.

**Noise** The change in detector response over a period of time in the absence of analyte. This consists of two components, namely the short-term random

---

[†] As the majority of readers will confirm, this observation regularly holds true in all walks of life. Also known under various other names!

variation in signal intensity, and the drift, i.e. the increase or decrease in the average noise level over a period of time.

**Normal-phase HPLC** An HPLC system in which the mobile phase is less polar that the stationary phase.

**Packed column** An HPLC column containing particles of inert material of typically 5 μm diameter on which the stationary phase is coated.

**Peptide** A linear chain of a small number of amino acids linked by peptide bonds. The number of amino acids which differentiates between a protein and a peptide remains a matter for discussion.

**Peptide mapping** The process of considering the amino acid sequence information from peptides obtained by enzyme digestion in an attempt to derive the (amino acid) sequence of the parent protein.

**Phase I metabolism** The introduction of a polar group, e.g. an hydroxyl group, into a parent drug structure prior to its elimination from the body.

**Phase II metabolism** The reaction of a phase I metabolite with an endogenous compound, e.g. glucuronic acid, to form a polar compound that is eliminated from the body.

**Plate height** The width of a theoretical plate.

**Plate number** The number of theoretical plates in a chromatographic column. This is a measure of the efficiency of the column.

**Polyimide belt** The continuous belt used in the moving-belt LC–MS interface.

**Post-source decay** The term used to describe the production of product-ion MS–MS spectra in a time-of-flight mass analyser.

**Post-translation modification** Changes that occur to proteins after peptide-bond formation has occurred, e.g. glycosylation and acylation.

**Precision** The closeness of replicate measurements on the same sample.

**Precursor-ion scan** An MS–MS scan in which those ions that fragment to a given product ion are detected.

**Product-ion scan** An MS–MS scan in which those ions obtained by fragmentation of a given precursor ion are detected.

**Protein** A linear chain of amino acids linked by peptide bonds.

**Q–ToF** The combination of quadrupole and time-of-flight mass analysers. This allows the $m/z$ ratios of ions produced during a product-ion scan to be measured accurately and the elemental composition of these ions to be determined.

**Quadrupole** A low-resolution mass analyser.

**Qualitative analysis** The analysis of a sample to determine the identity of any compounds present.

**Quantitative analysis** The analysis of a sample to determine the amount of an analyte present.

**Reagent gas** A gas used in chemical ionization to produce species which react with molecules of the analyte of interest to produce a molecular species.

**Rearrangement ion** An ion formed under certain ionization conditions in which the original molecular structure of the analyte has undergone some modification, i.e. has not been produced by simple bond scission.

**Reconstructed ion chromatogram** A plot of the intensity of an ion of chosen $m/z$ ratio as a function of analysis time. This is produced by computer analysis of mass spectral data acquired over an extended mass range.

**Reflectron** An ion lens used in the time-of-flight mass analyser to increase the distance travelled by an ion and thereby increase the resolution of the instrument.

**Repeatability** The closeness of a set of measurements carried out by a single analyst on a single instrument within a narrow time-interval using the same reagents.

**Repeller** An electrode used in thermospray ionization to effect fragmentation of molecular species.

**Reproducibility** The closeness of a set of measurements carried out by a number of analysts on a number of instruments over an extended period.

**Resolution** A term which indicates the ability of a device/technique to separate/distinguish between closely related signals. In chromatography, it relates to the ability to separate compounds with similar retention characteristics, and in mass spectrometry to the ability to separate ions of similar $m/z$ ratios.

**Retention index** The parameter used in gas chromatography to measure the retention of an analyte.

**Reverse-geometry double-focusing mass spectrometer** A double-focusing mass spectrometer in which the magnetic analyser precedes the electrostatic analyser.

**Reversed-phase HPLC** An HPLC system in which the mobile phase is more polar than the stationary phase.

**Selected-decomposition monitoring** An MS–MS scan in which the first stage of mass spectrometry is set to transmit a selected ion and the second to transmit only a selected product ion. This technique increases the selectivity of the analysis.

**Selected-ion monitoring** A technique in which the mass spectrometer is used to monitor only a small number of ions characteristic of the analyte of interest.

**Selected-ion recording** (*see* Selected-ion monitoring)

**Selected-reaction monitoring** (*See* Selected-decomposition monitoring)

**Selective detector** A detector which responds only to compounds containing a certain structural feature.

**Selectivity** The ability to determine the analyte of interest with accuracy and precision in the presence of other materials.

**Separation factor** (*see* Chromatographic selectivity)

**Sequence tagging** The use of MS–MS to investigate the amino acid sequence of a peptide.

**Sequencing** The determination of the order in which the repeating units occur in a biopolymer, e.g. amino acids in a protein, sugar residues in a carbohydrate, etc.

**Signal enhancement** The increase in analyte signal intensity brought about by the presence of extraneous materials in the sample.

**Signal-to-noise ratio** The ratio of the intensity of the analytical signal to that of the noise. This is used in determining the limits of detection and quantitation.

**Sodium dodecyl sulfate–polyacrylamide gel electrophoresis (SDS–PAGE)** An electrophoretic technique used for the separation of proteins.

**Soft ionization technique** An ionization technique that produces molecular species with few, if any, fragment ions.

**Solute-property detector** A detector which monitors a property of the analyte, e.g. the UV detector.

**Solvent-property detector** A detector which monitors a property of the HPLC mobile phase which is perturbed when an analyte elutes from the chromatographic column.

**Spray deposition** A method used to apply HPLC eluate in later versions of the moving-belt interface to provide a uniform layer of mobile phase on the belt and thus minimize the production of droplets.

**Standard additions** A method of relating the intensity of signal from an analyte measured in an 'unknown' to the amount of analyte present. This technique is designed to take matrix effects into account.

**Stationary phase** That part of the chromatographic system with which the analytes interact, over which the mobile phase flows.

**Suppression effects** The decrease in analyte signal intensity brought about by the presence of extraneous materials in the sample.

**Tandem mass spectrometry** An alternative term for MS–MS.

**Tandem technique** An alternative term for the combination of two or more analytical techniques.

**Theoretical plate** In plate theory, the chromatographic column is viewed as a series of narrow layers, known as theoretical plates, within each of which equilibration of the analyte between mobile and stationary phases occurs.

**Thermally labile compound** A compound that decomposes under the influence of heat.

**Thermospray** The process whereby a liquid stream is broken up into droplets by the action of a high temperature.

**Thermospray ionization** The formation of ions from droplets produced by the thermospray process.

**Three-dimensional quadrupole** (*see* Ion-trap)

**Time-of-flight mass analyser** A mass analyser in which ions are separated 'in time' as they drift through a field-free flight tube.

**Total-ion-current trace** A plot of the total number of ions reaching the mass spectrometry detector as a function of analysis time.

**Transformation** The mathematical process of changing a 'raw' electrospray spectrum containing a number of multiply charged ions into a mass spectrum plotted on a true mass scale.

**Triple quadrupole** A mass spectrometer consisting of three sets of quadrupole rods in series, which is used extensively for studies involving MS–MS.

**Tri-sector mass spectrometer** A mass spectrometer consisting of an electrostatic analyser (ESA), a magnetic sector and a second ESA in series.

**Universal detector** An alternative term for a general detector.

**Western blotting** A means of transferring protein bands from an electrophoresis gel onto a fixing medium for further analysis.

**Z-spray** An electrospray source in which ions are extracted into the mass spectrometer at 90° to the direction in which the spray is produced.

*MS Glossary on the Web*
[http://chemed.chem.purdue.edu/analyticalreview/mass_spec/msglossary.htm]

# SI Units and Physical Constants

## SI Units

The SI system of units is generally used throughout this book. It should be noted, however, that according to present practice, there are some exceptions to this, for example, wavenumber ($cm^{-1}$) and ionization energy (eV).

**Base SI units and physical quantities**

| Quantity | Symbol | SI Unit | Symbol |
|---|---|---|---|
| length | $l$ | metre | m |
| mass | $m$ | kilogram | kg |
| time | $t$ | second | s |
| electric current | $I$ | ampere | A |
| thermodynamic temperature | $T$ | kelvin | K |
| amount of substance | $n$ | mole | mol |
| luminous intensity | $I_v$ | candela | cd |

**Prefixes used for SI units**

| Factor | Prefix | Symbol |
|---|---|---|
| $10^{21}$ | zetta | Z |
| $10^{18}$ | exa | E |
| $10^{15}$ | peta | P |
| $10^{12}$ | tera | T |
| $10^{9}$ | giga | G |
| $10^{6}$ | mega | M |
| $10^{3}$ | kilo | k |

*(continued overleaf)*

**Prefixes used for SI units** *(continued)*

| Factor | Prefix | Symbol |
|---|---|---|
| $10^2$ | hecto | h |
| 10 | deca | da |
| $10^{-1}$ | deci | d |
| $10^{-2}$ | centi | c |
| $10^{-3}$ | milli | m |
| $10^{-6}$ | micro | μ |
| $10^{-9}$ | nano | n |
| $10^{-12}$ | pico | p |
| $10^{-15}$ | femto | f |
| $10^{-18}$ | atto | a |
| $10^{-21}$ | zepto | z |

**Derived SI units with special names and symbols**

| Physical quantity | SI unit | | Expression in terms of base or derived SI units |
|---|---|---|---|
| | Name | Symbol | |
| frequency | hertz | Hz | $1\ Hz = 1\ s^{-1}$ |
| force | newton | N | $1\ N = 1\ kg\,m\,s^{-2}$ |
| pressure; stress | pascal | Pa | $1\ Pa = 1\ Nm^{-2}$ |
| energy; work; quantity of heat | joule | J | $1\ J = 1\ Nm$ |
| power | watt | W | $1\ W = 1\ J\,s^{-1}$ |
| electric charge; quantity of electricity | coulomb | C | $1\ C = 1\ A\,s$ |
| electric potential; potential difference; electromotive force; tension | volt | V | $1\ V = 1\ J\,C^{-1}$ |
| electric capacitance | farad | F | $1\ F = 1\ C\,V^{-1}$ |
| electric resistance | ohm | Ω | $1\ \Omega = 1\ V^{-1}$ |
| electric conductance | siemens | S | $1\ S = 1\ \Omega^{-1}$ |
| magnetic flux; flux of magnetic induction | weber | Wb | $1\ Wb = 1\ V\,s$ |
| magnetic flux density; magnetic induction | tesla | T | $1\ T = 1\ Wb\,m^{-2}$ |
| inductance | henry | H | $1\ H = 1\ Wb\,A^{-1}$ |
| Celsius temperature | degree Celsius | °C | $1°C = 1\ K$ |
| luminous flux | lumen | lm | $1\ lm = 1\ cd\,sr$ |

**Derived SI units with special names and symbols** (*continued*)

| Physical quantity | SI unit | | Expression in terms of base or derived SI units |
|---|---|---|---|
| | Name | Symbol | |
| illuminance | lux | lx | $1\ \text{lx} = 1\ \text{lm}\,\text{m}^{-2}$ |
| activity (of a radionuclide) | becquerel | Bq | $1\text{Bq} = 1\ \text{s}^{-1}$ |
| absorbed dose; specific energy | gray | Gy | $1\ \text{Gy} = 1\ \text{J}\,\text{kg}^{-1}$ |
| dose equivalent | sievert | Sv | $1\ \text{Sv} = 1\ \text{J}\,\text{kg}^{-1}$ |
| plane angle | radian | rad | 1[a] |
| solid angle | steradian | sr | 1[a] |

[a]rad and sr may be included or omitted in expressions for the derived units.

# Physical Constants

**Recommended values of selected physical constants**[a]

| Constant | Symbol | Value |
|---|---|---|
| acceleration of free fall (acceleration due to gravity) | $g_n$ | $9.806\,65\ \text{m}\,\text{s}^{-2}$[b] |
| atomic mass constant (unified atomic mass unit) | $m_u$ | $1.660\,540\,2(10) \times 10^{-27}\ \text{kg}$ |
| Avogadro constant | $L, N_A$ | $6.022\,136\,7(36) \times 10^{23}\ \text{mol}^{-1}$ |
| Boltzmann constant | $k_B$ | $1.380\,658(12) \times 10^{-23}\ \text{J}\,\text{K}^{-1}$ |
| electron specific charge (charge-to-mass ratio) | $-e/m_e$ | $-1.758\,819 \times 10^{11}\ \text{C}\,\text{kg}^{-1}$ |
| electron charge (elementary charge) | $e$ | $1.602\,177\,33(49) \times 10^{-19}\ \text{C}$ |
| Faraday constant | $F$ | $9.648\,530\,9(29) \times 10^{4}\ \text{C}\,\text{mol}^{-1}$ |
| ice-point temperature | $T_{ice}$ | $273.15\ \text{K}$[b] |
| molar gas constant | $R$ | $8.314\,510(70)\ \text{J}\,\text{K}^{-1}\ \text{mol}^{-1}$ |
| molar volume of ideal gas (at 273.15 K and 101 325 Pa) | $V_m$ | $22.414\,10(19) \times 10^{-3}\ \text{m}^3\ \text{mol}^{-1}$ |
| Planck constant | $h$ | $6.626\,075\,5(40) \times 10^{-34}\ \text{J}\,\text{s}$ |
| standard atmosphere | atm | $101\,325\ \text{Pa}$[b] |
| speed of light in vacuum | $c$ | $2.997\,924\,58 \times 10^{8}\ \text{m}\,\text{s}^{-1}$[b] |

[a]Data are presented in their full precision, although often no more than the first four or five significant digits are used; figures in parentheses represent the standard deviation uncertainty in the least significant digits.

[b]Exactly defined values.

# The Periodic Table

Key:

| 3 0.98<br>**Li**<br>6.941 | |
|---|---|
| 0.98 | Pauling electronegativity |
| 3 | Atomic number |
| Li | Element |
| 6.941 | Atomic weight ($^{12}C$) |

| | |
|---|---|
| 1 2.20<br>**H**<br>1.008 | 2<br>**He**<br>4.003 |

d transition elements: Group 3 to 12

| Group 1 | Group 2 | 3 | 4 | 5 | 6 | 7 | 8 | 9 | 10 | 11 | 12 | Group 13 | Group 14 | Group 15 | Group 16 | Group 17 | Group 18 |
|---|---|---|---|---|---|---|---|---|---|---|---|---|---|---|---|---|---|
| 3 0.98<br>**Li**<br>6.941 | 4 1.57<br>**Be**<br>9.012 | | | | | | | | | | | 5 2.04<br>**B**<br>10.811 | 6 2.55<br>**C**<br>12.011 | 7 3.04<br>**N**<br>14.007 | 8 3.44<br>**O**<br>15.999 | 9 3.98<br>**F**<br>18.998 | 10<br>**Ne**<br>20.179 |
| 11 0.93<br>**Na**<br>22.990 | 12 1.31<br>**Mg**<br>24.305 | | | | | | | | | | | 13 1.61<br>**Al**<br>26.98 | 14 1.90<br>**Si**<br>28.086 | 15 2.19<br>**P**<br>30.974 | 16 2.58<br>**S**<br>32.064 | 17 3.16<br>**Cl**<br>35.453 | 18<br>**Ar**<br>39.948 |
| 19 0.82<br>**K**<br>39.102 | 20 1.00<br>**Ca**<br>40.08 | 21<br>**Sc**<br>44.956 | 22<br>**Ti**<br>47.90 | 23<br>**V**<br>50.941 | 24<br>**Cr**<br>51.996 | 25<br>**Mn**<br>54.938 | 26<br>**Fe**<br>55.847 | 27<br>**Co**<br>58.933 | 28<br>**Ni**<br>58.71 | 29<br>**Cu**<br>63.546 | 30<br>**Zn**<br>65.37 | 31 1.81<br>**Ga**<br>69.72 | 32 2.01<br>**Ge**<br>72.59 | 33 2.18<br>**As**<br>74.922 | 34 2.55<br>**Se**<br>78.96 | 35 2.96<br>**Br**<br>79.909 | 36<br>**Kr**<br>83.80 |
| 37 0.82<br>**Rb**<br>85.47 | 38 0.95<br>**Sr**<br>87.62 | 39<br>**Y**<br>88.906 | 40<br>**Zr**<br>91.22 | 41<br>**Nb**<br>92.906 | 42<br>**Mo**<br>95.94 | 43<br>**Tc**<br>(99) | 44<br>**Ru**<br>101.07 | 45<br>**Rh**<br>102.91 | 46<br>**Pd**<br>106.4 | 47<br>**Ag**<br>107.87 | 48<br>**Cd**<br>112.40 | 49 1.78<br>**In**<br>114.82 | 50 1.96<br>**Sn**<br>118.69 | 51 2.05<br>**Sb**<br>121.75 | 52 2.10<br>**Te**<br>127.60 | 53 2.66<br>**I**<br>126.90 | 54<br>**Xe**<br>131.30 |
| 55 0.79<br>**Cs**<br>132.91 | 56 0.89<br>**Ba**<br>137.34 | 57<br>**La**<br>138.91 | 72<br>**Hf**<br>178.49 | 73<br>**Ta**<br>180.95 | 74<br>**W**<br>183.85 | 75<br>**Re**<br>186.2 | 76<br>**Os**<br>190.2 | 77<br>**Ir**<br>192.22 | 78<br>**Pt**<br>195.09 | 79<br>**Au**<br>196.97 | 80<br>**Hg**<br>200.59 | 81 2.04<br>**Ti**<br>204.37 | 82 2.32<br>**Pb**<br>207.19 | 83 2.02<br>**Bi**<br>208.98 | 84<br>**Po**<br>(210) | 85<br>***At***<br>(210) | 86<br>**Rn**<br>(222) |
| 87<br>**Fr**<br>(223) | 88<br>**Ra**<br>226.025 | 89<br>**Ac**<br>227.0 | 104<br>**Rf**<br>(261) | 105<br>**Db**<br>(262) | 106<br>**Sg**<br>(263) | 107<br>**Bh** | 108<br>**Hs** | 109<br>**Mt** | 110<br>**Uun** | 111<br>**Uuu** | 112<br>**Unb** | | | | | | |

| | | | | | | | | | | | | | |
|---|---|---|---|---|---|---|---|---|---|---|---|---|---|
| 58<br>**Ce**<br>140.12 | 59<br>**Pr**<br>140.91 | 60<br>**Nd**<br>144.24 | 61<br>**Pm**<br>(147) | 62<br>**Sm**<br>150.35 | 63<br>**Eu**<br>151.96 | 64<br>**Gd**<br>157.25 | 65<br>**Tb**<br>158.92 | 66<br>**Dy**<br>162.50 | 67<br>**Ho**<br>164.93 | 68<br>**Er**<br>167.26 | 69<br>**Tm**<br>168.93 | 70<br>**Yb**<br>173.04 | 71<br>**Lu**<br>174.97 |
| 90<br>**Th**<br>232.04 | 91<br>**Pa**<br>(231) | 92<br>**U**<br>238.03 | 93<br>**Np**<br>(237) | 94<br>**Pu**<br>(242) | 95<br>**Am**<br>(243) | 96<br>**Cm**<br>(247) | 97<br>**Bk**<br>(247) | 98<br>**Cf**<br>(249) | 99<br>**Es**<br>(254) | 100<br>**Fm**<br>(253) | 101<br>**Md**<br>(253) | 102<br>**No**<br>(256) | 103<br>**Lw**<br>(260) |

# Index

Page numbers given in italics indicate a reference to a figure or table in the text.